Lecture Notes in Mechanical Engineering

Lecture Notes in Mechanical Engineering (LNME) publishes the latest developments in Mechanical Engineering—quickly, informally and with high quality. Original research reported in proceedings and post-proceedings represents the core of LNME. Volumes published in LNME embrace all aspects, subfields and new challenges of mechanical engineering. Topics in the series include:

- Engineering Design
- Machinery and Machine Elements
- Mechanical Structures and Stress Analysis
- Automotive Engineering
- Engine Technology
- Aerospace Technology and Astronautics
- Nanotechnology and Microengineering
- Control, Robotics, Mechatronics
- MEMS
- Theoretical and Applied Mechanics
- Dynamical Systems, Control
- Fluid Mechanics
- Engineering Thermodynamics, Heat and Mass Transfer
- Manufacturing
- Precision Engineering, Instrumentation, Measurement
- Materials Engineering
- Tribology and Surface Technology

To submit a proposal or request further information, please contact the Springer Editor of your location:

China: Ms. Ella Zhang at ella.zhang@springer.com
India: Priya Vyas at priya.vyas@springer.com
Rest of Asia, Australia, New Zealand: Swati Meherishi at swati.meherishi@springer.com
All other countries: Dr. Leontina Di Cecco at Leontina.dicecco@springer.com

To submit a proposal for a monograph, please check our Springer Tracts in Mechanical Engineering at http://www.springer.com/series/11693 or contact Leontina.dicecco@springer.com

Indexed by SCOPUS. All books published in the series are submitted for consideration in Web of Science.

More information about this series at http://www.springer.com/series/11236

Sadhan Kumar Ghosh · Koushik Ghosh ·
Santanu Das · Pranab Kumar Dan · Arijit Kundu
Editors

Advances in Thermal Engineering, Manufacturing, and Production Management

Select Proceedings of ICTEMA 2020

Springer

Editors
Sadhan Kumar Ghosh
Department of Production Engineering
and Management
Jadavpur University
Kolkata, West Bengal, India

Koushik Ghosh
Department of Thermal Engineering
Jadavpur University
Kolkata, West Bengal, India

Santanu Das
Kalyani Government Engineering College
Kalyani, West Bengal, India

Pranab Kumar Dan
Department of Production Management
Indian Institute of Technology Kharagpur
Kharagpur, West Bengal, India

Arijit Kundu
Department of Thermal Engineering
Jalpaiguri Government Engineering College
Jalpaiguri, West Bengal, India

ISSN 2195-4356 ISSN 2195-4364 (electronic)
Lecture Notes in Mechanical Engineering
ISBN 978-981-16-2346-2 ISBN 978-981-16-2347-9 (eBook)
https://doi.org/10.1007/978-981-16-2347-9

This Springer imprint is published by the registered company Springer Nature Singapore Pte Ltd.
The registered company address is: 152 Beach Road, #21-01/04 Gateway East, Singapore 189721, Singapore

Organizing Committee

Patron

Prof. Amitava Ray, Principal

Organizing Secretary

Dr. Arijit Kundu

Joint Organizing Secretary

Md. Naim Hossain

Executive Members

Prof. S. Bhattacharya
Prof. G. Panda
Prof. P. K. Saha
Prof. Arghadeep Biswas
Prof. Arindam Saha
Prof. Dhiman mandal
Prof. Jishan Mehedi
Prof. Mirwize Rahaman
Prof. Nripen Mondal
Prof. Samir Das

Prof. Sandipan Gunguly
Prof. Santanu Das
Prof. Suman Konar
Prof. Suman Mondal

Treasurer

Prof. Madhab Mandal
Smt. Elora Rakshit

Advisory Committee

Prof. A. Mani, IIT Madras
Prof. Arup Kumar Das, IIT Roorkee
Prof. Balbir Singh, SMVDU, Jammu
Prof. D. B. Jani, Gujarat Technological University, Ahmedabad
Prof. Dilip Sharma, MNIT Jaipur
Prof. Gautam Biswas, IIT Guwahati
Prof. Gautam Majumdar, Jadavpur University
Prof. I. V. Singh, IIT Roorkee
Prof. K. K. Shukla, NIT Jamshedpur
Prof. Kausik Ghosh, Jadavpur University
Prof. M. P. Maiya, IIT Madras
Prof. P. K. Das, IIT Kharagpur
Prof. Pallab S. Mahapatra, IIT Madras
Prof. Pankaj Biswas, IIT Guwahati
Prof. Ravi Kumar, IIT Roorkee
Prof. Sadhan Ghosh, Jadavpur University
Prof. Saptarshi Basu, IISc Bangalore
Prof. Shankar Singh, SLIET, Longowal
Prof. Suman Chakraborty, IIT Kharagpur
Prof. Swarnendu Sen, Jadavpur University
Prof. T. V. V. L. N. Rao, SRM Institute of Science and Technology, Kattankulathur
Prof. Vineet V. Tyagi, SMVDU, Jammu
Prof. Amaresh Dalal, IIT Guwahati
Prof. Ramesh Gupta, SNU, GB Nagar, Uttar Pradesh

International Advisors

Prof. Gherhardt Ribatski, University of São Paulo, Brazil
Prof. H. K. Sachidananda, Manipal Academy of Higher Education, Dubai
Prof. J. Paulo Davim, University of Aveiro, UA
Prof. Ravi P. Agarwal, Texas A&M University-Kingsville, Texas
Prof. Biswa Datta, Northern Illinois University, Illinois, USA
Prof. Irem Deniz, Manisa Celal Bayar University, Manisa, Turkey
Prof. Thanh Bui, Ho Chi Minh City University of Technology, Vietnam
Prof. Z. A. B. Zakaria, Universiti Teknologi Malaysia, Malaysia
Prof. Dr. Bassim H. Hameed, Universiti Sains Malaysia, Malaysia

Session Chairs

Dr. Nitesh Dutt, COER, Dehradun
Dr. D. B. Jani, Gujarat Technological University, Ahmedabad
Dr. Arun Balodi, Atria Institute of Technology, Bangalore
Dr. Susovon Chatterjee, Coochbehar Government Engineering College
Dr. Ranjib Biswas, MCKV Institute of Engineering, Liluah
Dr. Sanjay Singh, BIT Sindri
Dr. Arijit Dutta, Kalyani Government Engineering College
Dr. Sanchayan Mukherjee, Kalyani Government Engineering College
Dr. Jagabondhu Shit, Purulia Government Engineering College
Dr. Vijay kumar Dalla, NIT Jamshedpur
Dr. Mukesh Sharma, BIT Mesra

Preface

International Conference on Thermal Engineering and Management Advances (ICTEMA 2020) took place on December 19–20, 2020, in Jalpaiguri Government Engineering College, West Bengal, India. The conference was organized by the Department of Mechanical Engineering, Jalpaiguri Government Engineering College, West Bengal, India. The conference was really a large-scaled and international. The program committee and reviewers selected more than 90 research papers. The conference participants submitted papers reflecting recent advances in the field of thermal engineering, renewable energy, manufacturing, and production management. The conference was organized in ten sections, including.

Part 1: Thermal Engineering Research and Development of Machines and Mechanisms (Dynamics of Machines and Working Processes; Friction, Wear, and Lubrication in Machines; Design and Manufacturing Engineering of Industrial Facilities; Surface Transport and Technological Machines; Equipment and Technology of Welding Production).

Part 2: Renewable Energy Materials Engineering and Technologies for Production and Processing (New Functional Materials and Technologies; Innovation and Cost-Effective Use of Resources of Metallurgy Industry).

Part 3: Production Management Control and Automation Systems for Manufacturing in the Industrial Production Areas (Control Systems and Their Industrial Application; Industrial Mechatronics, Automation, and Robotics; Electric Power Systems and Renewable Energy Sources; Power Electronics, Electrical Machines, and Drives; Signal Processing and Real-Time Embedded Control; Modeling and Computer Technologies; Theory and Applications of Dynamical Measurements).

Part 4: Manufacturing Engineering Control and Automation Systems for Manufacturing in the Industrial Production Areas (Control Systems and Their Industrial Application; Industrial Mechatronics, Automation, and Robotics; Electric Power Systems and Renewable Energy Sources; Power Electronics, Electrical Machines, and Drives; Signal Processing and Real-Time Embedded Control; Modeling and Computer Technologies; Theory and Applications of Dynamical Measurements).

The delegates represented several international universities. The plenary lectures bridged the gap between the different fields of mechanical engineering making it possible for non-experts to gain insight into new areas. We hope to learn new ideas from each other, which we could be adopted to further improve our work in the important areas of thermal engineering, renewable energy, manufacturing, and production management. Many good experiences have been shared, and good lesson has been learned. The present volume gathers 42 peer-reviewed papers, and these papers were selected by the editors for publication in Springer book series "Lecture Notes in Mechanical Engineering."

The organizing committee would like to express its sincere gratitude to everybody who has contributed to the conference. Heartfelt thanks are due to authors, reviewers, participants, and all the team of organizers for their support and enthusiasm which granted success to the conference.

Jalpaiguri, India

Dr. Arijit Kundu
Organizing Secretary and Corresponding Editor

Contents

About the Editors

Dr. Sadhan Kumar Ghosh is currently Professor in Department of Mechanical Engineering, Jadavpur University, Kolkata, India. He obtained his B. Tech (Mechanical Engineering), M. Tech and Ph.D. from Jadavpur University, Kolkata, India. His major areas of research interests include waste management, circular economy, environmental protection, quality environment safety and energy management systems, green manufacturing, supply chain management, SME sustainability, and total quality management. He has published 03 patents, 42 peer-reviewed journal articles, 167 conference proceedings, 11 books, 02 book chapters and principal investigator of 32 projects. Dr. Ghosh delivered 56 Lectures in various countries. Recently, he was an editorial board member of the Waste Management and Resource Efficiency, Proceedings of 6th IconSWM 2016, Sustainable waste management: Policies & Case Studies: Proceedings of 7th IconSWM , Waste Water Recycling and Management: Proceedings of 7th IconSWM 2017.

Dr. Koushik Ghosh is currently Professor in the Department of Mechanical Engineering, Jadavpur University, Kolkata, India. He obtained his B.E (Mechanical Engineering) from North Bengal University, India, M. Tech and Ph.D. from Jadavpur University, Kolkata, India. His major areas of research interests include heat transfer, multiphase flow, boiling and condensation and transport in porous media. He has published 31 journal papers, 52 international and national conferences, 06 book chapters. Currently, he is a Co-Editor of Two-Phase Flow for Automotive and Power Generation Sectors (Book), Energy, Environment, and Sustainability, Springer, 2019; and member of reviewer panel of international journals (Applied Thermal Engineering (Elsevier), Applied Energy (Elsevier) Engineering Applications of Computational Fluid Mechanics (Taylors and Francis), Heat and Mass Transfer (Springer), Journal of Thermal Science and Engineering Applications (ASME).

Dr. Santanu Das is currently Professor and Head of the Department of Mechanical Engineering, Kalyani Government Engineering College, India. He obtained his B.Tech (Mechanical Engineering) and M. Tech from Jadavpur University, Kolkata, India, and Ph.D. from Indian Institute of Technology, Kharagpur, West Bengal, India. His major areas of research interests include manufacturing technology: machining, grinding, welding, weld cladding, coating, production and inventory management, etc. He has published 132 journal papers, 80 international conferences, 100 national conferences/seminars, 16 national magazines, 10 workshops, 09 short term training course papers, 07 books edited, 07 books reviewed. He has received the 'Shiksha Ratna Award' in 2018 as an 'Outstanding Teacher' from Education Department, Government of West Bengal. Currently, he is an Editor, Journal of The Association of Engineers, India; Member of Editorial Board, Indian Welding Journal; Chief Editor, Indian Welding Journal; and Editor-in-Chief, Bulletin of the Association of Machines and Mechanisms.

Dr. Pranab Kumar Dan is currently Associate Professor, Indian Institute of Technology, Kharagpur, in the Rajendra Mishra School of Engineering Entrepreneurship, Kharagpur, West Bengal, India. He obtained his B.Tech (Mechanical Engineering) and M.Tech from the Institute of Engineering Science and Technology (IIEST), Shibpur, West Bengal, India, and Ph.D. from Jadavpur University, Kolkata, India. His major research interests include production management and industrial Engineering, engineering design process and product innovation and development. He has published 34 journal papers, 12 international conferences, 08 books chapters, 08 national journals, 12 national conferences and 01 patent filed. Currently, he is a reviewer of International Journal of Advanced Manufacturing Technology (Springer) and Journal of Engineering Manufacture (Sage).

Dr. Arijit Kundu is currently Associate Professor in the Department of Mechanical Engineering, Jalpaiguri Government Engineering College, Jalpaiguri, West Bengal, India. He obtained his B.Tech (Mechanical Engineering) and M.Tech (Heat Power) from Jadavpur University, Kolkata, India, and Ph.D. (Refrigeration and Air Conditioning; Thermal Engineering) from the Indian Institute of Technology, Roorkee, India. His research interests include heat transfer, multiphase flow, boiling and condensation of eco-friendly refrigerants in re-entrant tubes, waste management, environmental protection, energy management systems, green manufacturing, sustainability, etc. He has published 05 international journals, 18 international and national conferences and running one project sanctioned from the Department of Higher Education, Science and Technology and Biotechnology, Government of West Bengal. Currently, he is an editorial board member of the Proceedings of International Conference on Multidisciplinary Innovation in Academic Research (ICMIAR) 2019.

Numerical Analysis of Heat Transfer Characteristics Under Single-Jet Air Impingement

Gourab Banerjee, Achintya Mukhopadhyay, Swarnendu Sen, Pranibesh Mandal, and Sourav Sarkar

1 Introduction

Cooling via jet impingement are commonly adopted solutions in various industrial applications. Impinging jets are known to yield a high heat transfer rate via forced convection. A large group of industrial processes such as the enhancement of cooling efficiency in laser and electronic components, cooling of turbine blades, quenching of steel, tempering of glass products, drying of tissue, paper, textiles—employs impingement cooling. As reported by Ligrani [1], any typical application of impinging jets during a cooling process results in a higher heat transfer coefficient (2–10 times) than that of a conventional cross-circulation dryer [2]. Due to a wide base of application and complicated fluid dynamics, extensive research has been carried out with jet impingement heat transfer.

To obtain an optimal technique for heat transfer characteristics, effects of several design variables such as nozzle geometry, nozzle size, nozzle-to-impinged surface spacing, nozzle-to-nozzle spacing, impingement surface motion, and operating variables such as jet velocity, cross flow, etc., need to be studied. A majority of the industrial applications like printing boards, production of foodstuffs, drying of continuous metal sheets, cooling of turbine aerofoils in the space industry, are concerned about the turbulent regime of the flow, downstream of the nozzle. It is a complicated job to accurately model the turbulent flow and predict the impingement heat transfer for both single jet and multi-jet impingement.

Several authors have simulated the case of impingement flows involving different operating parameters and turbulence models however there are limited works that compare the capability and applicability of the different turbulence models. Among the available turbulence models, standard k-ε model has been the predominant model of choice over the years for the researchers [3–8].

G. Banerjee (✉) · A. Mukhopadhyay · S. Sen · P. Mandal · S. Sarkar
Department of Mechanical Engineering, Jadavpur University, Kolkata, West Bengal 700032, India

S. K. Ghosh et al. (eds.), *Advances in Thermal Engineering, Manufacturing, and Production Management*, Lecture Notes in Mechanical Engineering,
https://doi.org/10.1007/978-981-16-2347-9_1

In the works of Shi et.al [9] results were presented for a semi-confined turbulence with rectangular slot jet. Their study incorporated the effects of near wall treatments, turbulence intensity, Reynolds number and different thermal boundary conditions on heat transfer rate, using two different viscous models. Inadequacy of both standard k-ε and Reynolds stress model (RSM) was seen in their simulation results. This inadequacy in the prediction of heat transfer rates is more prominent for the case of low nozzle-to-surface spacing.

Thakare and Joshi [10] also used standard k-ε and Reynolds stress model under low Reynolds number. They used several sets of Reynolds number for computation. Comparative analysis between the two models for the prediction of Nusselt number showed better consistency for the standard k-ε model over the RSM model. However, for the prediction of turbulence Prandtl number the Reynolds stress model was more favourable.

A comprehensive study by Polat et.al [11] highlighted the problems associated with the computation of impingement heat transfer under high Reynolds number. Works of [12] showed that the predictions of impingement heat transfer using the RSM model to be more consistent over the standard k-ε and RNG k-ε models. Simulation results indicated the importance of considering near wall treatment and variation of turbulent Prandtl number for evaluating heat transfer coefficients.

Morris et al. [13] studied the effects of near wall functions. They used the standard wall function for a Reynolds number ranging between 2000~13,000 and H/W = 2~4 and the non-equilibrium wall function for Reynolds number ranging above 13,000. They didn't include a direct comparison between the wall functions for predicting the heat transfer rate.

Zuckerman and Lior [14] studied the governing physics of fluid flow and calculated the heat transfer characteristics using empirical correlations. They studied the flow for a single jet impinging on a flat surface. Results indicated the fact that reduction in jet-to-surface distance and a simultaneous increase in Reynolds number, increased the local Nusselt number.

In this work, a numerical study of a single free jet impinging normally on a flat surface is presented. A powerful and versatile CFD tool package—ANSYS FLUENT 14.6 [15] is used to carry out the numerical calculations of the heat transfer characteristics. Results from these simulations are validated with existing experimental data. To justify the adaptability of the turbulence models, a comprehensive study is done by varying certain parameters such as near wall functions and turbulent Prandtl number.

2 Numerical Analysis

2.1 Mathematical Model

While simulating the fluid flow for the impinging jet, the following conservation equation in Orthogonal coordinates are solved:

Assuming the fluid (air) to be steady and incompressible,

1. Mass Balance:

$$(\partial U_i)/(\partial x_i) = 0 \quad (1)$$

2. Momentum Balance:

$$-\partial P/(\partial x_i) + \partial/(\partial x_j)$$
$$\{\mu[(\partial U_i)/(\partial x_j) + (\partial U_j)/(\partial x_i)] - \rho(u_i\, u_j)\} = 0 \quad (2)$$

3. Energy Balance:

$$\partial/(\partial x_j)\{\mu/\mathrm{Pr} - \partial T/(\partial x_j) - \rho(u_j\, T)\} = 0 \quad (3)$$

FLUENT 14.6 has five turbulence models which can be utilized to solve the governing Eqs. (1)–(3). The models available in the FLUENT database are: (i) standard k-ε model, (ii) RNG (Re-Normalisation Group) k-ε model, (iii) realizable k-ε model, (iv) RSM (Reynolds Stress Model), and (v) Large Eddy Simulation model.

There are fewer numerical works and documentation involving the realizable k-ε model and LES model since these are a new addition to the FLUENT database. In this paper, simulation is limited to the standard k-ε model and RSM model. The capability of each model, to predict the flow field and heat transfer rate, is evaluated. A deep review of the previous works has led to the omission of the RNG k-ε model in the scope of this work. The RNG k-ε model doesn't take into consideration the turbulent Prandtl number function, which is however seen to have a significant effect on the impingement heat transfer rate [10, 12].

There are 3 near wall functions in the FLUENT database: (i) standard wall functions, (ii) non-equilibrium wall functions and, (iii) enhanced wall functions. In this study, the first two wall functions are chosen for computation. The standard wall function in FLUENT is the default near wall treatment. It based on the work of Launder and Spalding [16]. They considered the turbulent kinetic energy and its dissipation to be equal in the wall-adjacent cells. This is the most widely used wall function involving industrial flows.

Non-equilibrium wall function is a two-layer based function given by Kim and Choudhury [17]. It employs the concept that the wall-adjacent cells consist of a linear viscous sub-layer ($y+ \leq 11.0$) and a fully turbulent logarithmic layer ($11 < y+ < 400$) [18]. The logarithmic region is very sensitive to large pressure gradients; thus, this makes the non-equilibrium wall function to be more suitable for use in complex flows experiencing high-pressure gradients.

2.2 Geometry

A schematic of the physical geometry is shown in Figs. 1 and 2. The computational domain is filled with air. Due to the symmetric nature of the geometry, half of the domain is used to save computation time. Proper choice of grid distribution (meshing) is of great importance while solving a numerical problem. In this work, a non-uniform orthogonal meshing is used to simulate the flow field more accurately. The grid layout and physical domain are chosen from the works of Shi et al. [9].

Meshing within the impingement region is kept much finer than that in the rest of the domain (Fig. 3). The grid size is gradually increased along the x-direction. Similarly, the meshing is gradually decreased along the y-direction i.e., from the impinged surface to the semi-confined surface. A grid density of 20×80 in the impingement region is selected from the grid-independent results. Such mesh adaptation is capable enough to capture the near wall interaction of the flow downstream of the nozzle and predict the heat transfer characteristics.

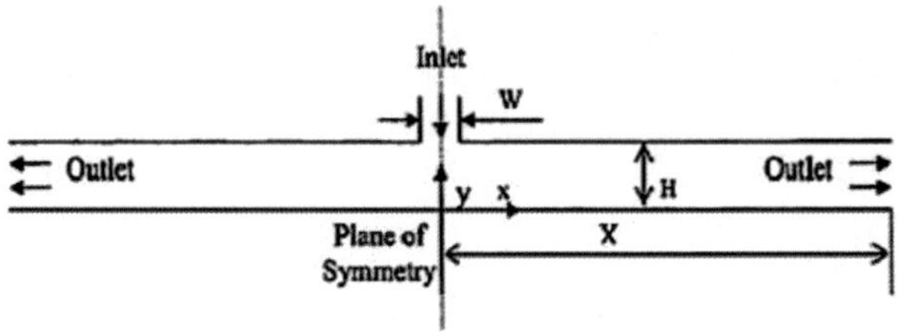

Fig. 1 Geometry of physical domain

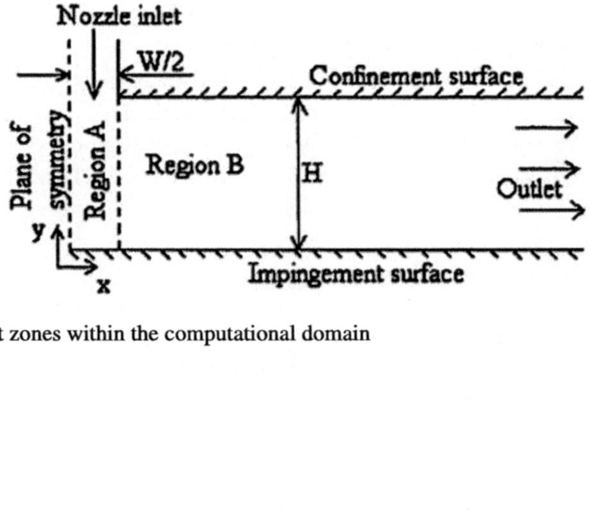

Fig. 2 Different zones within the computational domain

Fig. 3 Grid structure of the domain

2.3 Solver

In ANSYS FLUENT a finite volume code with a pressure-based solver is used to simulate the case. The following boundary conditions and solution methods are considered during computation:

1. Inlet: velocity inlet (depending on the Re)
2. Outlet: outflow condition
3. Impinged surface: isothermal wall
4. Semi-confined surface: adiabatic wall
5. Plane of symmetry: symmetry condition
6. Turbulent intensity: 2%
7. Length scale: 0.07D (hydraulic diameter)
8. Solution scheme: SIMPLEC algorithm (for pressure-velocity coupling).

The second-order spatial discretization scheme is used for the pressure; first-order upwind discretization is used for momentum, turbulence kinetic energy, turbulent dissipation rate, and energy. The under-relaxation factors and convergence criteria are taken as specified by default in the FLUENT database unless otherwise required for special cases. The convergence control is set at 1000 iterations and can be changed if convergence is not attained.

3 Results and Discussion

3.1 Effect of Turbulence Models

Two turbulence viscous models (standard k-ε and RSM) are used in this computation. Obtained results are compared with the experimental data from the work of van-Heiningen [19]. The work of van-Heiningen is chosen for comparison out of the numerous available data in the literature, considering the similarity of their experiment set-up, operating procedure and boundary conditions with the present simulation. The operating parameters are noted in Table 1. The local heat transfer coefficient

Table 1 Operating parameters for the test cases

Case	W (mm)	H/W	Re	T_i (K)	T_s (K)
1	14.1	2.6	10,400	310	348
2	6.2	6	11,000	303	338

(h_x) obtained from the CFD simulation of the impinged surface is computed in the form of local Nusselt number (Nu_x) as:

$$Nu_x = (h_x W)/k \tag{4}$$

$$h_x = q/(T_s - T_i) \tag{5}$$

Local Nusselt number distribution over the entire domain is compared with the experimental results. Figure 4 shows the comparison between the experimental data and the simulation results using the standard k-ε model and RSM model for case 1. The nature of Nusselt number distribution over the domain, as predicted by the viscous turbulence models are similar to the experimental data however both the models slightly overpredict the Nusselt number. Two peaks are noticed in the Nusselt number distribution curve as calculated by van-Heiningen. The first peak in Nusselt number is seen around the stagnation region of the impinging jet (X/W ~ 0) while the second peak is seen at X/W ~ 6. Although the turbulence models are seen to be capable in predicting the peaks in Nusselt number distribution, the standard k-ε model fails to predict the correct numerical value of either of the peaks while the RSM model does not predict the position of the secondary peak correctly.

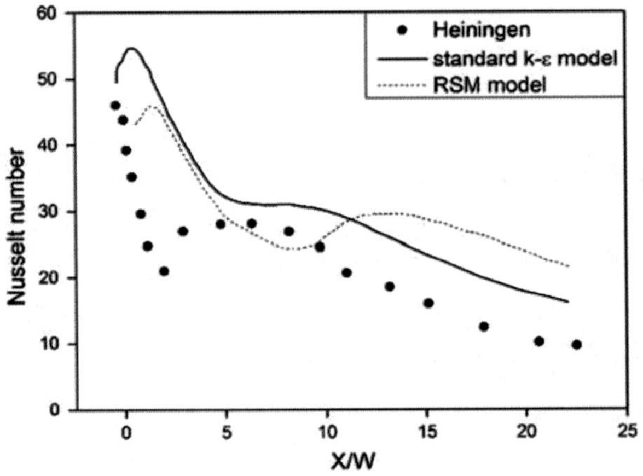

Fig. 4 Comparison of Nusselt number between experimental and simulation data for case 1 using two turbulence model

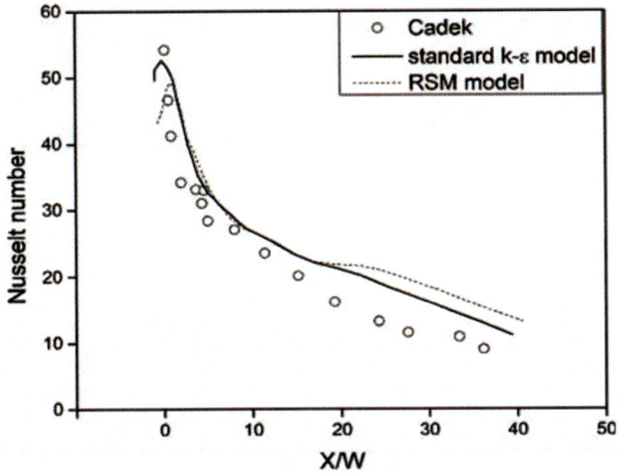

Fig. 5 Comparison of Nusselt number between experimental and simulation data for case 2 using two turbulence model

Over prediction of the Nusselt number via the first peak can be attributed to the high level of turbulence of the impinging jet in the stagnation zone. Formation of a secondary peak can be appreciated by the fact that in a semi-confined flow, vortices are formed within the shear layer near the confinement surface. These vortical flows are broken down into miniaturized turbulence (bubbles) which constricts the channel area, enhances the vertical fluctuations and hence increases the local heat transfer coefficient. This effect is prominent only for a smaller nozzle-to-surface spacing (case 1).

For a larger nozzle-to-surface spacing (case 2), no such secondary peak is formed in Fig. 5. The standard k-ε model shows a better prediction of local Nusselt number when compared with the experimental data of Cadek [20] for a larger nozzle-to-surface spacing.

3.2 *Effect of Turbulent Prandtl Number*

In this section, the effect of turbulent Prandtl number (Pr_t) on the Nusselt number is computed for case 1 by the standard k-ε model and RSM model respectively. Three sets of turbulent Prandtl number is chosen: *0.85* (default); *1.1*; *1.5* and the calculated data is compared with van-Heiningen [19]. Figures 6 and 7 show the computed results for the standard k-ε and RSM model respectively. The standard k-ε model overpredicts the Nusselt number in the stagnation region. The predicted data fit well with an increase in turbulent Prandtl number and in the downstream region.

Computation by the RSM model predicted the Nusselt number peaks more accurately than the standard k-ε model, however, the position of the secondary peak is

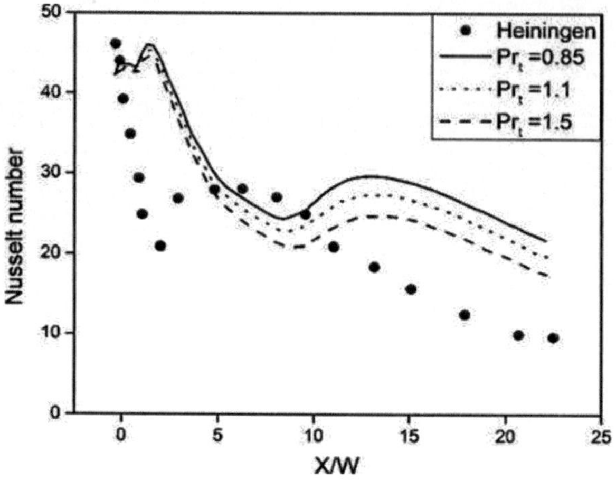

Fig. 6 Comparison of Nusselt number between experimental and simulation data under increasing turbulent Prandtl number using standard k-ε model

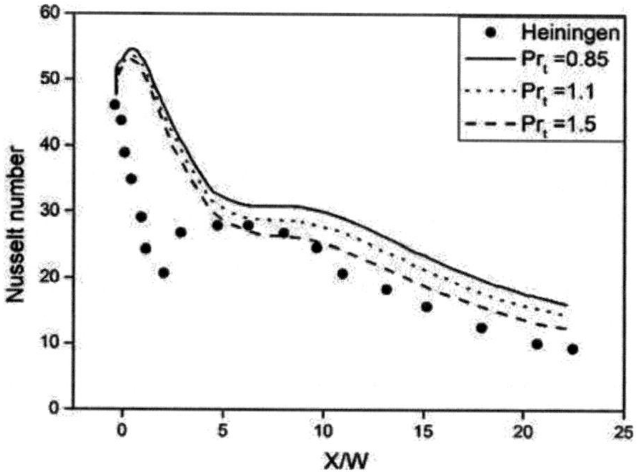

Fig. 7 Comparison of Nusselt number between experimental and simulation data under increasing turbulent Prandtl number using RSM model

predicted inaccurately. Contribution of the turbulent Prandtl number for the RSM model and standard k-ε model is seen to be more significant in the region away from the impingement zone. The results for standard k-ε model with high turbulent Prandtl number shows better consistency with the experimental data away from the impingement zone.

Table 2 Operating parameters for effect of near wall functions

Case	W (mm)	H/W	Re	T_i (K)	T_s (K)
3	14.1	2.6	71,300	310	348

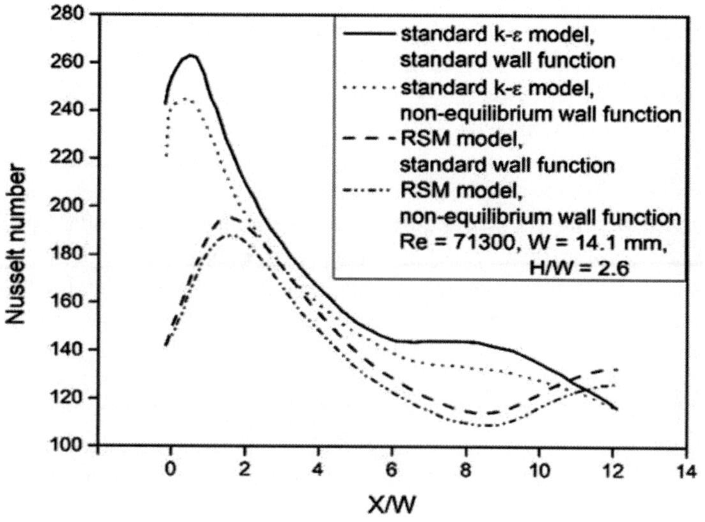

Fig. 8 Effect of near wall functions on Nusselt number distribution

3.3 *Effect of Near Wall Functions*

The effect of various wall functions available in FLUENT is used to calculate the impingement heat transfer. The operating parameters and boundary conditions for this simulation are given in Table 2.

Figure 8 shows the Nusselt number distribution under the influence of four near wall functions. The prediction of Nusselt number by the non-equilibrium wall functions is slightly lower than the standard wall functions for both the turbulence models. The numerical value of the Nusselt number is much higher for the k-ε models than the RSM models within the stagnation zone.

A general trend is seen from the previous results that the prediction of Nusselt number distribution by the turbulence models are generally higher in comparison to the experimental results. So, the use of non-equilibrium wall functions will give better results in comparison to the experimental data. However, the convergence rate as seen from the simulation is much stable and faster for the standard wall functions than the non-equilibrium wall functions. Thus, for high Reynolds number, it is best suited to choose the non-equilibrium wall functions but for a low Reynolds number, the standard wall functions prove marginally better.

4 Conclusion

Results from the CFD simulation, reveals the fact that the turbulence models slightly over predict the local Nusselt number, although the nature of the distribution being accurate in most of the cases. The simulation results for higher nozzle-to-surface spacing complements the experimental data but the case for lower nozzle-to-surface spacing draws attention for further improvement.

Unlike the RSM model, the standard k-ε model fails to predict the secondary peak effectively for a smaller value of H/W. However, the position of the peak in the RSM model deviates from the one as given in experimental results. No secondary peak is present for the case with higher H/W. It is more critical to choose the correct model for low nozzle-to-surface spacing due to the generation of a secondary peak.

The variance of turbulent Prandtl number is seen to have a prominent effect on the Nusselt number distribution. There are noticeable effects of near wall functions on the Nusselt number. Prediction by the standard wall function is slightly higher than the non-equilibrium wall function for both the turbulence models. The non-equilibrium wall function with the RSM model yields a superior result due to the consideration of the effect of high-pressure gradients and flow separation for a turbulent flow.

The simulation results are appreciably good for a semi-confined impinging jet. The model is capable enough to simulate the turbulent characteristics of the jet flow and calculate the heat transfer rate. This CFD analysis encourages further computation in this domain by taking into consideration other operating parameters affecting the impingement heat transfer.

References

1. Chandramohan P, Murugesan SN, Arivazhagan S (2017) Heat transfer analysis of flat plate subjected to multi-jet air impingement using principal component analysis and computational technique. J Appl Fluid Mech 10:293–306
2. Seyedein SH, Hasan M, Mujumdar AS (1995) Turbulent flow and heat transfer from confined multiple impinging slot jets. Numerical Heat Transf Part A
3. Agarwal RK, Bower WW (1982) Navier-stokes computations of turbulent compressible two-dimensional impinging jet flow fields. AIAA J 20(5):577
4. Amano RS, Brandt H (1984) Numerical study of turbulent axisymmetric jets impinging on a flat plate and flowing into an axisymmetric cavity. J Fluids Eng 106:410
5. Seyedein SH, Hasan M, Mujumdar AS (1994) Modelling of a single confined turbulent slot jet impingement using various k-ε model. Turbulent Models Mach Model 18:526
6. Hosseinalipour SM, Mujumdar AS (1995) Comparative evaluation of different turbulence models for confined impinging and opposing jet flows. Numer Heat Transf Part A 28:647
7. Behnia M, Parneix S, Durbin PA (1998) Prediction of heat transfer in an axisymmetric turbulent jet impinging on a flat plate. Int J Heat Mass Transf 41(12):1845
8. Cziesla Z, Biswas G, Chattopadhyay H, Mitra NK (2001) Large-eddy simulation of flow and heat transfer in an impinging slot Jet. Int J Heat Fluid Flow
9. Shi YL, Ray MB, Mujumdar AS (2002) Computational study of impingement heat transfer under a turbulence slot jet. Ind Eng Chem Res 41:4643–4651

10. Thakare SS, Joshi JB (2000) CFD modeling of heat transfer in turbulence pipe flows. AIChE J 46:1798–1812
11. Polat S, Huang B, Mujumdar AS, Douglas WJM (1989) Numerical flow and heat transfer under impinging jets. Annu Rev Numer Fluid Mech Heat Transf 2:157
12. Morris GK, Garimella SV, Amano RS (1996) Prediction of jet impingement heat transfer using a hybrid wall treatment with different turbulent Prandtl number functions. Trans ASME 118:562
13. Morris GK, Garimella SV, Fitzgerald JA (1999) Flow-field prediction in submerged and confined jet impingement using the Reynold stress model. J Electron Packaging 121:255
14. Zuckerman N, Lior N (2006) Jet impingement heat transfer: physics, correlations, and numerical modeling. Adv Heat Transf 39:565–631
15. (2012) FLUENT 14.5 User's guide. Ansys Inc
16. Launder BE, Spalding DB (1974) The numerical computation of turbulent flows. Comput Methods Appl Mech Eng 3:269–289
17. Kim SE, Choudhury D (1995) A near wall treatment using wall functions sensitized to pressure gradient. ASME 217
18. Chieng CC, Launder BE (1980) On the calculation of turbulent heat transport downstream from an abrupt pipe expansion. Numerical Heat Transf 3:189
19. van Heiningen ARP (1982) Heat transfer under an impinging slot jet. Ph.D. thesis, McGill University
20. Cadek FFA (1968) Fundamental investigation of jet impingement heat transfer. Ph.D. thesis, University of Cincinnati

Effect of Liquid Bridge Volume on Cohesive Sediment Motion

Debasish Biswas, Arijit Dutta, Sanchayan Mukherjee, and Asis Mazumdar

Nomenclature

R_s	Sediment grain radius (mm);
D_s	Inter-granular distance (mm);
V	Inter-granular liquid bridge volume (nl);
α_s	Angular acceleration of grain (rad/s^2);
P_{ws}	Pore water pressure (N/m^2);
m_s	Mass of the spherical particle (kg);
m_w	Entrapped water mass (kg);
φ	Contact angle (rad);
β	Water content index angle (rad);

1 Introduction

Due to dynamic nature of the river system, its shape continuously changes with time as the water level changes from time to time. As bank erosion is associated with the land loss and it weakens the flood defence, it leads the researchers to study in detail over a long period. Complexity nature of the river bank system takes huge time to study through experimentally. Sometimes it takes more than a year which involves huge cost investment as well. And these experimental analyses provide a fair idea for a specific river system to a great extent. However, these kinds of experimental or macroscopic analysis have certain limitations as well. These macroscopic analyses are very much specific for that particular river system. And the results obtained by these types of analyses are difficult to use for another river system, if not feasible.

D. Biswas (✉) · A. Dutta · S. Mukherjee · A. Mazumdar
Department of Mechanical Engineering, Kalyani Government Engineering College, Kalyani, India

S. K. Ghosh et al. (eds.), *Advances in Thermal Engineering, Manufacturing, and Production Management*, Lecture Notes in Mechanical Engineering,
https://doi.org/10.1007/978-981-16-2347-9_2

These facts lead the researchers to study the bank erosion in microscopic level for obtaining a generalized concept for every river system. However, very little progress in this regard has been made till date.

Macroscopic approach for bank failure has been studied by many previous researchers [1–3]. Results obtained in such a manner are not applicable for the other river system. The present analysis deals with "Truncated Pyramid Model (TPM)," for the grains frame-up in microscopic level which is devised by Mukherjee and Mazumdar [4]. Mukherjee [5] has studied the effect of grains size variation on separation speed of the grains in microscopic level. Goncharov [6] developed the concept of sediment threshold velocity which is the least average velocity for which individual grains dislodging from the bank surface continually. Pore water pressure and hydrodynamic static pressure have been correlated with the bank failure by Osman and Throne [7] and Darby and Throne [8]. Tokaldany and Darby [9] have analysed the hydrostatic pressure due to water level in the river system along with both the positive and negative pore water pressure. Duan [10] has devised a new analytical model to measure the bank erosion rate.

Soulie et al. [11] have suggested a series of equations for the account of the capillary cohesion between two grains as a function of the water entrapment between the grains as follows:

$$F_{s_c} = \pi\sigma \times R_s \times \left[\exp\left\{b_s + a_s \times \left(\frac{D_s}{R_s}\right)\right\} + c_s\right] \quad (1)$$

Here, the force of cohesion acting between two sediment grains of radius R_s is denoted by F_{s_c} with of D_s indicates inter-granular distance, and σ is the surface tension. The variables a_s, b_s and c_s are the functions of liquid bridge volume, angle of contact and radius of the grains as a following manner:

$$a_s = -1.1\left(V/R_s^3\right)^{-0.53} \quad (2a)$$

$$b_s = 0.48 + \left(-0.148\ln\left(V/R_s^3\right) - 0.96\right)\varphi^2 - 0.0082\ln\left(V/R_s^3\right) \quad (2b)$$

$$c_s = 0.078 + 0.0018\ln\left(V/R_s^3\right) \quad (2c)$$

where, V denotes inter-particle liquid bridge volume and (φ) denotes the angle of contact.

Likos and Lu [12] have been modelled the pore pressure force between two grains and developed a number of equations to describe the pore pressure force in microscopic level as follows:

$$F_{s_P} = P_{ws} \times \pi \times r_{s_2}^2 + 2 \times \sigma \times \pi \times r_{s_2} \quad (3)$$

Here, P_{ws} is the pore water pressure, and

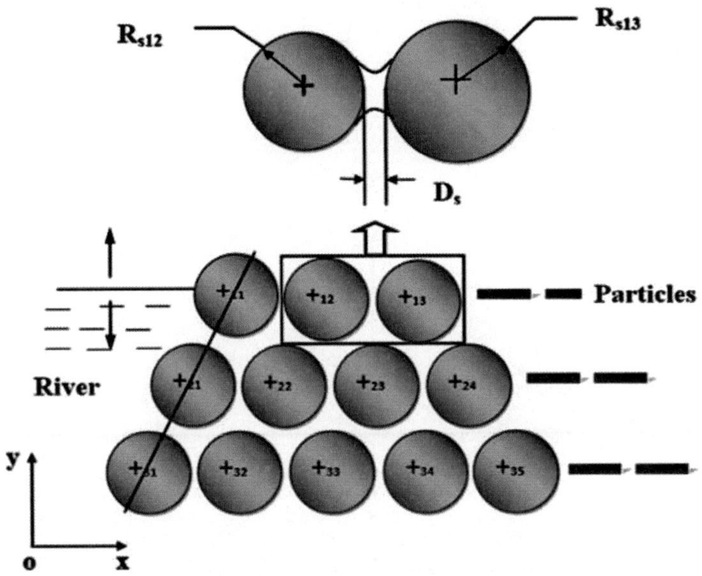

Fig. 1 Grains distribution in TPM with enlargement view of two consecutive grains

$$r_{s_2} = R_s \times [\tan \beta - \sec \beta + 1] \quad (4)$$

where, β is the water content index angle.

The effect of the above-mentioned two forces along with the submerged weight of the grain and hydrostatic force on TPM has been analysed here by varying grain size of 300, 400 and 500 μm and liquid bridge volume of 10, 20 and 30 nl for fully submerged grains for both case of water level rising and falling to account the dynamic effect of bank sediment grains. Conservation of moment of momentum principle has been implemented to determine escape velocity of the grains and the variation of the same has been plotted against inter-particle distance for three different sizes of the grain. Grains distribution in TMP has been shown in Fig. 1.

2 Model Frameworks and Freebody Diagram

Figure 2 shows the free-body diagram of the grain *11* with the different forces acting on the particle. The grain 11 has been chosen for the analysis as this being top most and left most is most vulnerable with respect to dislodgement from the bank surface under the action of different forces in microscopic level. The point of contact between the particles 11 and 21 is the instantaneous centre of rotation as it tends to turn up about that point. So this contact point (point A) has been chosen for the application of conservation of moment of momentum.

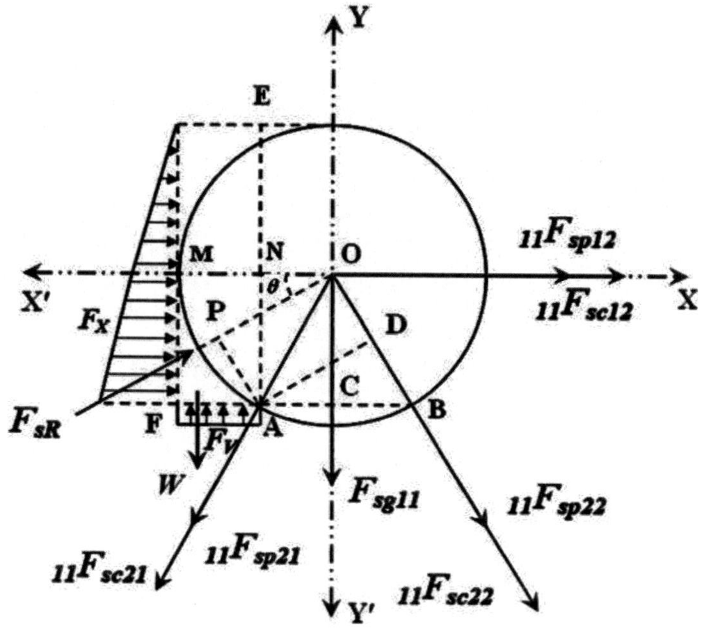

Fig. 2 Diagram of forces distribution on grain *11*

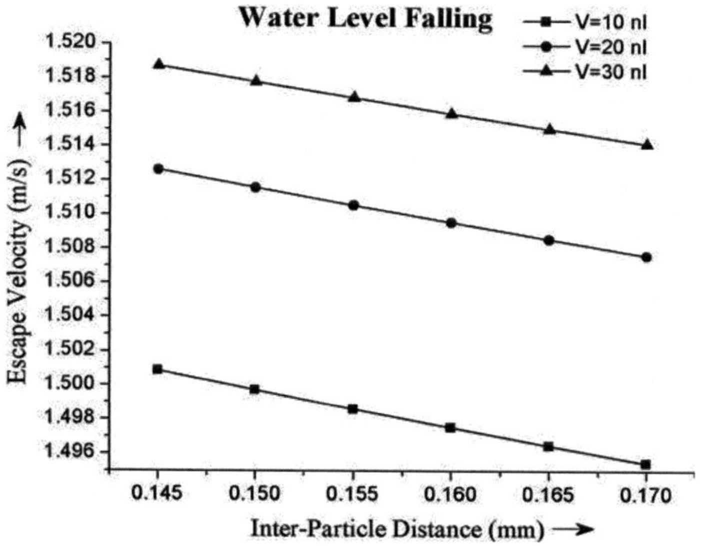

Fig. 3 Variation of $\vartheta_{\text{escape}}$ with D_s with water level dropping down for the radius of 300 μm

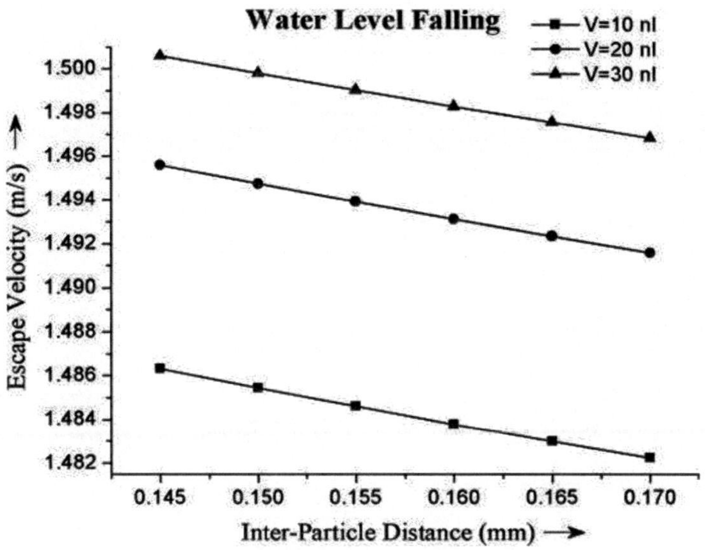

Fig. 4 Variation of ϑ_{escape} with D_s with water level dropping down for the radius of 400 μm

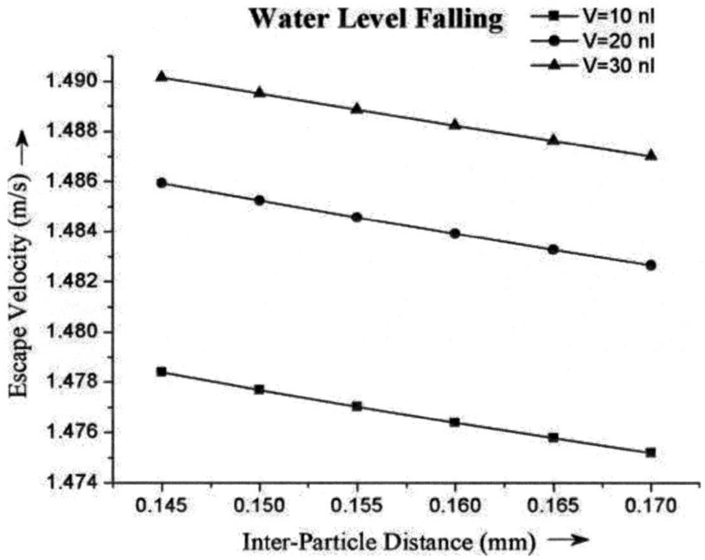

Fig. 5 Variation of ϑ_{escape} with D_s with water level dropping down for the radius of 500 μm

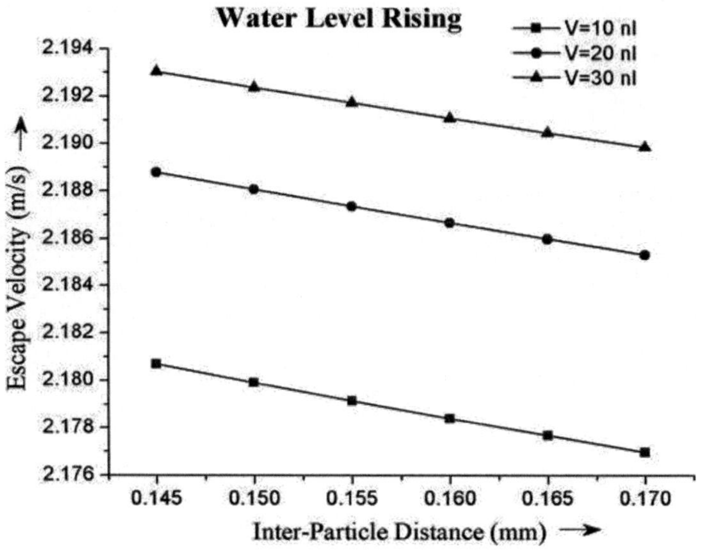

Fig. 6 Variation of ϑ_{escape} with D_s with water level increasing up for the radius of 300 μm

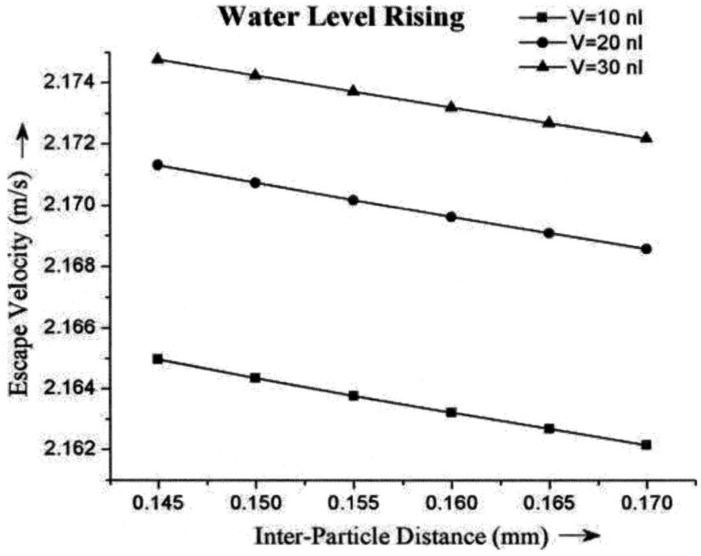

Fig. 7 Variation of ϑ_{escape} with D_s with water level increasing up for the radius of 400 μm

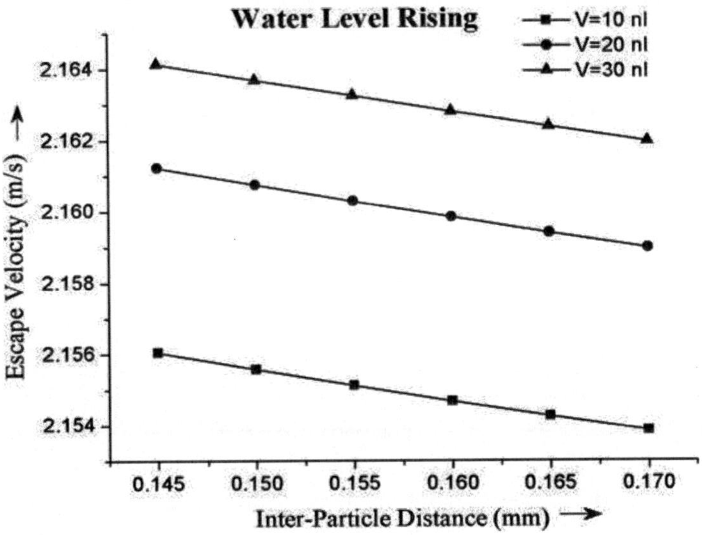

Fig. 8 Variation of ϑ_{escape} with D_s with water level increasing up for the radius of 500 μm

In this section we discuss how to format the title, authors and affiliations. Please follow these instructions as carefully as possible so all articles within a conference have the same style as the title page. This paragraph follows a section title so it should not be indented.

3 Escape Velocity Calculations

The 2-D planar analysis has been considered. The x-directional hydrostatic force is given by

$$F_X = F_H = P_{\text{avg}} \times A_X = \rho_w \times g \times h_c \times A_X \tag{5a}$$

$$= \rho_w \times g \times \frac{2}{3} \times \overline{\text{AE}} \times \overline{\text{AE}} \tag{5b}$$

The y-directional hydrostatic force (upward)

$$F_V = P \times A_Y = \rho \times g \times h \times A_Y \tag{6a}$$

$$F_V = \rho_w \times g \times \overline{\text{AE}} \times \overline{\text{AF}} \tag{6b}$$

The fluid block weight per unit length (downward)

$$W = m_w \times g = \rho \times g \times V \tag{7a}$$

$$= \rho_w \times g \times \left[\overline{\text{FM}} \times \overline{\text{AF}} - \frac{1}{2} \times \overline{\text{MN}} \times \overline{\text{AN}}\right] \tag{7b}$$

(AMN is assumed as triangle).

Net y-directional hydrostatic force (upward)

$$F_Y = F_V - W \tag{8}$$

Resultant hydrostatic force

$$F_{s_R} = \left[(F_X)^2 + (F_Y)^2\right]^{0.5} \tag{9}$$

And the direction of the resultant hydrostatic force with horizontal direction

$$\theta = \tan^{-1}\left(\frac{F_Y}{F_X}\right) \tag{10}$$

Now, conservation of moment of momentum principle has been applied about the instantaneous centre of rotation A. So taking moments about that point

$$F_{s_C} \times \left(\overline{\text{AN}} + \overline{\text{AD}}\right) + F_{s_R} \times \overline{\text{AP}} + F_{s_P} \times \left(\overline{\text{AN}} + \overline{\text{AD}}\right) + F_{s_g} \times \overline{\text{AC}} = I_A \times \alpha_s \tag{11a}$$

Here, I_A is the moment of inertia of the grains and α_s is the angular acceleration of the grains.

$$\Rightarrow F_{s_C} \times \left(\frac{\sqrt{3}R_s}{2} + \frac{\sqrt{3}R_s}{2}\right) + F_{s_R} \times R_s \sin(60 - \theta)$$
$$+ F_{s_P} \times \left(\frac{\sqrt{3}R_s}{2} + \frac{\sqrt{3}R_s}{2}\right) + F_{s_g} \times \frac{R_s}{2} = \frac{7}{5}\left(\frac{4}{3}\pi R_s^3 \rho_s\right) \times R_s^2 \times \alpha_s \tag{11b}$$

Solving Eq. (11b) angular acceleration comes out to be

$$\alpha_s = \frac{\left(F_{F_{s_C}} \times \sqrt{3}\right) + F_{s_R} \sin(60 - \theta) + \left(F_{s_P} \times \sqrt{3}\right) + \frac{F_{s_g}}{2}}{\frac{7}{5}\left(\frac{4}{3}\pi R_s^3 \rho_s\right) \times R_s} \tag{12}$$

Hence, the impending acceleration (in m/s^2) would be

$$f_s = \alpha_s \times R_s \tag{13}$$

So, the final expression of the grain escape velocity (in m/s) is

$$\vartheta_{\text{escape}} = (2 \times R_s \times f_s)^{0.5} \tag{14}$$

3.1 Input Parameters of the Model

Three values of radii are 300, 400 and 500 μm.

4 Results and Discussion

Influence of the inter-granular distance on the escape velocity is evident from all the figures for different sizes of the grains and for different volumes of water entrapped between the grains. All the figures clarify that with the increase in inter-granular distance escape velocity goes down linearly. It has been also seen that with the increase in the sizes of grains escape velocity falls down. For any size of the particle, as the volume entrapment of liquid increases from 10 to 30 nl, escape velocity increases continuously. Also the plot indicates that the escape velocity is the larger value in the case when the water level rises in comparison with that when the water level drops.

5 Conclusions

The conclusions which can be drawn from the present analysis are as follows:

- Inter-granular distance has an adverse effect on escape velocity owing to the rise in the gap between two grains weakening the bond between them.
- Due to less amount of binding force, larger size of grains have the lower escape velocity and they are more vulnerable with respect to separation from the bank surface.
- The entrapped volume raises the escape velocity; this proves the fact that the water is very good binding agent that binds the grains together. This is why a sand castle cannot be made by dry sand.
- It is seen that when water level falls, the escape velocity becomes less. The physics behind this fact is that when water level drops down, the momentum acts in opposite direction which helps the grains to detach from the bank surface, lowering the escape velocity. Due to this river systems appear to be more vulnerable in terms of bank erosion on macro-scale when water level starts falling.

Table 1 Input parameters

ρ_w	=	1000 kg/m^3
φ	=	0°
σ	=	0.073 N/m
V	=	10, 20 and 30 nl
ρ_s	=	2650 kg/m^3
β	=	45°
P_{ws}	=	10 kPa

References

1. El Kadi Abderrezzak K, Moran AD, Mosselman E, Bouchard JP, Habersack H, Aelbrecht D (2014) A physical, movable-bed model for non-uniform sediment transport, fluvial erosion and bank failure in rivers. J Hydro-environ Res 8(2):95–114
2. Odgaard AJ, Masconi CE (1987) Streambank protection by submerged vanes. J Hydraul Eng ASCE 113(4):520–536
3. Carroll RWH, Warwick JJ, James AI, Miller JR (2004) Modeling erosion and overbank deposition during extreme floods conditions on carson river Nevada. J Hydrol 297(1–4):1–21
4. Mukherjee S, Mazumdar A (2010) Study of effect of the variation of inter-particle distance on the erodibility of a riverbank under cohesion with a new model. J Hydro-environ Res 4(3):235–242
5. Mukherjee S (2011) Application of Truncated Pyramid model in determination of escape velocity of particles of different diameters in varying conditions. Int J Soft Comput Eng (IJSCE) 1(5):75–79
6. Goncharov VN (1964) Dynamics of channel flow. Israel Programme for Scientific Translation, Moscow, Russia
7. Osman AM, Thorne CR (1988) Riverbank stability analysis I: theory. J Hydraul Eng 114(2):134–150
8. Darby SE, Thorne CR (1996) Stability analysis for steep, eroding, cohesive riverbanks. J Hydraul Eng 122:443–454
9. Tokaldany EA, Darby SE (2006) A model for stability analysis of a multi-layered river bank (SMALR). J Agric Sci Technol 8:61–76
10. Duan JG (2005) Analytical approach to calculate rate of bank erosion. J Hydraul Eng 131(11):980–989
11. Soulie F, Youssoufi EI, Cherblanc FMS, Saix C (2006) Capillary cohesion and mechanical strength of polydisperse granular materials. Eur Phys J E 21:349–357
12. Likos JW, Lu N (2002) Hysteresis of capillary Cohesiom in unsaturated soils. In: 15th ASCE engineering mechanics conference, Columbia University, New York, NY, 2–5 June 2002

Boiling Behavior of Iso-Butane on a Horizontal Plain Tube

Ashok K. Dewangan and Sanjeev K. Sajjan

Nomenclature

d	Diameter (mm);
g	Acceleration due to gravity (ms^{-2});
h	Heat transfer coefficient (Wm^{-2} K^{-1})
k	Thermal conductivity (Wm^{-1} K^{-1})
M	Molecular mass of refrigerant (g mol^{-1})
P	Operating pressure (MPa)
P_c	Maximum (critical) pressure (MPa)
p_r	Pressure ratio, P/P_c (–)
q	Applied heat flux (Wm^{-2} K^{-1})
R_p	Roughness (mm)
T	Operating temperature (K)
T_r	Temperature ratio, T/T_c (–)
ΔT	Wall superheat (K)

Greek Letters

ϕ	Wetting angle (deg.)
ν	Momentum diffusivity (m^2 s^{-1})
ρ	Density of refrigerant (kg m^{-3})

A. K. Dewangan (✉)
Mechanical Engineering Department, National Institute of Technology Delhi, Delhi 110040, India
e-mail: ashokd@nitdelhi.ac.in

S. K. Sajjan
Mechanical Engineering Department, VCE Warangal, Warangal, Telangana 506005, India

S. K. Ghosh et al. (eds.), *Advances in Thermal Engineering, Manufacturing, and Production Management*, Lecture Notes in Mechanical Engineering,
https://doi.org/10.1007/978-981-16-2347-9_3

σ Surface tension of refrigerant (Nm^{-1})

Subscripts

c Maximum or critical state;
l Liquid;
sat Saturation;
v Vapor

1 Introduction

In refrigeration and air-conditioning industry, the Chlorofluorocarbons (CFCs) refrigerants were widely used. However, the production of this type of refrigerants have been prohibited by the international regulation (Montreal protocol [1]) due to its ozone depletion potential. Thus, CFCs have been phased out. In addition, the other group of refrigerants such as hydrochlorofluorocarbons (HCFCs) are also going to be phased out before 2030. Therefore, refrigeration and air-conditioning industry have stimulated to evolution of new alternative refrigerants due to environmental issues. Therefore, refrigeration industry has been focusing and developing a new alternative refrigerant that has better qualities than other refrigerants. Refrigerant R-600a (Iso-butane) has been proposed as a suitable replacement for CFC-12. Boiling of refrigerants (on evaporator) is an essential application in the refrigeration plants. The experimental evaluation is still vital to observe the behavior of bubble during boiling of refrigerant. The horizontal plain tube is the simplest geometrical configuration for scientific research. The plenty of boiling heat transfer experiments have been tested with horizontal tubes due to experimental simplicity. Nucleate is an evolving boiling heat transfer method which removes enormous quantities of heat from the heating surfaces with maintaining the lower temperature differences. Therefore, the refrigeration and air-conditioning industry focuses to research on improvement of boiling heat transfer. Many researches work with inconsistent pool boiling data for plain tubes are available [2–6]. The comparisons between experimental and predicted (existing correlations) results were also discussed for the nucleate pool boiling of refrigerants on plain tube. However, some empirical correlations were not suitable to estimate the heat transfer coefficients of alternative (Iso-butane) refrigerant [7–11]. The main aim of this work is to visualize bubble behavior on the plain heating surface and compare the present data with predicted data for understanding the boiling behavior of Iso-butane.

2 Experimental Apparatus

Figure 1 shows the pool boiling experimental arrangement. It is made of boiling vessel, test section and cooling circuit. The boiling vessel consists of stainless steel with inner diameter of 150 and 400 mm length. Two inspection windows were also involved in the opposite sides of the boiling vessel to observe the boiling

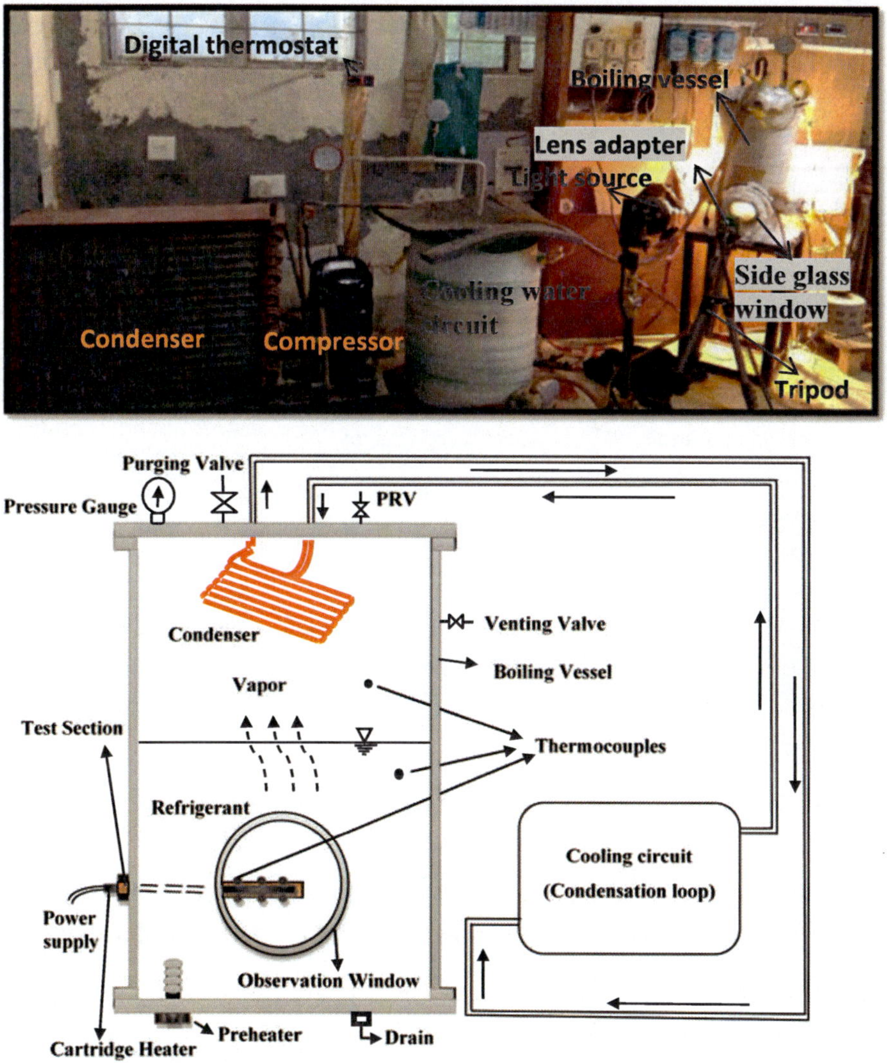

Fig. 1 Photographic view and schematic diagram of pool boiling setup

phenomenon. An internal condenser, a cartridge heater, an auxiliary heater, a pressure gauge, thermocouples, and a test section arrangement were also mounted in the vessel. The cooper tube was used as test tube of outer diameter of 25.4 and 200 mm length. The cartridge heater having an outside diameter of 16.6 mm and a length of 210 mm. In this experiment, the cartridge heater maintains uniformity in the heat flux throughout the heating surface. The saturation temperature of refrigerant and system pressure was controlled by an auxiliary heater. The temperatures of the heating surface and liquid pool were measured using K type thermocouples. The complete experimental procedure for R-600a was discussed by Dewangan et al. [12]. The behavior of bubbles was analyzed using a visualization study during pool boiling of R-600a over plain heating surface. The data were taken for heat flux varies between 10 and 50 kW m^{-2} at 12 °C saturation temperature. A photographic view of experimental facility with camera arrangement is also shown in Fig. 1. The transparent sight-glasses were mounted on the opposite side of the vessel to observe bubble behaviors during nucleate boiling.

The high-speed camera was used to capture the bubble formation on the heating surface. This camera is a device capable of capturing photographs with a frame rate of 1000 frames per second. The images were recorded at full resolution (1024 × 640 pixels) with maintaining a shutter speed of 1/9600 s. Videos were recorded from the tube surface positions (top and sides). From recorded videos the frames (single pictures) were extracted for the purpose of analysis. The bubble behavior was observed using the analysis of captured images.

3 Results and Discussions

The boiling heat transfer coefficient of refrigerant for each power input is calculated using convection heat transfer equation as:

$$h = \frac{q}{T_w - T_{\text{sat}}} \tag{1}$$

The average wall (surface) temperature of tube was calculated by considering the temperatures at top, bottom and two sides positions of the tube. The experiments were conducted with refrigerant R-600a at saturation temperature of 12 °C. Using REFPROP program [13], the properties of refrigerant were calculated and shown in Table 1. In order to verify experimental setup and temperature measure-

Table 1 Properties of refrigerant at 12 °C [2010]

Refri	T_{sat}	P	k_l	k_v	ν_l	ν_v	σ
	°C	MPa	Wm^{-1} K^{-1}	Wm^{-1} K^{-1}	cm^2 s^{-1}	cm^2 s^{-1}	Nm^{-1}
R600a	12	0.24	0.094	0.016	0.003	0.001	0.0012

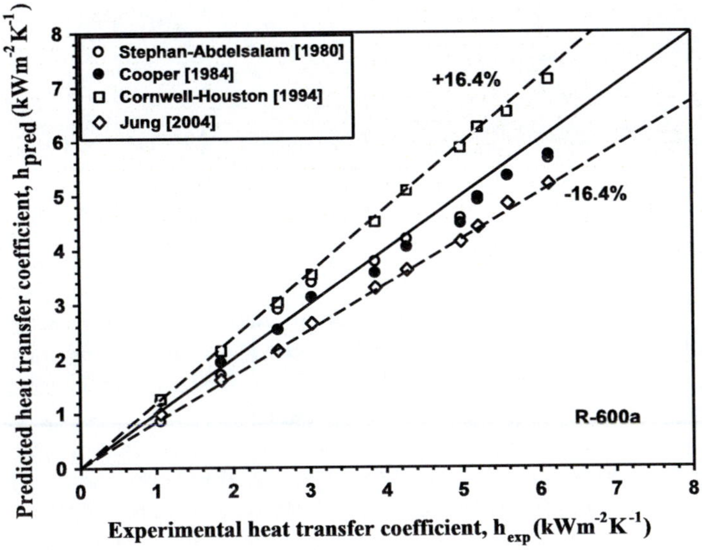

Fig. 2 Comparison of present result with predicted data

ment, the present data of refrigerant was compared with the predictive data. The result obtained from present work shows a good agreement with the predicted result and consequently this setup is reliable for given operating conditions and temperature measurement. The present heat transfer coefficient shows an error of $\pm 17\%$ with the predicted data and depicted in Fig. 2. The predicted results of Stephan and Abdelsalam [14], Cooper [15], Cornwell and Houston [16] and Jung [17] were compared with the experimental heat transfer coefficient of R-600a. The Stephan and Abdelsalam correlation is expressed as:

$$h = 207\left(\frac{k_l}{d}\right)\left(\frac{qd}{k_l}\right)^{0.745}\left(\frac{\rho_v}{\rho_l}\right)^{0.581}\left(\underset{l}{\mathrm{Pr}}\right)^{0.533} \tag{2}$$

where $d = 0.0146\emptyset\sqrt{2\sigma/g(\rho_l - \rho_v)}$, $\emptyset = 35°$.

The Cooper correlation explained the behavior of heat transfer from heating surface by using following equation:

$$h = 90(p_r)^{(0.12 - 0.2\log R_p)}(-\log p_r)^{-0.55}(M)^{-0.5}(q)^{0.67} \tag{3}$$

In this study, the roughness value (R_p) is taken as 0.4. The nucleate boiling heat transfer coefficient is calculated using Cornwell and Houston correlation:

$$h = \mathrm{AF}(p)\left(\frac{k_l}{d}\right)(\mathrm{Re}_b)^{0.67}\left(\underset{l}{\mathrm{Pr}}\right)^{0.4} \tag{4}$$

where $A = 9.7(p_c)^{0.5}$ $F(p) = 1.8(p_r)^{0.17} + 4(p_r)^{1.2} + 10(p_r)^{10}$.

And the Jung correlation is given as:

$$\frac{hd}{k_l} = 10\left(\frac{qd}{Ak_l T_{sat}}\right)^{c_1} (1 - T_r)^{-1.4} \left(\frac{\text{Pr}}{l}\right)^{-0.25} \tag{5}$$

where $c_1 = 0.855\left(\frac{\rho_v}{\rho_l}\right)^{0.309} (p_r)^{-0.437}$.

The present data indicate a good agreement with the predictions. For R-600a at 12 °C, Stephan and Abdelsalam and Cooper correlations give a good estimate for the given heat flux range. The prediction of Cornwell-Houston and Jung correlations are well with the experimental data for $q < 20$ kW m^{-2}. Stephan-Abdelsalam correlation underpredicted the heat transfer coefficients data of present study and showing a mean deviation of 13.6%. The heat transfer coefficients data of present study and showing a mean deviation of 13.6%. Cooper correlations also yielded a good agreement exhibiting a mean deviation of 14.2%. The other correlations (Cornwell-Houston and Jung) showed a little bit larger deviation for R-600a. The reason behind this large deviation might be that the data used for developing the correlation was inconsistent. The variations of heat transfer coefficients with wall superheats were observed. This may be due to the variation of size of bubbles. Generally, it offers the thermal resistance at high wall superheat.

This effect can be seen in Fig. 3. The lower amount of superheat is required to saturate refrigerant R-600a, thus higher heat transfer coefficient achieved. The visualization measurements were used to obtain the bubble characteristics for a given heat flux range. The captures high-speed images of bubble behaviors were analyzed

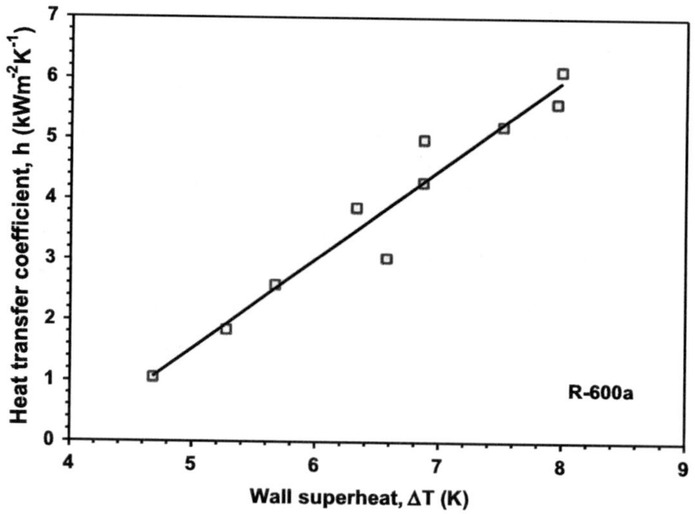

Fig. 3 Deviation of boiling heat transfer coefficient

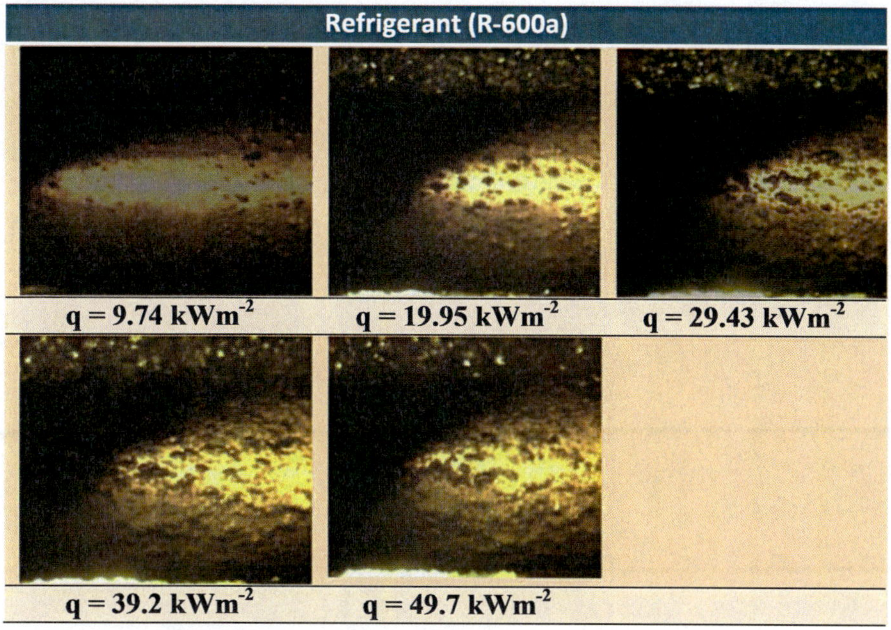

Fig. 4 Boiling over plain surface

with the help of Image-J software. This analysis enabled to obtain bubble parameters such as activated site density, departure bubble diameter and bubble frequency of plain surface. Figure 4 shows the bubble formation of refrigerant R-600a on a plain surface at 12 °C saturation temperature. The images are captured for different heat fluxes. This figure detects individual and coalesced bubble regimes in pool boiling of refrigerant. As the heat flux increases from 10 to 50 kW m^{-2} the bubbles merge to each other and form large bubbles on the surface. These bubbles are continuously escaping to liquid pool. At low heat flux $q = 10$ kW m^{-2}, the bubbles form at certain parts of surface and sites are activated over the whole surface. More number of active sites are visualized as the level of heat flux increases. The initial bubbles size on the top surface are larger than those on the side surface. The growth of the bubble's initiates from the bottom surface and merge with each other. These merging bubbles (larger bubble size) slide along the side surface and departed from the top surface. As a result, the large departure diameters are achieved for the R-600a. The bubble characteristics depend on the bubble parameters. The generation of active site density on the surface is dependent upon the applied heat input and conditions of the surface and it can be shown in Fig. 5. The trends of the curve are quite similar to Zuber model. The large deviations were observed between experimental and predicted results due to differences in operating/surface conditions and type of fluid used. The results reveal that the more bubbles formed on the surface as increases the applied heat flux. This occurs because the larger size cavity present on the surface gets activated with

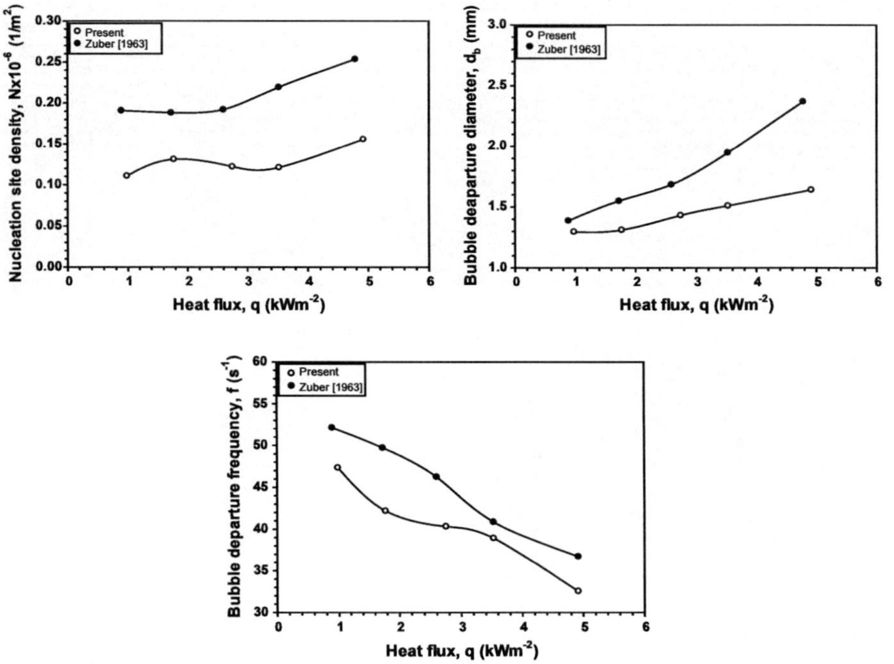

Fig. 5 Bubble parameters

heat flux increase. More small size of bubbles is nucleated on the heating surface at low heat flux.

As the heat flux increases bubbles merge each other and surface is covered with many bubbles. As a result, active site density gets reduced due to the coalescence and merging of bubbles. The bubble site density on the plain surface was also compared with predicted values by Zuber [1963]. The different site density behavior was detected for the heat flux levels of 1 to 3 kW m^{-2}. This discrepancy was observed between the present data and predicted result due to theirs operating and surface conditions. To obtain the bubble departure diameters, 500 recorded photographs of the heating surface were analyzed at heat flux of 1 to 5 kW m^{-2} and shown in Fig. 5. At each heat flux, the bubble characteristics were captured for 0.7 s. The bubble departure diameters varied with applied heat input on the heating surface. The surface tension of the refrigerant R-600a also affects the bubbles size. The experimentally measured bubble departure diameters were also compared to correlation proposed by Zuber [1963]. The same behavior was detected for 1–3 kW m^{-2} heat fluxes. Above this level, large variation in the departure diameter was found due to experimental conditions. The number of frames were observing between the consecutive bubble departures at a given site. This process gives the frequency of bubble departure. The number of frames was considered between 120 and 440 frames for 10 ms to get departure frequency. An average of the departure frequency was estimated at four active sites where same bubble size was considered. The frequency of departure bubble reduces rapidly with increasing heat flux and it can be represented in Fig. 5. Similarly, the bubble frequency data were compared with the predicted value of Zuber correlation [1959]. The similar behaviors between the predicted and present data values of the bubble departure frequency of R-600a are observed.

4 Conclusions

This work represents the boiling behavior of R-600a over the plain tube with maintaining the saturation temperature at 12 °C. The boiling behaviors were also visualized using a high-speed camera arrangement. The captured images were analyzed to detect the characteristics of bubble on the surface of test tube. The following conclusions were drawn based on the present study:

(1) For the boiling of R-600a over a plain horizontal tube, the present results were predicted within an error range of $\pm 17\%$. The agreement between predicted (Stephan and Abdelsalam and Cooper correlations) and present data is quite good. This satisfied the validity of experimental setup.
(2) The thermal performance (heat transfer coefficient) of refrigerant R-600a increases with increase of applied heat flux.
(3) The predicted results also satisfy the present data of experiments. Similar trends with large deviations were observed between present and predicted data. This is due to the working and surface conditions.
(4) The liquid–vapor exchange phenomena during boiling are visualized using high-speed camera arrangement. The captured photographs were analyzed to obtain the bubble parameters. The nucleation sites vary with increase of heat flux. The variations in departure diameters were also observed as heat flux increases. The departure frequency also decreases with increase of heat flux during boiling over the plain heating surface.

References

1. Molina MJ, Rowland FS (1974) Stratospheric sink for chlorofluromethane chlorine atom catalyzed destruction of ozone. Nature 249:810–812
2. Chiou C, Lu D, Wang C (1997) Pool boiling of R-22, R-124 and R-134a on a Plain Tube. Int J Heat Mass Transf 40(7):657–1666
3. Webb RL, Pais C (1992) Nucleate pool boiling data for five refrigerants on plain, integral-fin and enhanced tube geometries. Int J Heat Mass Transf 35(8):1893–1904
4. Kolev NI (1995) How accurate can we predict nucleate boiling. Eperimental Therm Fluid Sci 10:370–378
5. Cavallini A (1996) Working fluids for mechanical refrigeration. Int J Refrig 19(8):485–496
6. Gorenflo D (2001) State of the art in pool boiling heat transfer of new refrigerants. Int J Refrig 24(1):6–14

7. Barthau G, Hahne E (2004) Experimental study of nucleate pool boiling of R134a on a stainless-steel tube. Int J Heat Fluid Flow 25:305–312
8. Hsieh S, Hsu P (1994) Nucleate boiling characteristics of R-L 14, distended water (H_2O) and R-134a on plain and rib-roughened tube geometries. Int J Heat Mass Transf 37(10):1423–1432
9. Gorgy E, Eckels S (2012) Local heat transfer coefficient for pool boiling of R-134a and R-123 on smooth and enhanced tubes. Int J Heat Mass Transf 55:3021–3028
10. Zhang D, Zhu C, Yang X, Han T (2012) Experimental study of R134a pool boiling on single horizontal enhanced tubes. Adv Mater Res 550–553:3169–3172
11. Rocha S, Kannengieser O, Cardoso E, Passos J (2013) Nucleate pool boiling of R-134a on plain and micro-finned tubes. Int J Refrig 36:456–464
12. Ashok DK, Anil K, Ravi K (2017) Pool boiling of Iso-butane and quasi azeotropic refrigerant mixture on coated surfaes. Exp Therm Fluid Sci 85:176–188
13. Lemmon EW, Huber ML, McLinden MO (2010) NIST thermodynamic and transport properties of refrigerants and refrigerant mixtures. REFPROP version 9.0
14. Stephan K, Abdelsalam M (1980) Heat transfer correlations for natural convection boiling. Int J Heat Mass Transf 23:73–87
15. Cooper MG (1984) Saturation nucleate pool boiling—a simple correlation. Int Chem Eng Symp Ser 86:785–792
16. Cornwell K, Houston SD (1994) Nucleate pool boiling on horizontal tubes: a convection-based correlation. Int J Heat Mass Transf 37:303–309
17. Jung D, Kim Y, Ko Y, Song K (2003) Nucleate boiling heat transfer coefficients of pure halogenated refrigerants. Int J Refrig 26:240–248

MHD Thermal Convection of Nanofluid Saturated Porous Cavity Heated Linearly

Chitrak Mondal, Nirmalendu Biswas, and Nirmal K. Manna

Nomenclature

•

B	Magnetic field, tesla, (N/A·m^2)
Ha	Hartmann number
L	Length of the cavity/length scale, (m)
Nu	Average Nusselt number
P	Pressure, (Pa)
Pr	Prandtl number
Da	Darcy Number
Ra	Rayleigh number
T	Temperature, (K)
u, v	Velocity components, (m/s)
U, V	Dimensionless velocity components
x, y	Cartesian coordinates, (m)
X, Y	Dimensionless coordinates

Greek symbols

α	Thermal diffusivity, m^2/s

C. Mondal · N. K. Manna
Department of Mechanical Engineering, Jadavpur University, Kolkata 700032, India

N. Biswas (✉)
Department of Power Engineering, Jadavpur University, Salt Lake, Kolkata 700106, India

S. K. Ghosh et al. (eds.), *Advances in Thermal Engineering, Manufacturing, and Production Management*, Lecture Notes in Mechanical Engineering,
https://doi.org/10.1007/978-981-16-2347-9_4

β	Expansion coefficient of fluid, (K^{-1})
θ	Dimensionless temperature
ϕ	Concentration of nanoparticles
υ	Kinematic viscosity, (m^2/s)
ρ	Density, kg/m^3
κ	Electrical conductivity ($\mu S\ cm^{-1}$)
ψ	Dimensionless streamfunction
c, h	Cooling, heating

1 Introduction

Due to different applications of buoyancy-induced thermo-fluid flow and heat transfer such as in electronic chip cooling, biomedical application, and heat exchanger, it is still a relevant area of research. To enhance the heat transfer, nowadays nanoparticles with conductive materials are mixed with a base fluid [1–3]. By applying a magnetic field, the magnetohydrodynamic (MHD) fluid flow can be regulated more efficiently [4–6]. The investigation of convective heat transfer in a porous medium has also some practical relevance. Earlier, analysis of Cu-water nanofluid filled cavity is investigated by under different boundary conditions [7–10].

The medium of the domain is chosen based on a thorough literature survey. In earlier works, the Cu-water nanofluid is taken as the working medium as could be seen from the works of Rajarathinam et al. [7], Mansour et al. [8], Mahmoudi et al. [9], Bairi [10], Acharya et al. [11], Biswas et al. [12]. Mansour et al. [8] studied the influence of thermal buoyancy along with magnetic field to study the impact on heat transfer in a cavity. References [9, 10] considered different types of geometries and investigated the influence of a magnetic field inside the cavity. References [9, 12] considered the domain filled with a porous substance where the effect of porosity of the medium is included during their investigations.

From the extensive survey of the literature, it is observed that, though there are investigations related to MHD natural convection in the nanofluid-saturated porous domain, the buoyancy-induced convection effect in a cavity heated linearly is not studied yet extensively. Thus, the present work of MHD convection in a porous cavity containing Cu-water nanofluid is performed to fill the gap in this research area. The parameters of Ha, Da, Ra, and ϕ are varied to investigate the effects on heat transfer and fluid flow.

2 Problem Formulation

2.1 Problem Descriptions

The problem domain is presented schematically in Fig. 1 along with the boundary conditions. The computational domain is a square of length L. It is saturated with a porous medium and Cu-water nanofluid. To study the effect of MHD convection, an external magnetic field (of magnitude B) acted horizontally is considered here. The adiabatic wall is located at the bottom of the cavity. Whereas, the top wall is cold and acts as a heat sink. Both the sidewalls are linearly heated and the heating pattern is opposite in the two walls. For the left wall, the temperature decreases from bottom to top linearly, however, the opposite situation occurs for the right walls.

Many assumptions are imposed to eliminate some complexities, as the problem by definition involves multiphysics covering natural convection, magneto-hydrodynamics, nanofluid, and porous medium. The flow of fluid is thought as incompressible, homogeneous, Newtonian laminar flow and the properties are constant. The Hall effect, viscous dissipation, and Joule heating are neglected. The Boussinesq approximation is considered for the buoyancy effect. The flow is considered as steady. It yields the governing equations of continuity, momentum, and energy as given below.

$$\frac{\partial U}{\partial X}+\frac{\partial V}{\partial Y}=0 \tag{1}$$

$$\frac{1}{\varepsilon^2}\left(\frac{\partial U}{\partial X}+V\frac{\partial U}{\partial Y}\right)=-\frac{1}{\rho_r}\frac{\partial P}{\partial X}+\frac{\nu_r \Pr}{\varepsilon}\left(\frac{\partial^2 U}{\partial X^2}+\frac{\partial^2 U}{\partial Y^2}\right) -\left(\frac{\nu_r \Pr}{Da}+F_c\frac{\sqrt{U^2+V^2}}{\sqrt{Da\varepsilon^3}}\right)U \tag{2}$$

Fig. 1 Schematic diagram of the problem geometry, boundary conditions

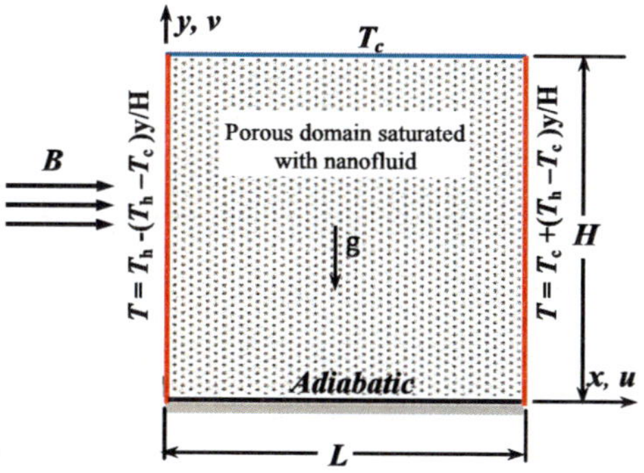

$$\frac{1}{\varepsilon^2}\left(U\frac{\partial V}{\partial X}+V\frac{\partial V}{\partial Y}\right)=-\frac{1}{\rho_r}\frac{\partial P}{\partial Y}+\frac{\nu_r \Pr}{\varepsilon}\left(\frac{\partial^2 V}{\partial X^2}+\frac{\partial^2 V}{\partial Y^2}\right)$$
$$-\left(\frac{\nu_r \Pr}{\text{Da}}+F_c\frac{\sqrt{U^2+V^2}}{\sqrt{Da\varepsilon^3}}\right)V$$
$$-\frac{1}{\rho_r}\frac{\kappa_{nf}}{\kappa_f}\text{Ha}^2 \Pr V+\beta_r Ra \Pr\theta \quad (3)$$

$$U\frac{\partial \theta}{\partial X}+V\frac{\partial \theta}{\partial Y}=\alpha_r\left(\frac{\partial^2 \theta}{\partial X^2}+\frac{\partial^2 \theta}{\partial Y^2}\right) \quad (4)$$

These non-dimensionalized equations consist of dimensionless variables of (X, Y) spatial coordinates, (U, V) velocity components, P pressure and θ temperature. The parameters involved are the Prandtl (Pr), Darcy (Da), Rayleigh (Ra), Hartmann (Ha) numbers. These are defined as

$$\begin{aligned}&(X,Y)=(x,y)/H,\ (U,V)=(u,v)/(\alpha/H)\\&P=p/\rho(\alpha/H)^2,\quad \theta=(T-T_c)/(T_h-T_c)\\&\text{Ra}=\frac{g\beta(T_h-T_c)H^3}{\alpha\upsilon},\quad \Pr=\frac{\upsilon}{\alpha},\quad \text{Ha}=\sqrt{\kappa_f/\mu_f}\end{aligned} \quad (5)$$

The modeling of this porous problem is performed by adopting the Brinkman-Forchheimer Darcy model (BFDM). This form accounts for the inertial frictional effect [13] when high velocity. It includes the Forchheimer coefficient (F_c) defined as

$$F_c = 1.75/\sqrt{150} \quad (6)$$

The physicochemical characteristics are given in Table 1.

In the governing equations, the properties of nanofluid (indicated by suffix *nf*) and base fluid (suffix *f*) are included in the ratio form. These ratios of density (ρ_r), thermal diffusivity (α_r), kinematic viscosity (ν_r), and volumetric expansion (β_r) are defined using a volumetric concentration of nanofluid (ϕ) and properties of solid nanoparticles.

$$\rho_r = \rho_{nf}/\rho_f \quad (7)$$

$$\rho_{nf} = (1-\phi)\rho_f + \phi\rho_s \quad (8)$$

Table 1 Physicochemical characteristics of base fluid and nanoparticles

	Density (kgm^{-3})	Thermal conductivity ($\text{Wm}^{-1}\text{K}^{-1}$)	Specific heat ($\text{Jkg}^{-1}\text{K}^{-1}$)	Thermal expansion (K^{-1})
Water	$\rho_f = 997.1$	$k_f = 0.613$	$C_{pf} = 4179$	$\beta_f = 21 \times 10^{-5}$
Cu	$\rho_s = 8933$	$k_s = 401$	$C_{ps} = 385$	$\beta_s = 1.67 \times 10^{-5}$

$$\nu_r = \nu_{nf}/\nu_f = (\mu_{nf}/\rho_{nf})/\nu_f \tag{9}$$

The nanofluid viscosity is computed by the correlation [14]

$$\mu_{nf} = \mu_f/(1-\phi)^{2.5}. \tag{10}$$

$$C_{pnf} = \{(1-\phi)C_{pf} + \phi C_{ps}\} \tag{11}$$

The thermal conductivity [15] is computed as

$$k_{nf} = k_f\left[\frac{(k_s + 2k_f) - 2\phi(k_f - k_s)}{(k_s + 2k_f) + \phi(k_f - k_s)}\right] \tag{12}$$

$$\alpha_r = \frac{\alpha_{nf}}{\alpha_f} = \frac{(k_{nf}/\rho_{nf}C_{pnf})}{(k_f/\rho_f C_{pf})} \tag{13}$$

The ratio of volumetric thermal expansion is expressed as

$$\beta_r = \frac{\beta_{nf}}{\beta_f} = \{(1-\phi)\beta_f + \phi\,\beta_s\} \tag{14}$$

From the problem description, the boundary conditions is obtained as given below

(a) $U = 0$, $V = 0$, $\theta = 0$ for the top cold wall ($Y = 1$)
(b) $U = 0$, $V = 0$, $\partial\,\theta/\partial Y = 0$ for the bottom wall ($Y = 0$)
(c) $U = V = 0$, $\theta = Y$ for the right cold wall ($X = 1$)
(d) $U = V = 0$, $\theta = 1 - Y$ for the left heating wall ($X = 0$)

The above-mentioned Eqs. (1–4) are solved using a FVM-based extensively validated in-house CFD code. The obtained results are finally processed into the streamlines and the Nusselt number (Nu). The Nu of the top wall is computed as an area-averaged Nusselt number as

$$\mathrm{Nu} = \int_0^1 \left(-\frac{\partial\theta}{\partial Y}\bigg|_{Y=1}\right) dX \tag{15}$$

The velocity components are used to generate the stream function (ψ) as under.

$$U = \frac{\partial\,\psi}{\partial X} \text{ and } V = -\frac{\partial\,\psi}{\partial Y} \tag{16}$$

The stream function value at the walls is considered zero due to the no-slip and no penetration condition.

2.2 Numerical Technique

The numerical simulation of the non-dimensionalized governing Eqs. (1–4) is performed using the Finite Volume Method (FVM) and SIMPLE algorithm [16]. This algorithm is incorporated in a well-validated [12, 17–19] in house CFD code. A second-order central differencing scheme for the diffusion terms, and a third-order upwind (QUICK) scheme for the advection terms are chosen. The residual values of the iterative scheme are set to 10^{-8} or below. Before finalizing the results, an extensive mesh-independent study is conducted which is omitted for the sake of brevity. Finally, a 160×160 grid size with non-uniform distribution is chosen for the entire simulation.

3 Result and Discussions

In this work, the influence of the magnetic field, porous medium, and the thermal convection are addressed by varying the Hartmann number (Ha), Darcy number (Da) and the Rayleigh number (Ra). The Hartmann number is varied in the range of 0–100, the Da value in the range of 10^{-5}–10^{-1}, and the Ra value as 10^5 and 10^6. For the proper visualization of the flow domain and the vortex structures, the streamlines are considered where the maximum and minimum streamlines values indicate the relative strength of the vortices. The isotherm contours are considered to determine the temperature field of the domain where the isotherm values are scaled to 1 and they are basically the non-dimensionalized value. The heat transfer is analyzed by considering the top-wall averaged Nusselt number while there is a variational study of Nu at the end of the section.

3.1 Effect of Different Rayleigh Number (Ra)

This study is conducted to study the effect of applying nanofluid in the linearly heated porous cavity by comparing streamline and isotherm contours with the base fluid case. The Nu value for the top wall is also considered to estimate the heat transfer improvement due to usage of nanofluid. Two values of Ra number are considered to capture the effect of thermal convection. For $Ra = 10^5$ situation of both the fluids, the flow field looks similar but the strengths of the vortices are different. For $Ra = 10^6$ case, in both the flow-fields, another corner vortex appears in the left top corner and the lower clockwise (CW) vortex has maximum strength. By comparing different Ra values, it is seen that the vortex strength increases drastically as Ra increases. The plot of isotherms shows the distribution of isotherms intuitively. As the circulation strength of $Ra = 10^5$ is lower the temperature distribution is very much similar to the boundary and not much interaction is there. However, for $Ra = 10^6$, the isotherms are deformed due to a strong circulation in the cavity and the higher temperature regions are suppressed to the walls. By comparing the Nu value, it is noticed that, with an increase in Ra, Nu increases and the addition of nanofluid further enhances the heat transfer (Fig. 2).

3.2 Effect of Darcy Number (Da)

Figure 3 considers the effect of Da in the nanofluid-filled linear heated cavity. The Darcy number suggests the relative effect of permeability with respect to its area and here the Da increases from 10^{-5} to 10^{-1}. The change in Da value does not change the relative position and the shape of the vortices, however, the maximum stream function increases as Da value increases. Due to this change in the strength of circulation, the shape of the isotherm contours is changing. As the vortices become stronger, the mixing of the hot and cold fluid is much more prominent. Therefore, the isotherms distort more with higher Da value. Moreover, with the increasing Da value as the permeability increases, the heat transfer also increases as indicated by the Nu values (though the changes are insignificant).

3.3 Impact of Hartmann Number (Ha)

Figure 4 presents the effect of the Hartmann number. This number is basically the non-dimensionalized parameter which measures the relative importance of the electromagnetic force to the viscous force. Here, Ha values are varied from 0 to 100. The magnetic field direction is from left to right, and thus, Ha effects the V-component of flow velocity directly. Through the continuity equation, the effect is transferred to the U-component of velocity. So, the vortices are weakened with an increase in Ha, and the upper left-corner vortex is diminished as Ha increases. Due to the overall damping effect, the maximum value of the streamfunction is also decreasing as Ha increases. The isotherm pattern can be explained from the stream function values also. For Ha $= 0$, the isotherms are most distorted but as the Ha increases the distortion is decreasing. With the increase in Ha, associated heat transfer from the top cold wall is also decreasing which is evident from the above-mentioned values.

3.4 Effect of Nanoparticle Concentration (ϕ)

The effect of the nanoparticles volume fraction (ϕ) suspended in the base fluid is illustrated in Fig. 5 where ϕ is varied from 0.001 to 0.05. The ϕ value is chosen in such a way that the medium behaves as a homogeneous medium. The isotherm contours for different ϕ are almost similar in shape. However, the increase in ϕ

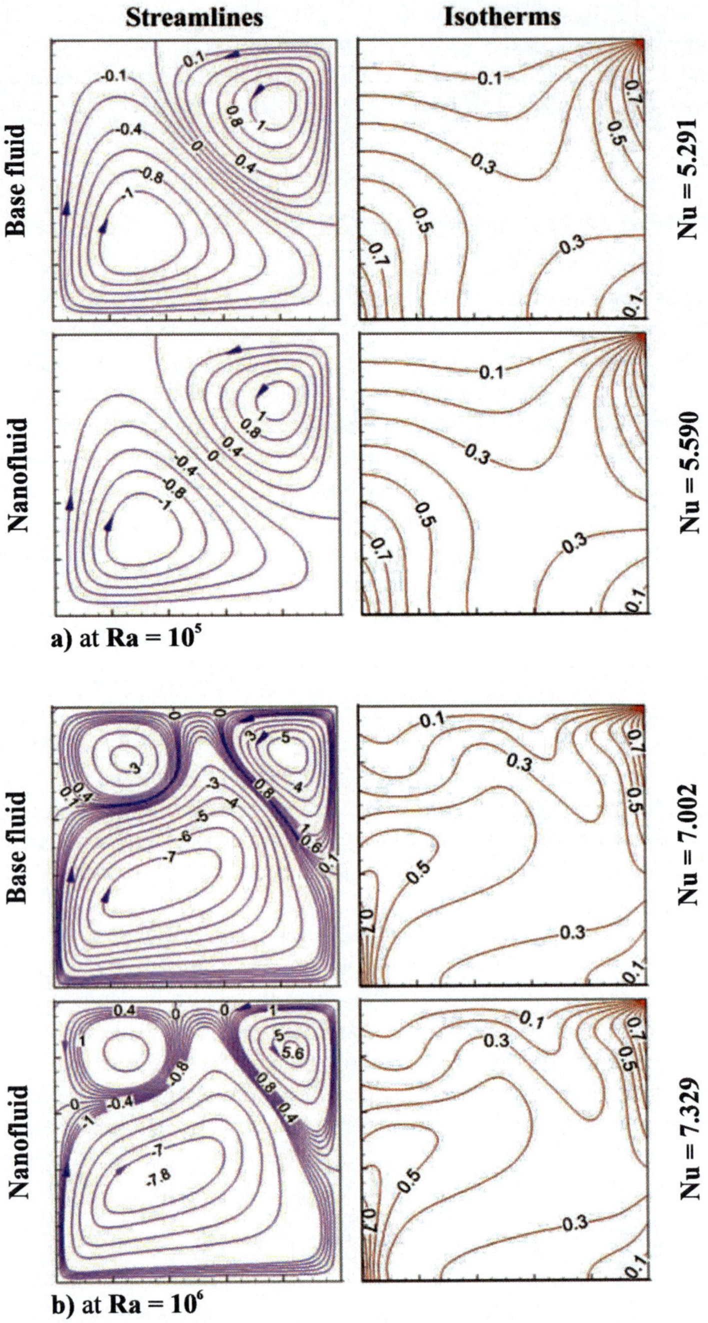

Fig. 2 Typical analysis of flow structure under base fluid and nanofluid saturated porous media at $\mathrm{Ra} = 10^5, 10^6$, $\mathrm{Ha} = 30$, $\phi = 0.02$, $\mathrm{Da} = 10^{-3}$, $\varepsilon = 0.6$

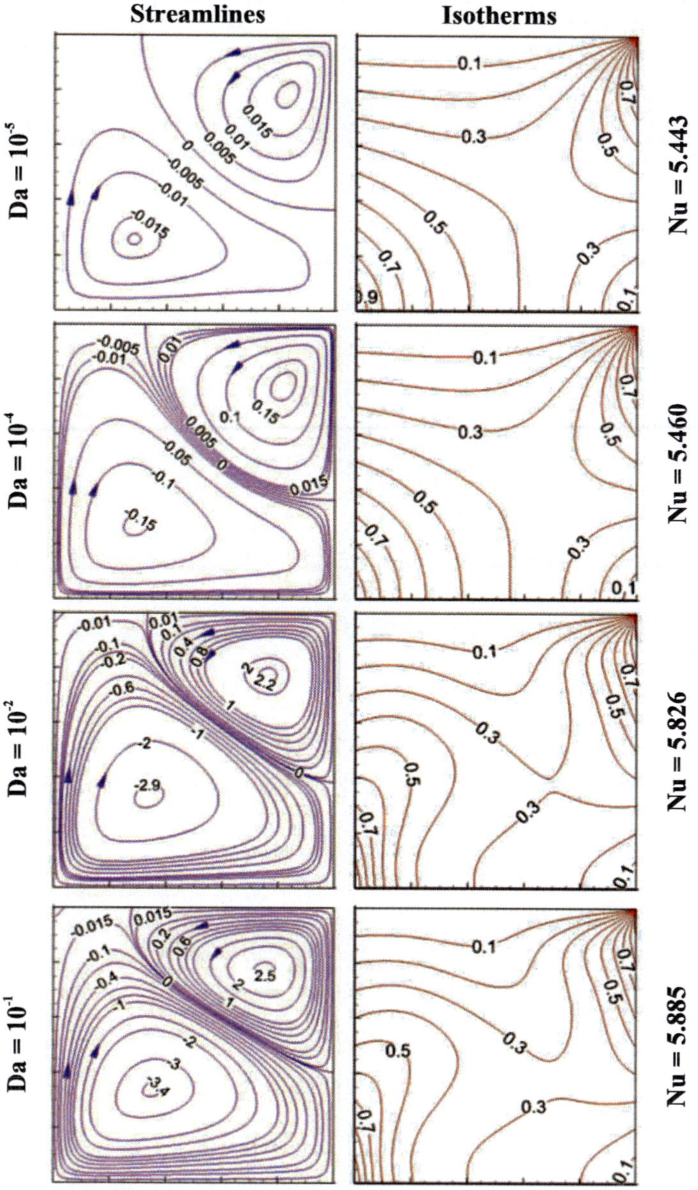

Fig. 3 Effect of Da on fluid flow and thermal characteristics at $\mathrm{Ra} = 10^5$, $\mathrm{Ha} = 30$, $\phi = 0.02$, $\varepsilon = 0.6$

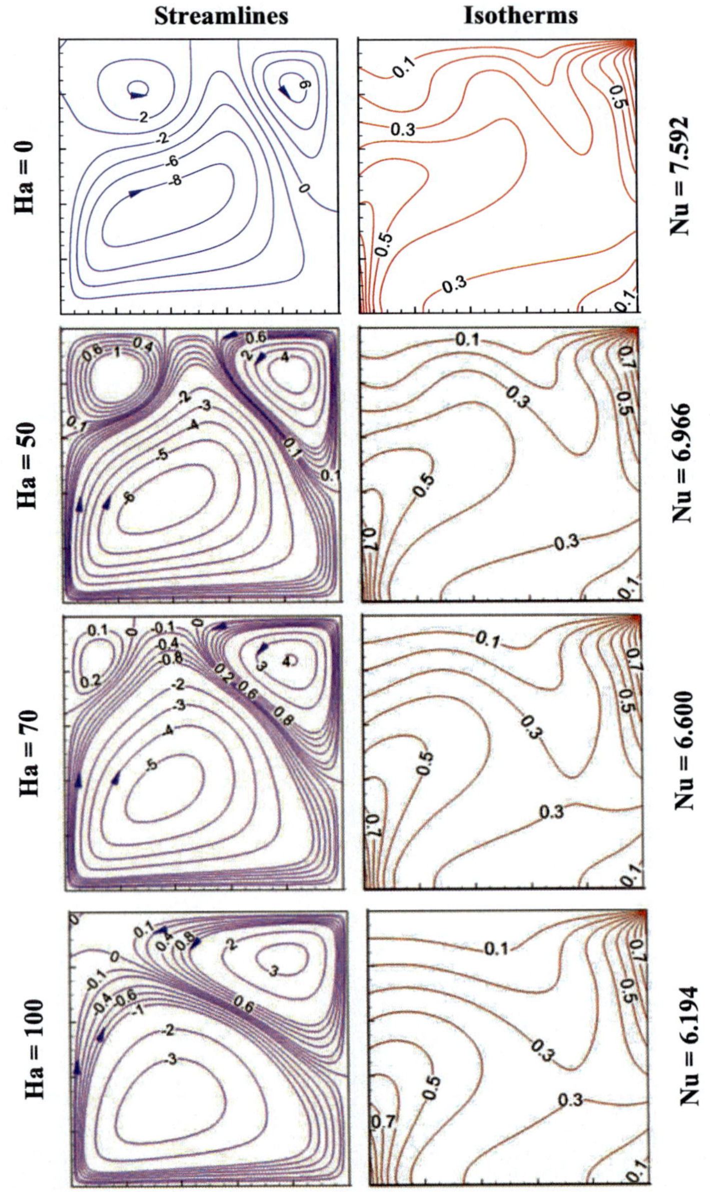

Fig. 4 Impact of Hartmann number on fluid flow and thermal characteristics on the linearly heated cavity at $Ra = 10^6$, $Da = 10^{-3}$, $\varepsilon = 0.6$, $\phi = 0.02$

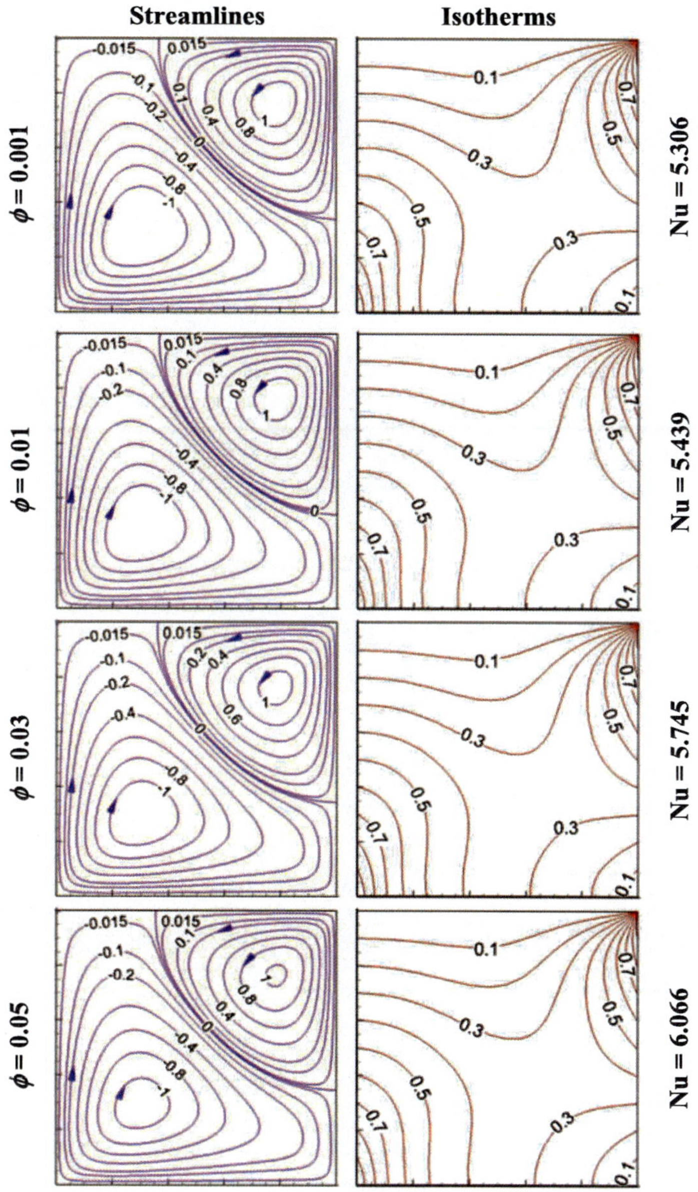

Fig. 5 Effect of ϕ on fluid flow and thermal characteristics on the linearly heated cavity at $Ra = 10^6$, $Da = 10^{-3}$, $\varepsilon = 0.6$, $Ha = 30$

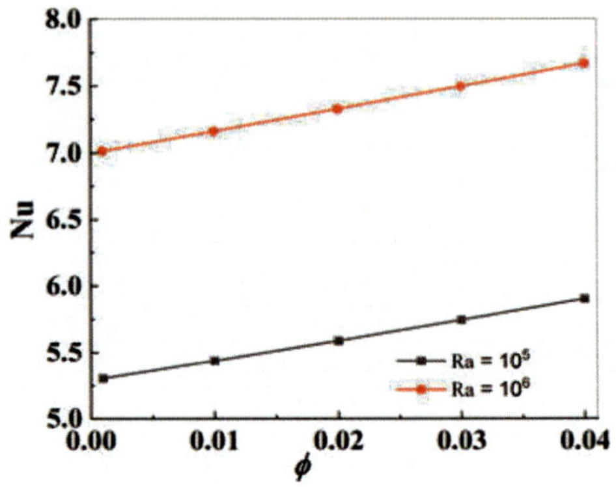

Fig. 6 Heat transfer characteristics (Nu) of the top wall at $Ra = 10^5, 10^6$, $Ha = 30$, $Da = 10^{-3}$

shows an increment in heat transfers through the top wall. This may be described due to the enhanced thermal conduction due to the presence of highly conductive Cu-nanoparticles.

3.5 *The Heat Transfer Characteristics (Nu)*

The changes in the Nu value of the top wall are considered with respect to the concentration of nanoparticles of the medium and the trend is reported in Fig. 6. The figure shows an almost linear variation of Nu with respect to ϕ. As the Ra increases, corresponding Nu increases significantly. The increment in heat transfer can be recognized due to the enhancement of convective current at higher Ra and ϕ for the porous substance of $Da = 10^{-3}$.

4 Conclusions

The impacts of the magnetic field, Darcy number, and the Cu-nanoparticles on the characteristics of thermo-fluid flow are studied extensively and systematically using the streamlines, isotherms, and the average Nusselt number. The variations of the pertinent parameters are significant. Some major observations are mentioned below:

- As Ra value increases, the strength of the vortices increases and the heat transfer increases. Due to the incorporation of the nanofluid, the heat transfer is heightened from the base fluid situation.
- The effect of the porous medium is encountered by changing the Da value. The study shows that, due to an increase in permeability, the flow is strengthened and the heat transfer enhances.
- The effect of the magnetic field is encountered with varying Ha value. The magnetic field has an overall dampening effect on the flow and heat transfer

characteristics. Depending upon the direction of the application and the relative strength of the buoyancy force, the flow structure is dictated accordingly.

- The effect of change in concentration of the nanofluid (ϕ) is studied by varying ϕ. The addition of nanofluid has an enhancement effect on heat transfer through the concentration, which studied shows similar patterns of fluid-flow and temperature structures within the cavity.
- Finally, the effect on heat transfer with suspension of nanoparticle concentration in the base fluid is studied. The concentration is within the permissible limit so that the fluid remains homogeneous. The Nu-curve is almost a linearly increasing trend. So with the addition of nanoparticles in the base fluid, the heat transfer is improved.

References

1. Santra AK, Sen S, Chakraborty N (2018) Study of heat transfer augmentation in a differentially heated square cavity using copper–water nanofluid. Int J Thermal Sci 47:1113–1122
2. Mehryan SAM, Izadpanahi E, Ghalambaz M, Chamkha AJ (2019) Mixed convection flow caused by an oscillating cylinder in a square cavity filled with Cu–Al_2O_3/water hybrid nanofluid. J Thermal Anal Calorim 137:965–982
3. Mansour MA, Siddiqa S, Gorla RSR, Rashad AM (2018) Effects of heat source and sink on entropy generation and MHD natural convection of Al_2O_3-Cu/water hybrid nanofluid filled with square porous cavity. Thermal Sci Eng Prog 6:57–71
4. Ozoe H (2005) Magnetic convection. Imperial College Press, Singapore
5. Matt CFT, Quaresma JNN, Cotta RM (2017) Analysis of magnetohydrodynamic natural convection in closed cavities through integral transforms. Int J Heat Mass Transfer 113:502–513
6. Al-Balushi LM, Rahman MM (2019) Convective heat transfer utilizing magnetic nanoparticles in the presence of a sloping magnetic field inside a square enclosure. J Thermal Sci Eng Appl 11:041013-1–041013-19
7. Rajarathinam M, Nithyadevi N, Chamkha AJ (2018) Heat transfer enhancement of mixed convection in an inclined porous cavity using Cu-water nanofluid. Adv Powder Technol 29:590–605
8. Mansour MA, Bakier MAY (2015) Influence of thermal boundary conditions on MHD natural convection in square enclosure using Cu-water nanofluid. Energy Rep 1:134–144
9. Mahmoudi AH, Pop I, Shahi M, Talebi F (2013) MHD natural convection and entropy generation in a trapezoidal enclosure using Cu-water nanofluid. Comp Fluids 72:46–62
10. Bairi A (2018) Natural convection between concentric and inclined hemispherical cavities filled with Cu-water nanofluid. J Mol Liquids 249:1263–1270
11. Acharya S, Dash SK (2019) Natural convection in a cavity with undulated walls filled with water-based non-Newtonian power-law CuO–water nanofluid under the influence of the external magnetic field. Numer Heat Transfer A 76(7):552–575
12. Biswas N, Manna NK, Datta P, Mahapatra PS (2018) Analysis of heat transfer and pumping for bottom-heated porous cavity saturated with Cu-water nanofluid. Powder Technol 326:356–369
13. Nield DA, Bejan A (2006) Convection in porous media, 3rd edn. Springer, New York
14. Brinkman HC (1952) The viscosity of concentrated suspensions and solutions. J Chem Phys 20:571
15. Maxwell J (1904) A treatise on electricity and magnetism. Oxford University Press, Cambridge, UK
16. Patankar SV (1980) Numerical heat transfer and fluid flow. Taylor and Francis

17. Biswas N, Sarkar UK, Chamkha AJ, Manna NK (2020) Magneto-hydrodynamic thermal convection of Cu–Al_2O_3/water hybrid nanofluid saturated with porous media subjected to half-sinusoidal nonuniform heating. J Thermal Anal Calorim. https://doi.org/10.1007/s10973-020-10123-0
18. Biswas N, Chamkha AJ, Manna NK (2020) Energy-saving method of heat transfer enhancement during magneto–thermal convection in typical thermal cavities adopting aspiration. SN Applied Sci 2–1911:1–25
19. Biswas N, Chamkha AJ, Manna NK, (2020) Effects of half-sinusoidal nonuniform heating during MHD thermal convection in Cu–Al_2O_3/water hybrid nanofluid saturated with porous media. J Thermal Anal Calorim. https://doi.org/10.1007/s10973-020-10109-y

Natural Convection of Copper-Water Nanofluid in a Square Enclosure with an Isothermal Protruding Heater

Dipayan Sanpui, Sourav Sarkar, and Swarnendu Sen

Nomenclature

•

c_p	Specific heat, (J/kg-K)
g	Gravitational acceleration, (m/s^2)
k	Thermal conductivity, (W/mK)
L	Length of the enclosure, (m)
D	Distance from the left wall to the center of the protruding heater, (m)
W	Width of protruding geometry, (m)
H	Height of protruding geometry, (m)
Pr	Prandtl number
Ra	Rayleigh number
p	Pressure, (N/m^2)
m	Fluid behaviour index, (N sec^n/m^2)
n	Flow consistency index
P	Dimensionless pressure
Nu	Nusselt number
$\overline{Nu}$	Average Nusselt number
t	Time, (s)
T	Temperature, (K)
u, v	Velocity components, (m/s)
U, V	Dimensionless velocity components

D. Sanpui
School of Materials Science and Nanotechnology, Jadavpur University, Kolkata 700032, India

S. Sarkar (✉) · S. Sen
Department of Mechanical Engineering, Jadavpur University, Kolkata 700032, India

S. K. Ghosh et al. (eds.), *Advances in Thermal Engineering, Manufacturing, and Production Management*, Lecture Notes in Mechanical Engineering,
https://doi.org/10.1007/978-981-16-2347-9_5

x, y	Cartesian co-ordinates
X, Y	Dimensionless cartesian coordinates

Greek Symbols

α	Thermal diffusivity, (m/s^2)
β	Thermal expansion co-efficient, (1/K)
ϕ	Solid volume fraction
ψ	Dimensionless stream function
ρ	Density, (kg/m^3)
μ	Dynamic viscosity, (kg/ms)
υ	Kinematic viscosity, (m^2/s)
θ	Dimensionless temperature
ω	Vorticity
ε	Parameter used for error analysis

Subscripts

c	Condition for cold wall
s	Solid particle
f	Base fluid
h	Condition for hot wall
l	Local values
nf	Nanofluid

1 Introduction

The enhancement of heat transfer techniques, its adequacy and necessity is an integral part of a wide class of technological and industrial applications such as thermal shielding, cooling of industrial systems, and several others. Nanofluids is a well dispersed solution of solid nanoparticles in base fluids i.e. water, ethylene glycol (EG), acetone or oil found to be an emerging solution for heat transfer enhancement [1]. The chances of applying them as heat transfer medium for various nanotechnology based cooling applications, such as nuclear machineries, microchannels, MEMS devices and tribological applications i.e. nano-lubricants, hydraulic fluids, cutting fluids, nano-refrigerants is currently under consideration.

The numerical studies related to heat transfer enhancement using nanofluid are abundant in literature [1]. Now, in most of the numerical studies [2–6] nanofluid has been considered as of Newtonian behavior, in those studies, in most of the cases, the rate of heat transfer increases with increasing volume fraction, Rayleigh number and time steps. When we are considering the nanofluid as non-Newtonian the shear strain rate and shear stress components associated comes into the frame. Also, it is to be noted that, the constant term of viscosity has been replaced by an apparent or changing viscosity term which inhibits flow circulation as the volume fraction increases. Khanafer et al. [7] numerically studied the natural convection of Al_2O_3-water nanofluid inside rectangular cavities. They experienced an increase in heat transfer with increasing volume fraction. Natural convection inside a differentially heated square enclosure filled with nanofluid has been numerically performed by Ho et al. [8] and Santra et al. [9]. Separate mathematical models for viscosity and effective thermal conductivity have been considered by Ho et al. [8]. The Ostwald-de Waele model (power law fluid model, proposed in 1925) for a shear-thinning, non-Newtonian nanofluid has been used by Santra et al. [9]. It has been found that they used two rheological indexes fluid behaviour index (m) and flow consistency index (n), incorporated within the mathematical model and finally observed a decrease in average Nusselt number with increasing Rayleigh numbers at different volume fractions. In another study, Santra et al. [10] have also performed a numerical investigation on copper water nanofluid flowing through two isothermally heated parallel plates. In that problem, they experienced an increase in average Nusselt number with increasing Rayleigh numbers at a constant volume fraction in case of non-Newtonian nanofluid compared to Newtonian nanofluid. These two [9, 10] numerical studies can be considered as the numerical solutions of its first kind regarding natural and forced convection of non-Newtonian nanofluid. Loenko et al. [11] used the Ostwald-de Waele power law model with FDM to numerically study the natural convection of a non-Newtonian fluid in a square cavity in the presence of local heat-generating element. They observed that as the value of rheological index increases the fluid becomes more viscous which inhibits the heat transfer and as the fluid becomes shear-thinning it enhances the heat removal from the energy source. They also studied the dependency of thermal conductivity ratio on time steps and showed that as the thermal conductivity ratio increases, the number of time steps required to reach steady-state increases and heat transfer also increases simultaneously. Yigit et al. [12] studied the natural convection of power-law fluids in a square cavity where the bottom of the chamber was heated locally and the vertical walls were kept at comparatively less temperature. They observed with increasing n, the convection inside the cavity increases; but the rate of increase lowers as the value of n increases. The same scenario can be encountered for vertical velocity profile; which reflects the decrease in vertical velocity with an increase in n. That means as the flow inside the cavity becomes viscous; though the average Nusselt number value increases, but the rate of increment lowers. Khezzar et al. [13] performed numerical simulation on free convection of power law fluid inside a cavity at different angular orientations. The angle of inclination was gradually increased from 0 to 90° and the isotherms, streamlines patterns have been compared for both Newtonian and non-Newtonian nanofluids. Also, correlations were presented between the average Nusselt number and power law index, volume fractions at different Rayleigh numbers and varying aspect ratios of the geometry.

The present research work shows the effect of copper-water nanofluid for investigation of the heat removal from a square cavity with protruding heater at the bottom wall; where, the entire upper wall and the non-heated portions of the bottom wall are adiabatic. Symmetrical cooling has been done along the vertical walls of the cavity. Patel et al. [14] proposed a model to study the effective thermal conductivity which includes the parameters, Brownian motion and temperature dependency of the nanofluid. The viscosity of the nanofluid has been modelled according to the Ostwald-de Waele model. The nanofluid has been considered as of shear-thinning behavior. To the best of knowledge, no other numerical study on heat transfer performance analysis with a central isothermal heater inside a square enclosure has been investigated for a non-Newtonian nanofluid has been performed.

2 Mathematical Formulation

The problem under consideration (Fig. 1) is to numerically study natural convection of Cu-water nanofluid inside a square cavity; including a protruding isothermal heater on the center of the bottom wall. The nanofluid inside the enclosure is assumed to be of non-Newtonian behavior. The geometry under consideration and dimensions are shown in the figure. A heater of height H and width W whose temperature is maintained to be at T_h and the midpoint of the heater is kept at a distance of D from the left wall. The vertical sidewalls of the cavity are maintained at a less temperature

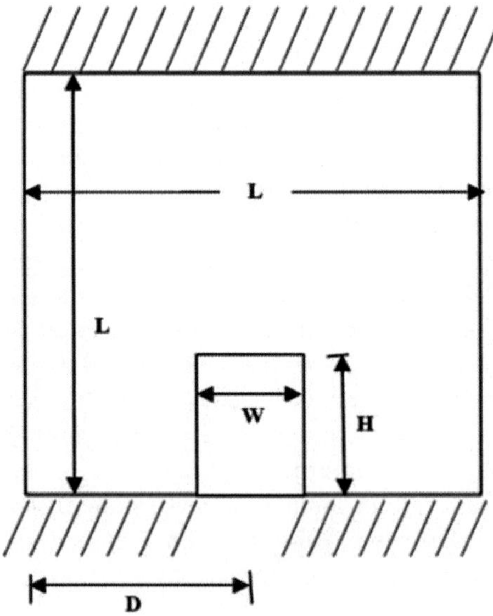

Fig. 1 Geometry of the problem

of T_c. Where the top horizontal wall is entirely insulated and the non-heated portions of the bottom horizontal wall are insulated. The nanofluid is incompressible in nature and consists of copper nanoparticles of diameter 100 nm. The thermophysical properties of the nanofluid are assumed constant, except for the apparent viscosity which changes accordingly with the change in shear rate. It has been assumed that density of the nanofluid varies only with the body force which is treated according to the Boussinesq approximation. It is to be noted that, the length (L) of the cavity and all the other dimensions i.e. D, W, H are in meter (m).

By culminating the respective assumptions; the conservation equations can be written as follows [9]:

$$\frac{\partial u}{\partial x}+\frac{\partial v}{\partial y}=0 \tag{1}$$

$$\frac{\partial u}{\partial t}+\rho_{nf}\left(u\frac{\partial u}{\partial x}+v\frac{\partial u}{\partial y}\right)=-\frac{\partial p}{\partial x}-\left[\frac{\partial \tau_{xx}}{\partial x}+\frac{\partial \tau_{yx}}{\partial y}\right] \tag{2}$$

$$\begin{aligned}\frac{\partial v}{\partial t}+\rho_{nf}\left(u\frac{\partial u}{\partial x}+v\frac{\partial u}{\partial y}\right)&=-\frac{\partial p}{\partial y}+\left(\phi\rho_p\beta_p+\rho_f\beta_f(1-\phi)\right)\left(\frac{\partial^2 u}{\partial x^2}+\frac{\partial^2 v}{\partial y^2}\right)\\&+\left(\frac{(\rho\beta)_{nf}}{\rho_{nf}}\right).g.(T-T_c)-\left[\frac{\partial \tau_{xy}}{\partial x}+\frac{\partial \tau_{yy}}{\partial y}\right]\end{aligned} \tag{3}$$

$$\frac{\partial T}{\partial t}+u\frac{\partial T}{\partial x}+v\frac{\partial T}{\partial y}=\alpha_{nf}\left[\frac{\partial^2 T}{\partial x^2}+\frac{\partial^2 T}{\partial y^2}\right] \tag{4}$$

where,

$$\alpha_{nf}=(k_{nf})/(\rho C_p)_{nf} \tag{5}$$

The interrelation between the shear stress and rate of shear strain in case of 2-D Cartesian system according to the Ostwald-de Waele power law model is:

$$\tau=-m\left[\left|\sqrt{\frac{1}{2}(\dot{\gamma}\cdot\dot{\gamma})}\right|^{(n-1)}\right]\dot{\gamma} \tag{6}$$

where,

$$\frac{1}{2}(\dot{\gamma}\cdot\dot{\gamma})=2\left\{\left(\frac{\partial u}{\partial x}\right)^2+\left(\frac{\partial v}{\partial y}\right)^2\right\}+\left(\frac{\partial v}{\partial x}+\frac{\partial u}{\partial y}\right)^2 \tag{7}$$

The stress tensors of Eqs. (2) and (3) can be expressed as

$$\tau_{xx} = -2\left\{m\left|\left[2\left\{\left(\frac{\partial u}{\partial x}\right)^2 + \left(\frac{\partial v}{\partial y}\right)^2\right\} + \left(\frac{\partial v}{\partial x} + \frac{\partial u}{\partial x}\right)^2\right]^{\frac{1}{2}}\right|^{(n-1)}\right\}\left(\frac{\partial u}{\partial x}\right) \tag{8}$$

$$\tau_{yx} = \tau_{xy} = -2\left\{m\left|\left[2\left\{\left(\frac{\partial u}{\partial x}\right)^2 + \left(\frac{\partial v}{\partial y}\right)^2\right\} + \left(\frac{\partial v}{\partial x} + \frac{\partial u}{\partial y}\right)^2\right]^{\frac{1}{2}}\right|^{(n-1)}\right\} \left(\frac{\partial u}{\partial y} + \frac{\partial v}{\partial x}\right) \tag{9}$$

$$\tau_{yy} = -2\left\{m\left|\left[2\left\{\left(\frac{\partial u}{\partial x}\right)^2 + \left(\frac{\partial v}{\partial y}\right)^2\right\} + \left(\frac{\partial v}{\partial x} + \frac{\partial u}{\partial y}\right)^2\right]^{\frac{1}{2}}\right|^{(n-1)}\right\}\left(\frac{\partial v}{\partial y}\right) \tag{10}$$

Here *m, n* are the two rheological indexes, depends on type of material used as nanoparticles for the preparation of nanofluid and solid volume fraction (ϕ). An experimental variation of shear stress with respect to shear strain has been found in Putra et al. [15], later which has been used by Santra et al. [9] for calculation of the values of rheological indexes i.e. *m* and *n* for 1 and 5% solid volume fraction. Proper interpolation and extrapolation have been done with keeping in mind that the nanofluid is showing shear thinning behavior with increasing volume fraction. The detailed values of *m* and *n* for different volume fractions are shown in Table 1. It is to be noted that if the value of the power-law index is 1 then the value of *m* becomes the viscosity of the base fluid.

The nanofluid effective density can be expressed as:

$$\rho_{nf} = (1 - \phi)\rho_f + \phi\rho_s \tag{11}$$

Table 1 Values of rheological indexes

Solid volume fraction (ϕ) (%)	m (N sec^n m^{-2})	n
0.5	0.00187	0.880
1.0	0.00230	0.830
1.5	0.00283	0.780
2.0	0.00347	0.730
2.5	0.00426	0.680
3.0	0.00535	0.625
3.5	0.00641	0.580
4.0	0.00750	0.540
4.5	0.00876	0.500
5.0	0.01020	0.460

Table 2 Thermophysical properties

Property	Fluid (Water)	Solid (Copper)
C_p (J/kg-K)	4181.80	383.1
ρ (kg/m^3)	1000.52	8954.0
k (W/m-K)	0.597	386.0
β (K^{-1})	210.0×10^{-6}	51.0×10^{-6}

Similarly, the formulation of effective heat capacity of the nanofluid is expressed as:

$$(\rho C_p)_{nf} = (\rho C_p)_f (1 - \phi) + \phi (\rho C_p)_s \tag{12}$$

The above Eqs. (11) and (12) have been taken from Mahian et al. [1].

Solid spherical copper nanoparticles of 100 nm diameter have been taken with base fluid i.e. water; as the nanofluid. The thermophysical properties of the nanofluid has been extracted from Santra et al. [9] (Table 2):

For formulation of effective thermal conductivity, Patel et al. [14] model has been used:

$$\frac{k_{nf}}{k_f} = 1 + \frac{k_p A_p}{k_f A_f} + C.k_p.Pe.\frac{A_p}{k_f.A_f} \tag{13}$$

where,

$$\frac{A_p}{A_f} = \frac{d_f}{d_p} \frac{\phi}{(1 - \phi)} \tag{14}$$

Here, $Pe = \frac{u_p d_p}{\alpha_f}$, where u_p is the velocity of the nanoparticles due to Brownian motion and is given by,

$$u_p = \frac{2 k_b T}{\pi \mu_f d_p^2} \tag{15}$$

The governing equations Eqs. 1–4 can be expressed in dimensionless form using the parameters given below [9]:

$$X = \frac{x}{h}, Y = \frac{y}{h}, U = \frac{uh}{\alpha_f}, V = \frac{vh}{\alpha_f}$$

$$\theta = \frac{(T - T_c)}{(T_h - T_c)}, \tau = \frac{\alpha_f}{L_y^2} t \text{ and } P = \frac{(p - p_0)h^2}{(\rho_{nf} \alpha_f^2)}$$

The governing equations Eqs. 1–4 takes the forms:

$$\frac{\partial U}{\partial X} + \frac{\partial V}{\partial Y} = 0 \tag{16}$$

$$\frac{\partial U}{\partial \tau} + U\frac{\partial U}{\partial X} + V\frac{\partial U}{\partial Y} = -\frac{\partial P}{\partial X} - \left(\frac{\mu_{app}}{\rho_{nf}\alpha_f}\right)\left[\frac{\partial^2 U}{\partial X^2} + \frac{\partial^2 U}{\partial Y^2}\right] \tag{17}$$

$$\frac{\partial V}{\partial \tau} + U\frac{\partial V}{\partial X} + V\frac{\partial V}{\partial Y} = -\frac{\partial P}{\partial Y} + Ra.Pr.\frac{\rho_f}{\rho_{nf}}\left(1 - \phi + \phi\frac{\rho_s\beta_s}{\rho_f\beta_f}\right)\theta + \left(\frac{\mu_{app}}{\rho_{nf}\alpha_f}\right)\left(\frac{\partial^2 V}{\partial X^2} + \frac{\partial^2 V}{\partial Y^2}\right) \tag{18}$$

$$\frac{\partial \theta}{\partial \tau} + U\frac{\partial \theta}{\partial X} + V\frac{\partial \theta}{\partial Y} = \frac{k_{nf}}{k_f}\frac{(\rho C_p)_f}{(\rho C_p)_{nf}}\left[\frac{\partial^2 \theta}{\partial X^2} + \frac{\partial^2 \theta}{\partial Y^2}\right] \tag{19}$$

The apparent viscosity of the nanofluid can be written as,

$$\mu_{app} = m\left(\frac{\alpha_f}{h^2}\right)^{(n-1)}\left|\left[2\left\{\left(\frac{\partial u}{\partial x}\right)^2 + \left(\frac{\partial v}{\partial y}\right)^2\right\} + \left(\frac{\partial v}{\partial x} + \frac{\partial u}{\partial x}\right)^2\right]^{\frac{1}{2}}\right|^{(n-1)} \tag{20}$$

The stream function-vorticity formulation has been used for solving this problem. The stream function ψ is defined by,

$$U = \frac{\partial \psi}{\partial Y} \tag{21}$$

$$V = -\frac{\partial \psi}{\partial X} \tag{22}$$

Vorticity $\omega = \frac{\partial V}{\partial X} - \frac{\partial U}{\partial Y}$ can be written in a simplified form:

$$\omega = -\nabla^2\psi \tag{23}$$

Again, the above Eqs. 15–18 are transformed using the stream-function (ψ)-vorticity (ω) formulation. These equations are as follows:

$$\frac{\partial^2 \psi}{\partial \mathrm{x}^2} + \frac{\partial^2 \psi}{\partial \mathrm{y}^2} = -\omega \tag{24}$$

$$\frac{\partial \omega}{\partial \tau} + \frac{\partial \psi}{\partial Y}\frac{\partial \omega}{\partial X} - \frac{\partial \psi}{\partial X}\frac{\partial \omega}{\partial Y} = \left(\frac{\mu_{app}}{\rho_{nf}\alpha_f}\right)\left(\frac{\partial^2 \omega}{\partial X^2} + \frac{\partial^2 \omega}{\partial Y^2}\right) + Ra.Pr.\frac{\rho_f}{\rho_{nf}}\frac{\partial \theta}{\partial x} \tag{25}$$

$$\frac{\partial \theta}{\partial \tau} + \frac{\partial \psi}{\partial Y}\frac{\partial \theta}{\partial X} - \frac{\partial \psi}{\partial X}\frac{\partial \theta}{\partial Y} = \left(\frac{k_{nf}}{k_f}\frac{(\rho C_p)_f}{(\rho C_p)_{nf}}\right)\left(\frac{\partial^2 \theta}{\partial X^2} + \frac{\partial^2 \theta}{\partial Y^2}\right) \tag{26}$$

After $\psi - \omega$ formulation, the term apparent viscosity also changes and can be written as,

$$\mu_{app} = m\left(\frac{\alpha_f}{h^2}\right)^{(n-1)}\left|\left[4\left(\frac{\partial^2\psi}{\partial X\partial Y}\right)^2 + \left(\frac{\partial^2\psi}{\partial Y^2} + \frac{\partial^2\psi}{\partial X^2}\right)^2\right]^{\frac{1}{2}}\right|^{(n-1)} \tag{27}$$

The dimensionless boundary conditions imposed for solving the equations Eqs. 24–26 are expressed as:

At $X = 0, L$ and $0 \leq Y \leq L$; along the walls $\psi = \theta = 0$

At $Y = L$ and $0 \leq X \leq L$; along the wall $\psi = 0$ and $\frac{\partial\theta}{\partial Y} = 0$

At $Y = 0$ and $0 \leq X \leq D - \frac{W}{2}$; along the wall portion $\psi = 0$ and $\frac{\partial\theta}{\partial Y} = 0$

At $Y = 0$ and $D + \frac{W}{2} \leq X \leq L$; along the wall portion $\psi = 0$ and $\frac{\partial\theta}{\partial Y} = 0$

On the protruding heater walls, $\psi = 0$ and $\theta = 1$.

Equations 24–26 were solved computationally; calculation of Nu_l (local Nusselt number) and $\overline{Nu}$ (average Nusselt number) has been done along the hot walls of the protruding heater.

$$Nu_l = -\frac{k_{nf}}{k_f}.\left(\frac{\partial\theta}{\partial y}\right)_{\text{hot walls}} \tag{28}$$

$$\overline{Nu} = \left(\frac{1}{H}\right)\int_0^H Nu_l.dY|_{\text{vertical hot walls}} + \left(\frac{1}{W}\right)\int_0^W Nu_l.dX|_{\text{horizontal hot wall}} \tag{29}$$

where, H is the length of the vertical hot walls and W is the length of the horizontal hot wall.

3 Numerical Scheme and Validation

Finite Difference Method has been used to solve the set of governing equations along with the boundary conditions. The domain for computation has been divided into 181 × 181 uniform grids. Besides, an under relaxation method has been used along with the convergence criteria:

$$\frac{\Sigma\left|\varepsilon_{i,j}^{n+1} - \varepsilon_{i,j}^{n}\right|}{\Sigma\left|\varepsilon_{i,j}^{n+1}\right|} \leq 10^{-5}$$

where ε represents the stream function ψ and dimensionless temperature θ.

Table 3 Grid independence test

Volume fraction (ϕ) (in %)	Rayleigh number (Ra)	No. of Grids	$\overline{Nu}$ (hot walls)
0.0	10^6	101 × 101	7.5966
0.0	10^6	141 × 141	7.3982
0.0	10^6	181 × 181	7.3467

A grid independence test has been performed for the problem under consideration at various grids, but after 181 × 181 grids, the value of average Nusselt number saturates and due to this reason, the 181 × 181 grid configuration has been taken as the converged grid. The data for grid independence has been given in Table 3. A validation has been performed in comparison with the results of Santra et al. [9] for Rayleigh numbers ranging from 10^4 to 10^6; the comparison has been shown in Figs. 2 and 3. The average Nusselt numbers also have been compared with the present work from which we have found less difference between the present work and Santra et al. [9] (given in Table 4).

4 Results and Discussion

Flow circulation and heat transfer characteristics for diverse Rayleigh numbers and volume fractions (ϕ) have been studied. Solid spherical copper nanoparticles of diameter 100 nm have been taken for the preparation of the nanofluid where water has been taken as the base fluid. Prandtl number of the base fluid i.e. water has been taken as 7.02 [9]. The effective thermal conductivity of nanofluid has been extracted from the model proposed by Patel et al. [14]. The value of constant "c" used in the mathematical model used by Patel et al. has been taken as 3.60×10^4. The values of the rheological indexes m and n have been calculated using the experimental data from Putra et al. [15]. Results have been presented for $Ra = 10^4$ to 10^6 and volume fraction ranged from 0, 1, 2.5 and 5%. Here the protruding heater is taken as an isothermal central heater. Now, the heat transfer has been considered strongly affected by the nature of flow circulation inside the enclosure. As the viscosity increases with an increase in solid volume fraction, the flow circulation gets inhibited due to addition of nanoparticles in the nanofluid. Different Nusselt numbers can be encountered with an increase in solid volume fraction and Rayleigh numbers.

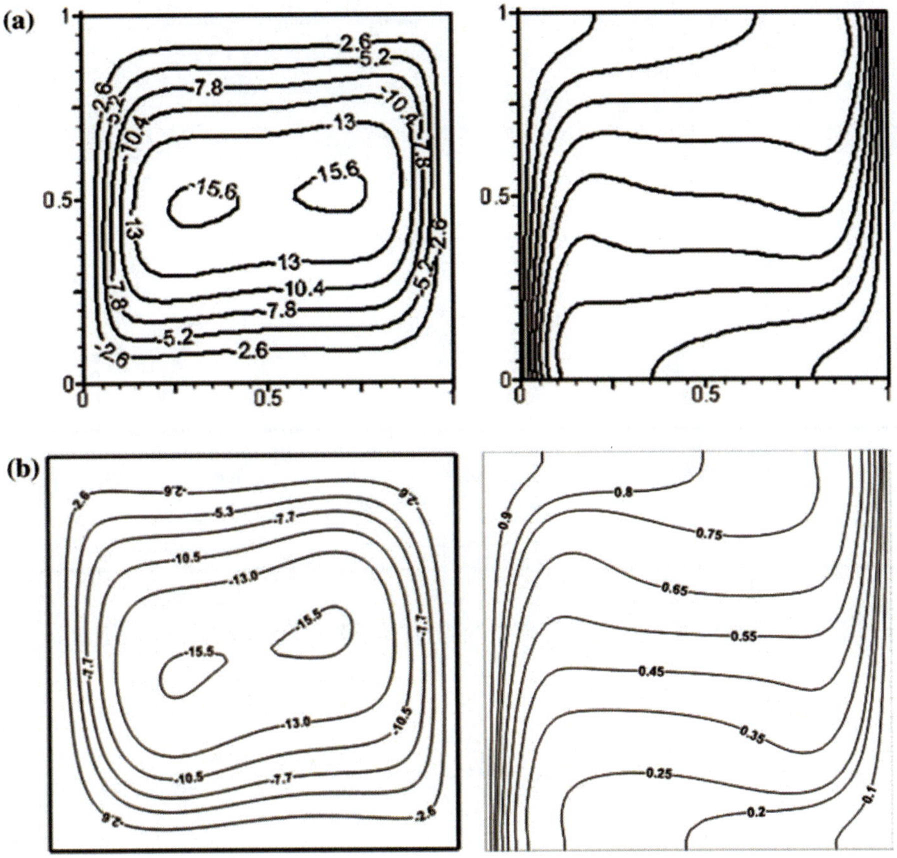

Fig. 2 Comparison of streamline (left) and isotherm (right) contours of **a** Santra et al. [9], with **b** Present work at $Ra = 10^6$ ($\phi = 2.5\%$)

4.1 Influence of Nanoparticle Volume Fraction on Vertical Velocity and Temperature Profiles

Figure 4 represents profiles of vertical velocities at the midplane of the cavity at $Y = 0.5$ for $1.0 \geq X \geq 0$, for $Ra = 10^6$ has been taken for different volume fractions, which shows that at higher volume fractions i.e. 5% volume fraction; the maximum v-velocity decreases. It means that, as the flow becomes more viscous, the flow circulation of nanofluid inside the cavity decreases. Hence, the buoyancy forces decreases and the vertical velocity profiles decreases with an increasing volume fraction. Also, it is to be noted that, the velocity boundary layers along the boundary walls appears to be thickened and the size of the symmetric static zones at the cores of counter-rotating flows decreases. The value of the stream function of the cores is extremely less and negative values which depict that the flow is almost static

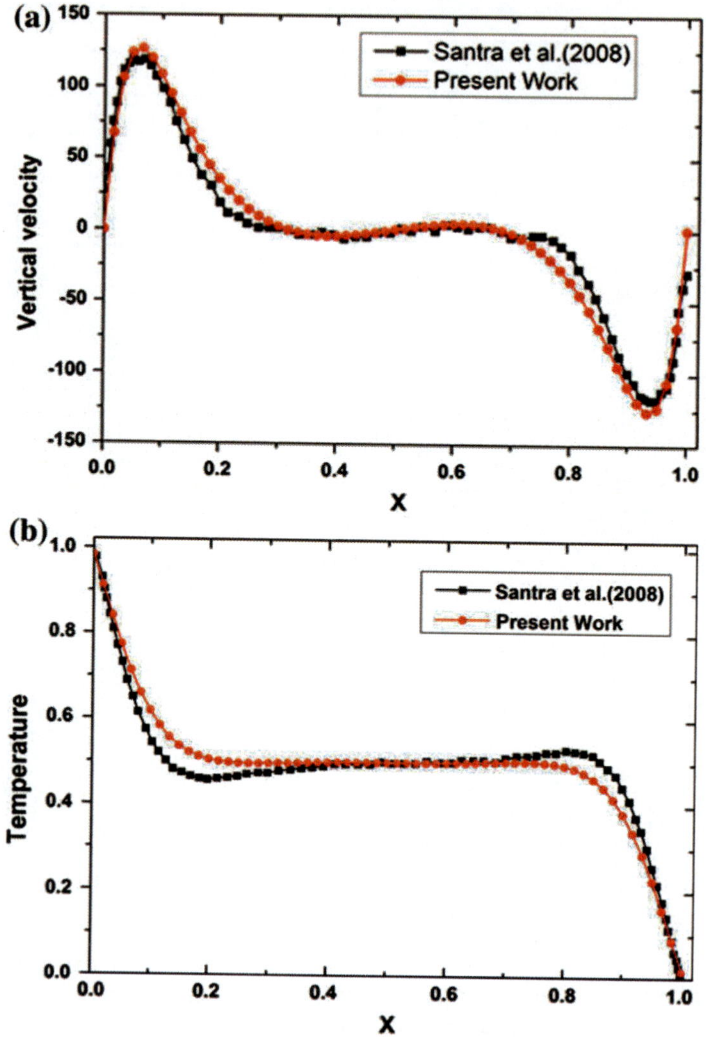

Fig. 3 Validation of **a** dimensionless vertical velocity, **b** dimensionless temperature profiles, at $Ra = 10^6$ ($\phi = 2.5\%$)

Table 4 Validation of Nusselt numbers

Rayleigh Number (Ra)	Volume fraction (ϕ)	$\overline{Nu}$ of Santra et al. [9]	$\overline{Nu}$ of present work
10^4	0.025	1.484	1.5813
10^5	0.025	3.426	3.8791
10^6	0.025	6.670	6.5321

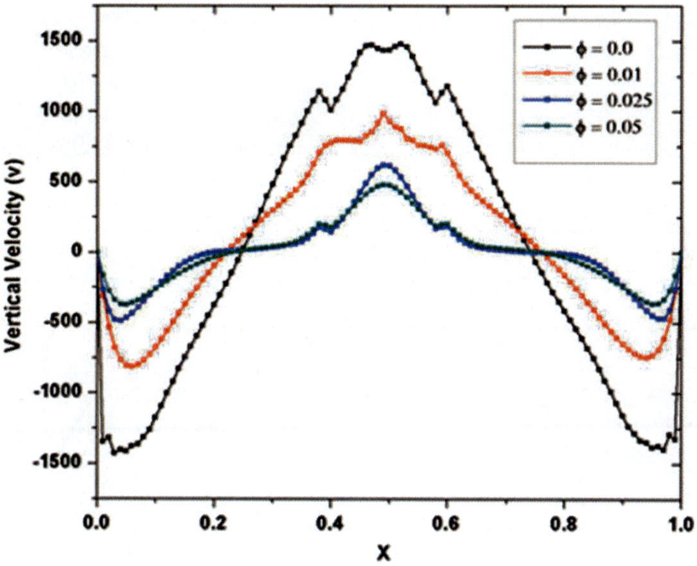

Fig. 4 Vertical velocity profiles at mid plane for different ϕ values at $Ra = 10^6$

at the cores. Figure 5 presents the temperature profile along the mid-plane of the cavity at different volume fractions at $Ra = 10^6$. Figure 5 shows that at $\phi = 0\%$ a uniform or dome shaped temperature profile can be seen, whereas, with increasing volume fraction, as the nanofluid gets more viscous, the nature of the distribution of temperature profiles changes. Besides, increasing effective thermal conductivity with increasing concentration of the nanofluid results in transport of heat better by diffusion, rather than advection at higher concentration of nanofluid.

4.2 Influence of Nanoparticle Volume Fraction on Streamline and Isotherm Contours

The contours of streamlines and isotherms for $Ra = 10^6$ for $\phi = 0\%$, 2.5 and 5% have been shown in Fig. 6. The figure illustrates that as Rayleigh number increases, the effect of buoyancy increases. As solid volume fraction increases inside the flow but that buoyancy effect has been overcome by the increasing effective thermal conductivity and viscosity of the nanofluid inside the cavity which increases the diffusive heat transfer with increasing volume fraction. But if the strength of circulation can be compared for Rayleigh number 10^6 at two different volume fractions; then a mild difference in the values of streamlines can be observed. Most probably this is due to the increasing viscous effect of the nanofluid with increase in volume fractions.

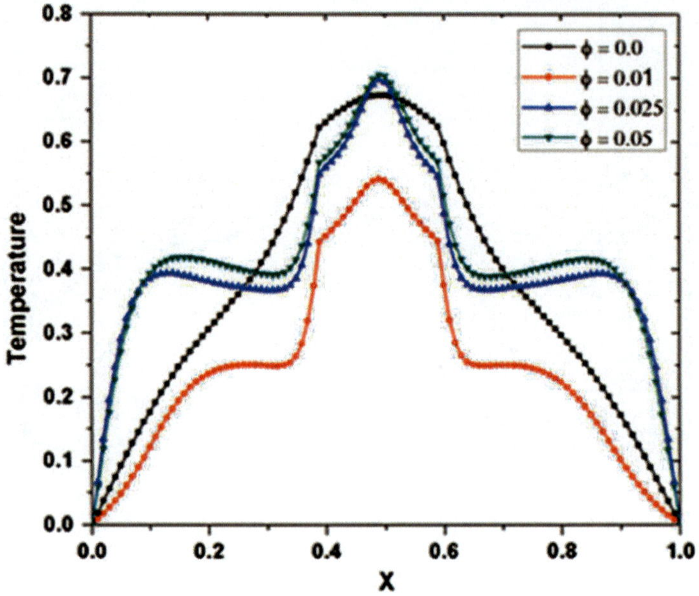

Fig. 5 Vertical velocity profiles at mid plane for different ϕ values at $Ra = 10^6$

Moreover, the apparent increase in thickness of the thermal boundary layer increases with increasing heat transfer; diffusive in nature.

Streamlines and isotherms for $\phi = 2.5\%$ for $Ra = 10^4$, 10^5 and 10^6 has been presented in Fig. 7. It has been observed that as the absolute value of stream function; which is a measure of flow circulation of fluid, increases quickly with an increase in *Ra*. As the buoyancy effect increases with the Rayleigh number, obviously the stagnant cores shift upwards and gets distorted. Hence the velocity boundary layer thickness increases, with an increase in Rayleigh number at the stagnant core regions and decreases along the symmetrically cooled vertical walls.

4.3 Influence of Nanoparticle Volume Fraction on Average Nusselt Number

The average Nusselt numbers $\left(\overline{Nu}\right)$ down the hot walls of the protruding heater have been calculated and presented for different ϕ and *Ra* in Fig. 8. The figure depicts the decrease of Nusselt numbers with an increase in solid volume fraction. The viscosity model of this work has been collected from Putra et al. [15], it is quite obvious that due to the mathematical model of the viscosity used here, revealed same results experimentally. Santra et al. [9] also found a decreasing trend of average Nusselt numbers with increasing volume fraction. It is also important that at $\phi = 2.5\%$

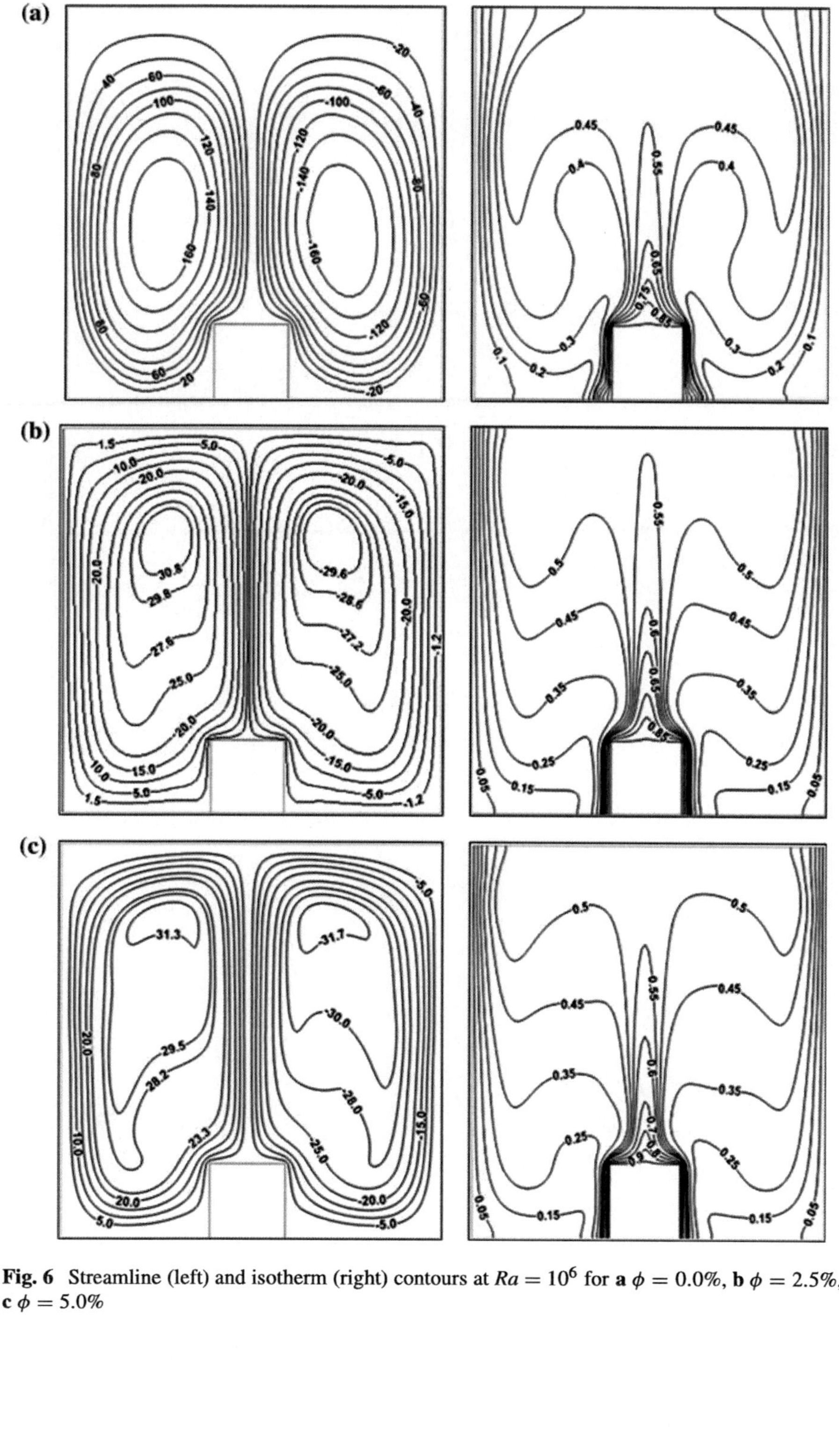

Fig. 6 Streamline (left) and isotherm (right) contours at $Ra = 10^6$ for **a** $\phi = 0.0\%$, **b** $\phi = 2.5\%$, **c** $\phi = 5.0\%$

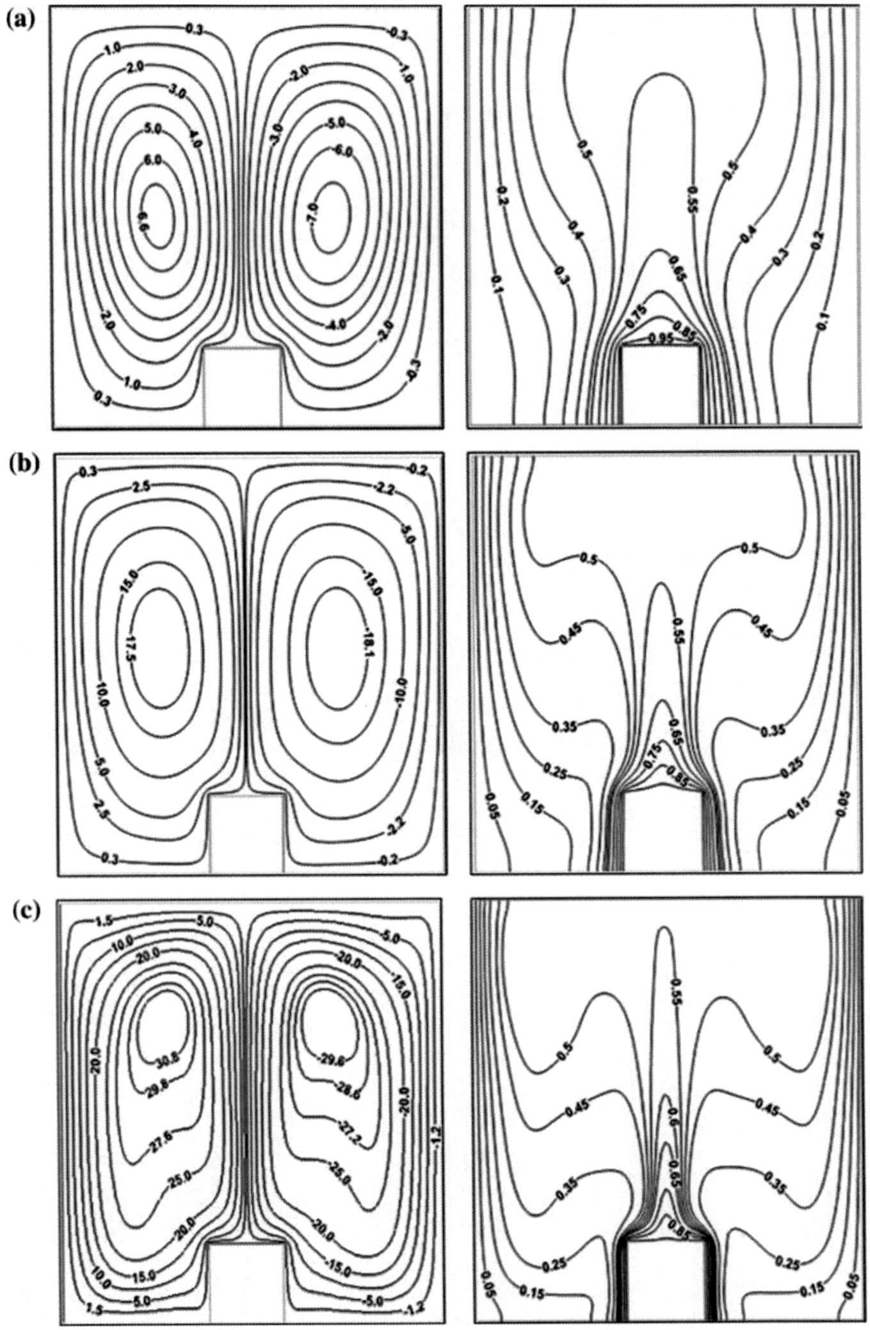

Fig. 7 Streamline (left) and isotherm (right) contours at $\phi = 2.5\%$ for **a** $Ra = 10^4$, **b** $Ra = 10^5$, **c** $Ra = 10^6$

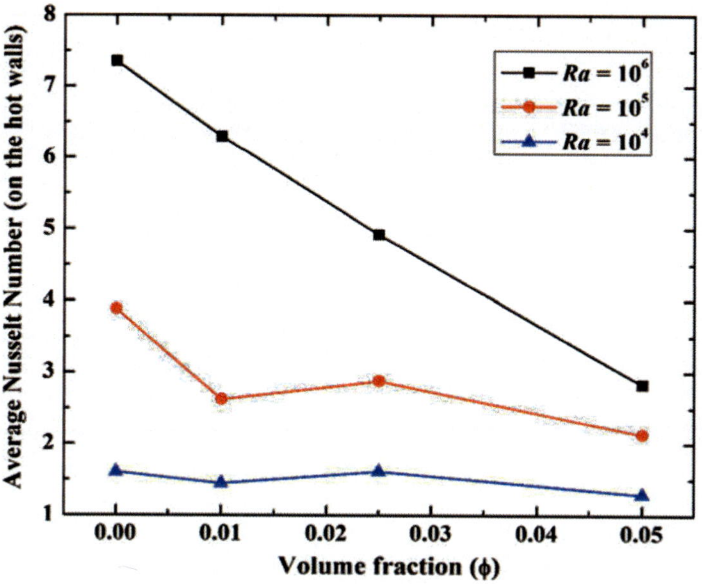

Fig. 8 Comparison of Nusselt number with solid volume fraction (ϕ) for different Rayleigh numbers (Ra)

and at both the Rayleigh numbers $Ra = 10^4$ and 10^5; the Nusselt number slightly increases but that increase is not sufficient for direct heat transfer enhancement and finally it decreases as the volume fraction increases up to 5%.

5 Conclusion

Enhancement of heat transfer utilizing copper-water nanofluid has been investigated in a two-dimensional square cavity with centrally installed isothermal heater inside has been studied numerically for Rayleigh numbers 10^4 to 10^6 and volume fraction of $5.0\% \geq \phi \geq 1.0\%$.The diameter of copper nanoparticles has been taken as 100 nm and Prandtl number for the base fluid i.e. water is taken as 7.02 [9]. The numerical computation has been done using the stream-function (ψ)-vorticity (ω) formulation. Ostwald-de Waele model has been used to formulate the shear stresses of the non-Newtonian nanofluid. The rheological indexes (m and n) have been calculated using the experimental data available from Putra et al. [15] and have been extracted from Santra et al. [9] The thermal conductivity of the nanofluid has been calculated using the model suggested by Patel et al. [14]. Final results reflected a decrease in Nusselt number for a constant volume fraction.

References

1. Mahian O, Kolsi L, Amani M, Estellé P, Ahmadi G, Kleinstreuer C, Marshall JS, Taylor RA, Abu-Nada E, Rashidi S, Niazmand H (2019) Recent advances in modelling and simulation of nanofluid flows—part II: applications. Phys Rep 791:1–59
2. Mahmoudi AH, Shahi M, Shahedin AM and Hemati N (2011) Numerical modeling of natural convection in an open cavity with two vertical thin heat sources subjected to a nanofluid. Int J Heat Mass Transf 38(1):110–118
3. Shahi M, Mahmoudi AH, Talebi F (2010) Numerical study of mixed convective cooling in a square cavity ventilated and partially heated from the below utilizing nanofluid. Int Commun Heat Mass Transf 37(2):201–213
4. Mahmoudi AH, Shahi M, Raouf AH (2011) Modeling of conjugated heat transfer in a thick walled enclosure filled with nanofluid. Int Commun Heat Mass Transf 38(1):119–127
5. Mahmoudi AH, Shahi M, Talebi F (2010) Effect of inlet and outlet location on the mixed convective cooling inside the ventilated cavity subjected to an external nanofluid. Int Commun Heat Mass Transf 37(8):1158–1173
6. AlAmiri A, Khanafer K, Pop I (2009) Buoyancy-induced flow and heat transfer in a partially divided square enclosure. Int J Heat Mass Transf 52(15–16):3818–3828
7. Khanafer K, Vafai K, Lightstone M (2003) Buoyancy-driven heat transfer enhancement in a two-dimensional enclosure utilizing nanofluids. Int J Heat Mass Transf 46(19):3639–3653
8. Ho CJ, Chen MW, Li ZW (2008) Numerical simulation of natural convection of nanofluid in a square enclosure: effects due to uncertainties of viscosity and thermal conductivity. Int J Heat Mass Transf 51(17–18):4506–4516
9. Santra AK, Sen S, Chakraborty N (2008) Study of heat transfer augmentation in a differentially heated square cavity using copper–water nanofluid. Int J Therm Sci 47(9):1113–1122
10. Santra AK, Sen S, Chakraborty N (2009) Study of heat transfer due to laminar flow of copper–water nanofluid through two isothermally heated parallel plates. Int J Therm Sci 48(2):391–400
11. Loenko DS, Shenoy A, Sheremet MA (2019) Natural Convection of Non-Newtonian Power-Law Fluid in a Square Cavity with a Heat-Generating Element. Energies 12(11):2149
12. Yigit S, Battu M, Turan O, Chakraborty N (2019) Free convection of power-law fluids in enclosures with partially heating from bottom and symmetrical cooling from sides. Int J Heat Mass Transf 145:
13. Khezzar L, Siginer D, Vinogradov I (2012) Natural convection of power law fluids in inclined cavities. Int J Therm Sci 53:8–17
14. Patel HE, Sundararajan T, Pradeep T, Dasgupta A, Dasgupta N, Das SK (2005) A micro-convection model for thermal conductivity of nanofluid. Pramana J Phys 65:863–869
15. Putra N, Roetzel W, Das SK (2002) Natural convection of nano-fluids. Heat and Mass Trans 39(8–9):775–784

Numerical Analysis of Micro-Scale Diffusion Flame Structure Using Methane Gas as Fuel

Tanumoy Banerjee and Nripen Mondal

Nomenclature

C_p	Heat capacity at constant pressure
D	Diffusion Coefficient
d	Burner diameter (mm)
Fr	Froude number
g	Gravitational Acceleration (m/s^2)
h	Flame height
n_i	Stoichiometric Coefficient based on mole for i-th species ($n_f = 1, n_o = 2$)
Pe	Peclet Number
p	Dynamic Pressure
q	Heat of formation per unit mass of methane
Re	Reynolds number
r	Radial direction
T	Stress Tensor
T	Temperature
u	Velocity vector with radial (=r) and axial (=x) components
u_d	Molecular diffusion velocity
u_e	Exit velocity of methane

T. Banerjee
Department of Mechanical Engineering, Jadavpur University, Kolkata 700032, West Bengal, India

N. Mondal (✉)
Department of Mechanical Engineering, Jalpaiguri Government Engineering College, Jalpaiguri 735102, West Bengal, India

S. K. Ghosh et al. (eds.), *Advances in Thermal Engineering, Manufacturing, and Production Management*, Lecture Notes in Mechanical Engineering,
https://doi.org/10.1007/978-981-16-2347-9_6

1 Introduction

Diffusion flames of laminar-jet types are broadly classified into three types: (i) Burke-Schumann flame (diffusion-controlled) [1], (ii) Roper flame (buoyancy controlled) [2], and (iii) Diffusion and convection controlled micro flame [3]. Recently the micro flame concept was investigated by Ban et al. [3] and Nakamura et al. [4]. The microscale flames are different from the observed candle-like diffusion-controlled flame, as they are found to be nearly spherical [2]. Buoyancy effect is neglected for Ropar diffusion-controlled flame, as it is mainly laminar diffusion flames. Due to these spatial characteristics, lots of studies are being conducted on microscale flames. Firstly, because of micro flame's spherical shape, it can be used as a model for microgravity (generally 10^{-5} g) [4] flames. Secondly, its small size attributes to minimum heat release from the flame, with extensive heat loss to the burner; indicating that every time at near-limiting state, this flame may be formed and thus giving a handful of tool to investigate the limiting behavior of extinction and stability of flame under different driven conditions. Thus, this shows that microscale flames are very important to investigate diffusion flames phenomena, extinction, and stability phenomenon. The micro flame has the following characteristics:

$$\mathrm{Fr}\left(=\frac{u_e^2}{gd}\right) > 1 \text{ and } \mathrm{Pe}(= u_e/u_d) = O(1) \tag{1}$$

here u_e is methane exit flow velocity from burner port, u_d denotes molecular diffusion velocity and d is the burner diameter. From scaling analysis, it can be defined that if a flame has $u_e = 0.1$ m/s, $d = 1$ mm, and Froude number (Fr) ≫1, it will show very little or no buoyancy effect. Fr ≫1 implies that characteristic acceleration time by buoyancy is of greater significance as compared to a characteristic flow time. Peclet number of the order 1, i.e., O (1) [1] indicates that characteristic flow time is at per with diffusion time. The micro flame is characterized as 'diffusion-convection controlled flame' because of the usage of diffusion time [4] here. The burner port size being small, the jet volume flow rate stays small when methane flow rate is high.

2 Numerical Model and Adopted Numerical Scheme

2.1 *Numerical Model and Adopted Numerical Scheme*

The assumptions made for mathematical modeling while doing numerical analysis in CFD package are (Fig. 1 [4]):

- We have taken 2D axisymmetric model.
- Ideal gas laws are incorporated regarding densities.

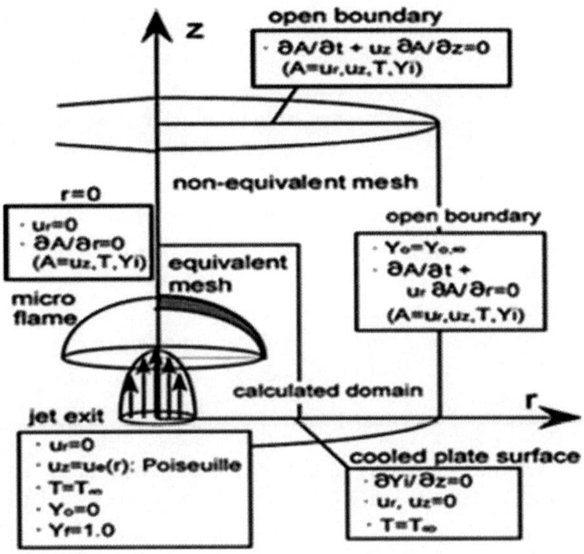

Fig. 1 A schematic of mathematical model and applied boundary conditions

- Temperature-dependent thermo-physical properties and transport properties for all the species (they are equal to those of air) using CHEMKIN reaction file are taken.
- For methane-air reaction model, we have considered one-step irreversible model [5].
- Laminar finite rate model is taken.
- Methane inlet is applied at the bottom part of the burner port.

We validated the numerical domain of the laminar jet diffusion flame by soling four partial differential equations of fluid dynamics (time-dependent continuity, momentum, energy, species) in finite volume method using commercial CFD software. We considered axisymmetric 2-D model as the jet is cylindrical. We injected pure methane from the bottom of the domain, called burner, into ambient atmosphere of 1 bar pressure. Initial temperature of numerical domain is mentioned as 300 K and initial chemical composition is taken as 21% oxygen and 79% nitrogen. Burner is modelled within the numerical domain so the properties are taken. Doing so, we have considered the backward species diffusion inside numerical domain. Considering gravity as an external force, at the open boundaries, far-field conditions are applied and on the burner surface, we have taken non-slip, non-catalytic, and fixed temperature (300 K) as boundary conditions. Later microgravity analysis is done. In order to meet the current experiments, inner and outer diameter of Bunsen burner, and volumetric flow rate are set. For this study, adaptation of complete transport and multi-step reaction mechanisms has been done for every species and all reversible reaction steps for methane–air mixture. We have used CHEMKIN properties file and incorporated SMOOKE's transport file in CFD package.

We have used finite-volume method to solve the equations numerically in a staggered grid system. Flux terms on the cell surface have been taken care of by using central difference scheme and while doing integration over time, Euler implicit method is used. At every time step, iterations have been done using method of line by line successive over-relaxation (SOR). SIMPLE [6] scheme is used to consider the two-way coupling between pressure and velocity in Navier Stokes equation. The numerical domain considered for calculation is taken up respectively as 40d and 60d along radial and axial direction, allocating the minimum fine grid size as d/10, nearest to the burner; coarser sized grids are created closer to the outer boundaries. At first, calculations were done for time step of (1.0×10^{-5}) s and it was continued until steady-state was reached. We calculated the same for different time steps from 10^{-5} to 10^{-7} s for time independence study. All the 3-time steps match with the results the same way. So, we took 10^{-5} time steps for our analysis to reduce computation time. In order to ensure that there is zero dynamic response in the system, solutions obtained from steady-state analysis are checked with the increment of time step size. Here the burner inner diameter was fixed at d = 0.4 mm and exit fuel flow velocity at the calculation zone has been varied from $u_e = 1{-}5.2$ m/s.

2.2 *Governing Equations*

The time-dependent fluid flow governing equations are solved numerically to get the steady-state result at each node. The four fluid dynamics equations are as follows (D/Dt is defined as non-dimensional substantial derivative):

2.2.1 Continuity Equation

$$\frac{D\rho}{Dt} + \rho\nabla.u = 0$$

2.2.2 Momentum Equation

$$\frac{\rho Du}{Dt} = -\nabla.\mathbf{T} - \nabla p + (\rho - \rho_\infty)g$$

$$\mathbf{T} = \left(p + \frac{2}{3}\mu\,\nabla.u\right)U - \mu\left[(\nabla.u) + (\nabla.u)^T\right]$$

2.2.3 Energy Conservation Equation

$$\rho c_P \frac{DT}{Dt} - \nabla.(\lambda \nabla T) = q\omega$$

2.2.4 Species Conservation Equation

$$\rho \frac{DY_i}{Dt} - \nabla.(\rho D \nabla Y_i) = -v_i \omega$$

2.3 *Results and Discussions*

2.3.1 Flame Height

Figure 2 shows a comparative plot for nondimensional flame height with respect to burner diameter (h/d) versus different methane exit flow velocities from 1 to 5.2 m/s. We defined the flame height where the flame front reaction rate is 5–6 times lower than the reaction rate at the burner locality, i.e., $\omega = 0.001$ g/cm^3.s; roughly one-fifth of the highest rate of reaction near burner locality and where emission of hydrocarbons is maximum. We found a linear increment of flame height with the

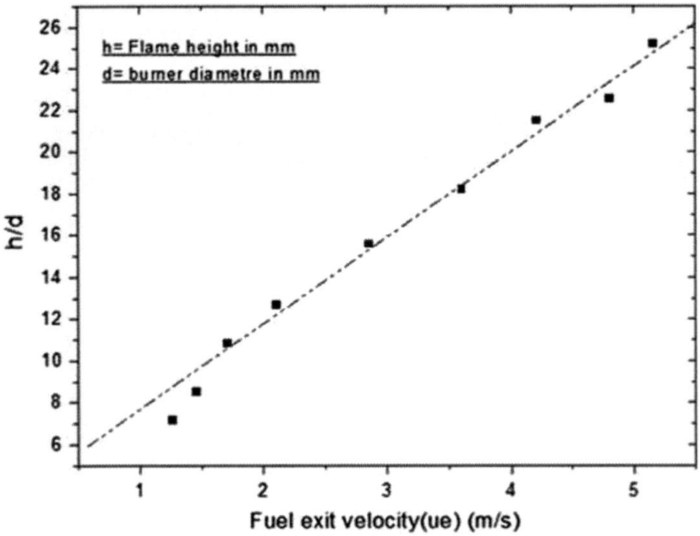

Fig. 2 Non-dimensional flame height with respect to burner diameter vs different exit methane flow velocity: u_e

increase of exit fuel flow velocity. We found almost a spherical flame shape for low Re flow at the burner tip and with an increase of velocity this shape changes to near candle and distorted candle shape which is due to the balancing between convection and diffusion velocity for methane. It has been shown that the height of the flame shows linear incremental relation with fuel (methane) exit velocity. This gives the indication that this type of flame has same feature with diffusion flames (Roper flame) characterized as "buoyancy controlled diffusion flame" [1]. Our numerical result also agrees with the same. This linear dependence of flame height with exit fuel velocity is comparable to previously defined Burke-Schumann and Roper flame type. We calculated the gradient from Fig. 2 between non-dimensional flame height (h/d) and fuel flow velocity and we found it to be 4.5 s/m which is slightly higher than Roper type diffusion flame which shows that this flame is slightly steeper than Roper flame when it comes to considering flame height.

2.3.2 Flame Structures and Characteristics

Figure 3 shows validated results for $u_e = 1$ and 2.5 m/s. We have compared several 2-D contour plots: (a) Vector plot of methane flow, (b) Methane concentration contour, and (c) Oxygen concentration contour. Our result successfully validates the results from Nakamura et al. [4]. We have successfully found out that temperature of outer skirt of flame is lower by at least 200 K than the adiabatic flame temperature, which was given by [4].

At the fully developed flame tip our temperature measures 800–850 K but at the flame edge temperature data shows 620–650 K. The difference in temperature between adiabatic flame temperature and temperature at outer skirt of the flame [7] of methane (roughly 2230 K) is more than 1400 K.

Here, for $u_e = 1$ m/s, temperature rises steadily along the burner axial direction and becomes maximum (842.5 K) at the flame tip, but for $u_e = 2.5$ m/s, after the exit from the burner temperature suddenly falls because of the initially exothermic reaction between methane and air, then increases steadily along both the axial and radial direction giving maximum flame tip temperature of 748 K at the flame edge along radial direction.

Figure 3 shows 2-D contour profile comparison of flow velocity vector and methane and oxygen concentration of micro flame for two different methane exit velocity. As the buoyancy-driven velocity is dependent on gravitational constant (g) and flow residing time inside channel, so very little effect of buoyancy can be seen on flow field when methane exit flow rate is either high or low. From Fig. 3a it can be seen the vertical velocity vector for low velocity is higher because of thermal expansion, low Re flow, and buoyancy effect. This accelerated flow disperses the species concentration balance given in figures (b), (c), showing that these properties change lot radially than axially. But the flame is controlled by only convection-diffusion balance.

In Fig. 3, we can see at higher flow rate, air flows near the quenching region of the flame for which the vertical displacement of velocity vector for $u_e = 2.5$ m/s is

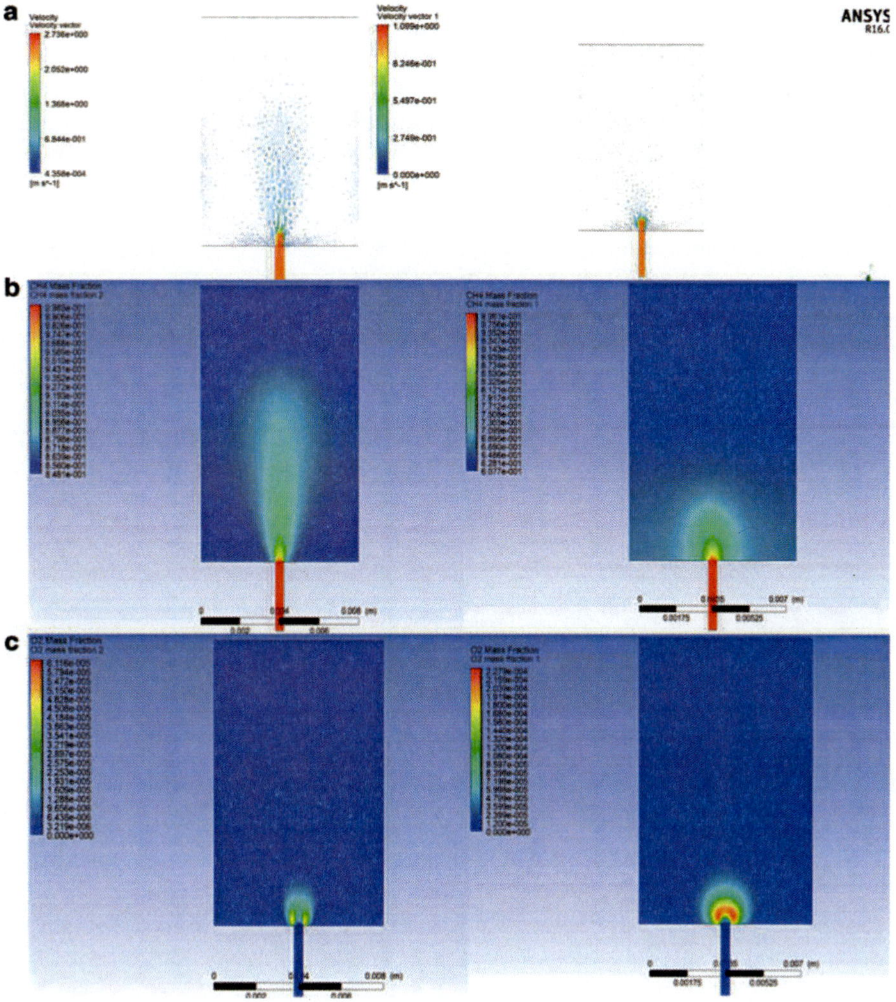

Fig. 3 Obtained 2-D profile contours under gravity: **(top)** flow velocity vectors, **(middle)** methane concentration contours, **(bottom)** oxygen concentration contours. Left one is for methane exit velocity: $u_e = 2.5$ m/s and right one is for flow velocity: $u_e = 1$ m/s

lower. Diffusion flame structure for high flow rate gets affected by ambient air due to flow entrainment. The flame characteristics, species concentration gradient can be greatly affected by this entrained air for low Re and high Re flow. Flow of entrained air can be minimized if we use flame sheet model in analyzing the mathematical model and also use laminar finite rate in rate reaction model.

A sharp increment of 35–40% of length of numerical domain is seen to have the oxygen concentration gradient for $u_e = 1$ m/s than for $u_e = 2.5$ m/s which is depicted in the Fig. 3c.

Figure 4 predicts the central axis structure of methane flame having two different methane velocity. We did comparative analysis of maximum temperate rise, temperature profile, and oxygen, methane concentration change with (Z/d), which defines non dimensional relative distance from reaction zone. Figure 5 clearly shows double-peak structure of CO mass concentration for the two conditions in this study, which means that inside the numerical domain two combustion zones are found and thus two peaks are there for produced CO mass concentration. As CO is produced due to non-stoichiometric combustion and this is the result of air entrainment near the quenching zone. We can say that any microscale flame cannot be categorized as only a premixed or non-premixed flame which is proved by two CO peaks. Quenching distance exists in all flames. Quenching distance means the distance along the axis of the burner, from burner surface up to the base of the flame produced. At low flow

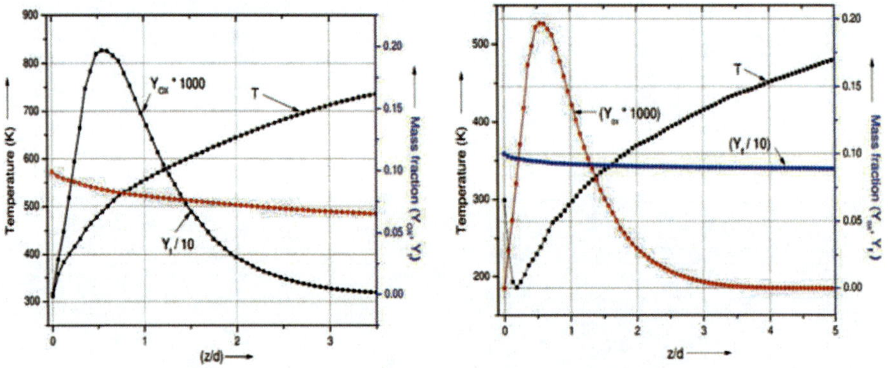

Fig. 4 Temperature (T), oxygen (Y_{ox}) and methane (Y_f) concentration versus (z/d) graph along $r = 0$ (center axis of the flame). Left one is for methane exit velocity, $u_e = 1$ m/s and right one is for $u_e = 2.5$ m/s

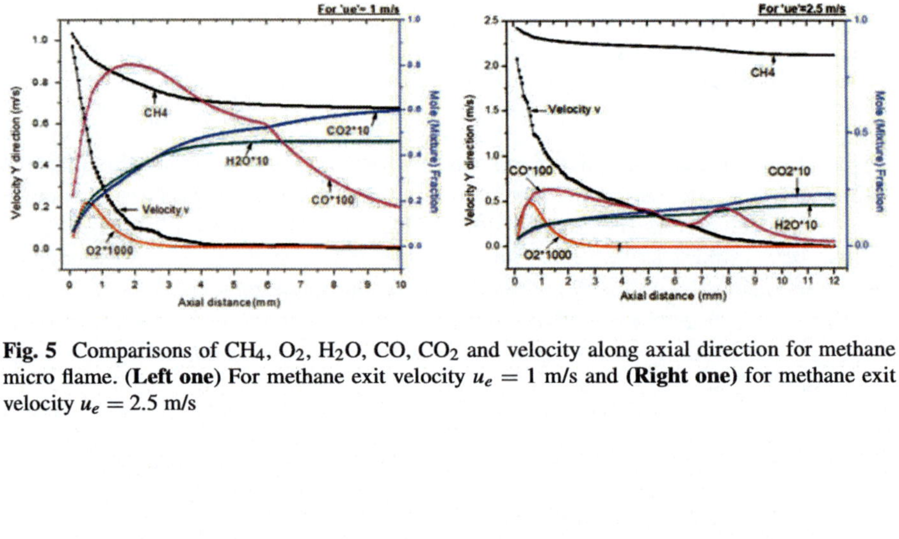

Fig. 5 Comparisons of CH_4, O_2, H_2O, CO, CO_2 and velocity along axial direction for methane micro flame. **(Left one)** For methane exit velocity $u_e = 1$ m/s and **(Right one)** for methane exit velocity $u_e = 2.5$ m/s

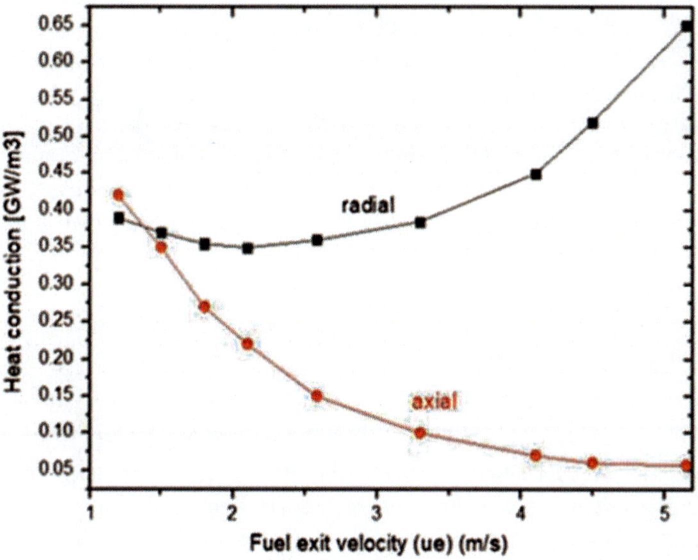

Fig. 6 Centre line heat conduction in both axial and radial direction vs fuel exit velocity comparison for different methane flow velocity

rates, the two CO peaks are closer but at high flow rates they tend to disperse showing that at outer part of flame non-premixed combustion is occurring.

Figure 6 shows heat conducted axially and radially from the energy equation at the center axis with the change in methane exit flow velocity. At the initial stage, axial heat conducted is larger than heat conducted radially at around $u_e = 1$ m/s but with increase of fuel exit flow velocity the radial heat conduction surpasses axial heat conduction due to wall heat transfer and wall quenching effect on the micro-flame. With the increase of velocity, the difference between radial and axial heat diffusion rate also increases in a quadratic fashion. The axial diffusion transport is less at high flow velocity due to thermal expansion.

From Fig. 4 it can also be seen that maximum temperature attained for $u_e =$ 1 m/s is at least 200 K higher in non-dimensional axial flame distance. This is due to the axial diffusion for higher fuel exit flow rate. Oxygen concentration is also substantially higher for low Re flow at non dimensional axial distance ($Z/d = 2$) due to very little effect of buoyancy. From Fig. 5 it can also be seen that flame velocity for high Re flow is high in axial direction and methane concentration diffuses very flatly for high methane flow exit velocity which is due to sharply defined non premixed flame outer zone for high Re flow.

3 Conclusion

Numerical studies were conducted on ANSYS FLUID FLOW (FLUENT v16.0) and on ANSYS FLUID FLOW (CFX v16.0) and graphs were plotted on ORIGIN PRO 8. Time independent study is done and we have taken the data for 10^{-6} time steps with a fine grid at the burner wall and applying inflation layer of 3.

The following conclusions can be drawn from this study:

- When exit methane flow velocity is 1–2.5 m/s, the micro-flame does not show any buoyancy effect which can be seen by the semi-spherical shape of the flame.
- We have found a linear dependency of non-dimentionalized flame height with different methane exit flow velocity. As we considered the effect of axial diffusion transport to measure flame height, this linear coefficient function is steeper than Ropar flame as it does not consider diffusion.
- To establish the microscale diffusion flame structure, we incorporated laminar finite rate model with stiff chemistry solver at the burner outlet. This numerical calculation properly validates with the numerical results from Nakamura et al. [4]. But at flame edge due to wall heat transfer and detailed laminar finite rate chemical kinetic model the result differs from their experimental work.
- As thermal expansion predominates at higher methane flow rate so mass, energy transport due to axial diffusion minimises. From various chemical concentrations comparison and thermal analysis at separate fuel flow rate along axial and radial direction from centre axis we conclude that heat conducted faster with increasing fuel velocity in radial direction but temperature gradient is sharper for high methane flow rate though maximum temperature is lower for $u_e = 2.5$ m/s.

References

1. Burke SP, Schumann TEW (1928) Diffusion flames. Ind Eng Chem 20:998–1004
2. Ropar FG (1977) The prediction of laminar jet diffusion flame sizes: part I. Theoretical Model Combust Flame 29:219–226
3. Ban H, Venkatesh S, Saito K (1994) Convection-diffusion controlled laminar micro flames. J Heat Transfer 116:954–959

4. Nakamura Y, Ban H, Saito K, Takeno T (1997) Structure of micro (millimetre size) diffusion flames. In: Proceedings of 1997 meeting of the central states section/the combustion institute, point clear, AL U.S.A., Apr 1997, pp 160–163
5. Coffee TP, Kotlar AJ, Miller MS (1984) The overall reaction concept in premixed, laminar, steady-state flames. II. Initial temperatures and pressures. Combust Flame 58:59–67
6. Patankar SV (1980) Numerical heat transfer and fluid flow. McGraw-Hill Inc.
7. Saito K, Williams FA, Gordon AS (1986) Structure of laminar coflow methane-air diffusion flames. J Heat Transf 108:640–648

Effect of Different Surface Types of Loading Vessel Wall on Combustion Performance of a Domestic LPG Cook Stove

Mithun Das

1 Introduction

Most of the developed country, liquefied petroleum gas (LPG) has been extensively used as a fuel in the domestic cooking appliances. Because it has high heating rates and rapid combustion properties and, at the same time, it produces relatively less polluted products [1, 2]. Hence, thermal efficiency study is an essential part of cooking stoves. It mainly depends on the fuel flow rate (heat input), equivalence ratio of air/fuel mixture, fuel composition, loading height (the height of the pan from the burner top), and jet to jet spacing [3–5]. In a cooking stove, the rich air-fuel premixed mixture enters into the burners from mixing tube. The optimal value of the equivalence ratio is around 1.4 for getting maximum efficiency [4]. With increasing in loading height, at first, the efficiency reaches an extreme, then reduces. Ashman et al. [6] performed the experiments on a single cooktop burner to find out the effects of loading height and thermal input on its thermal efficiency and emissions. They found that the efficiency reduced with rising load height. For a given load height, thermal efficiency was higher for less thermal inputs. For low load heights (0–50 mm), both of NO_2 and CO emission reduced with increasing load height. Hou and Ko [7] investigated the effects of oblique angle and load height on the combustion characteristics (i.e. flame structure, temperature distribution, and thermal efficiency) of a Bunsen-type burner. At a fixed oblique angle, the thermal efficiency first increased with heating height then reduced gradually. They also found the optimum value of oblique angle (60°) and load height (12 mm) to achieve the highest thermal efficiency. Hou et al. [3] investigated a comparative study on a swirl flow burner and radial flow burner and found that the thermal efficiency of the swirl flow burner was higher than

M. Das (✉)
Department of Mechanical, IIT Madras, Chennai, Tamil Nadu 600036, India

School of Nuclear Studies and Application, Jadavpur University Salt Lake Campus, LB 8, Sector III, Kolkata 700106, India

S. K. Ghosh et al. (eds.), *Advances in Thermal Engineering, Manufacturing, and Production Management*, Lecture Notes in Mechanical Engineering,
https://doi.org/10.1007/978-981-16-2347-9_7

the conventional radial flow burner. Zhen et al. [8] also noticed that swirl type burner improved thermal efficiency and reduced pollutant emission. Both thermal efficiency and CO emission are increased with increasing jet to jet spacing of the burner holes [2]. Boggavarapu et al. [4] studied both experimentally and numerically to find out the thermal efficiency of a domestic cook stove burner using liquefied petroleum gas (LPG) and piped natural gas (PNG) fuels. They also improved thermal efficiency with the attachment of a circular insert and a radiant sheet. Thermal efficiency increased with increasing fuel flowrate for both existing and modified burners. Thermal efficiency improvement using the design modification was predicted about 5% for LPG at a maximum flow rate. They further predicted that thermal efficiency of PNG fuel was mainly affected by loading height rather than modifications. At an optimal load height (24 mm), the efficiency of PNG was increased by about 10% from the baseline loading height. Recently our group investigated a detailed numerical study on the flow characteristics of the fuel-air mixture in a self-aspirated mixing tube of a domestic LPG cook stove [9] and the burning characteristics of the mixture on the top of the burner [10].

In this work, we study the thermal performance of a domestic LPG cookstove using computational fluid dynamics (CFD) analysis and find out the effect of surface types of the loading vessel. Aluminum alloy and stainless steel with polished/oxidized/anodized surfaces are considered as loading vessel walls. The simulation work is done on a 3-D domain of 12° section of the cooktop burner.

2 Numerical and Experimental Methodology

Thermal efficiency is a very significant parameter for the performance study of a domestic cook stove. According to Indian standard (IS) 4246:2002 [11], the thermal efficiency of a burner is defined as the percentage of the heat input of the fuel which is transferred to the water of the loading vessel. Numerically efficiency prediction according to the definition is very difficult and time-consuming. As it is required to the modeling of the whole vessel with containing water including water circulation and evaporation at the top surface. For simplicity, instead of water and loading vessel wall, the outer wall of the vessel is only considered as boundary and given a constant wall temperature boundary condition in the present computational domain. Thermal efficiency is here predicted from the percentage of total heat input which is transfer through the vessel wall.

In a domestic cookstove, the injected fuel is mixed with primary air in a self-aspirated mixing tube and mixture burns over the burner top in the presence of surrounding air. The flames from the burner top impinge on the loading vessel that is placed over burner top with the help of pan supports. The external diameter and height of the vessel are considered as 260 mm and 140 mm, respectively, for LPG flow rate 71 L/h (L/h), following the Indian standard for thermal efficiency testing [11]. LPG is taken as a mixture of 40% propane and 60% butane by mass.

The current numerical modeling of the combustion and flow dynamics are performed using the commercial code ANSYS-Fluent 17.2. The computational domain and grids are developed in the ANSYS SpaceClaim and ICEM CFD, respectively. A 12° sector of the burner head including a flat-bottomed vessel, which is positioned over the burner head, is used as a periodic symmetry domain for CFD simulation (Fig. 1). The CutCell method is used to generate the meshes for this complicated geometrical domain. Finer mesh grids are employed at the combustion zone and at the vessel wall boundaries, where higher gradients are expected (Fig. 2).

The standard κ-ε model is used for the transport equations. A three-step reaction mechanism—two for propane and one for butane—including four reactions (forward

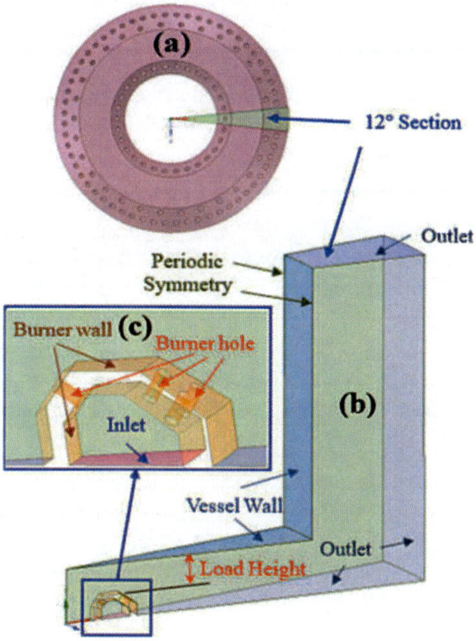

Fig. 1 Computational 3-D domain and boundary conditions for a 12° sector of the burner head: **a** the burner head, **b** numerical domain for cook stove and **c** enlarged view of the flow passages in the burner

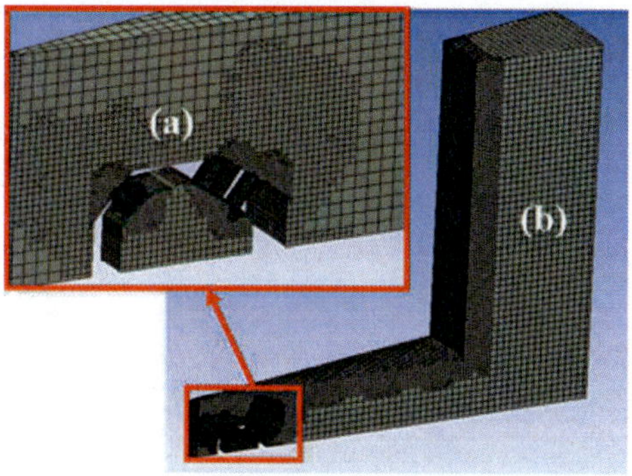

Fig. 2 Grid details of the computational domain: **a** 12° section and **b** enlarge view of the burner domain area

and backward) [10] is considered in the present study. The combustion model, which is primarily applied in the current computational work, follows the species transport along with finite-rate reaction model. The volumetric reaction and STIFF chemistry solver are enabled in Fluent setting.

All boundary condition (BC)s are shown in Fig. 1. Inlet BC is used for constant mass flow inlet of fuel-air mixture, while the pressure outlet BC is considered for the outlets. The burner walls are assumed as an adiabatic wall; however, a constant temperature (398 K) BC is set for both the bottom and side walls of the vessel. The discrete ordinates radiation model is used for radiation from the computational domain. Gravitational force is considered downward of the Y-axis to simulate the effect of the buoyancy forces around the vessel wall for the temperature gradients.

In combustion problem with considering buoyancy-driven flows, the solution is generally run using a two-step solution procedure. In first, the solution is obtained in cold flow conditions where the solution is iterated without reaction and then the ignition is patched near the combustion zone. The following values for the under-relaxation factor, 0.3 for pressure; 1.0 for density; 1.0 for body force: 0.7 for momentum; 0.8 for energy, and 1.0 for all species mass fractions, were used for converging the solution. The solution is considered converged when the heat flux through the vessel wall is almost unchanged.

The thermal efficiency test of an LPG stove is done also experimentally for validating numerical results. The schematic of the test setup is shown in Fig. 3. The thermal efficiency of the burner has been tested according to the standard. The LPG gas is supplied from the cylinder to the burner at a constant pressure of around 30 mbar. The water boil testing pan size and mass of water in the pan were considered as per standard. In the time of testing, the measured water of the pan is heated from initial temperature (T_1) to 90 °C (T_2) and noted down the time duration. During the testing period when the temperature of the water reaches ~80 °C, the water is continuously stirred until the test end. The efficiency (η) is calculated from the following expression:

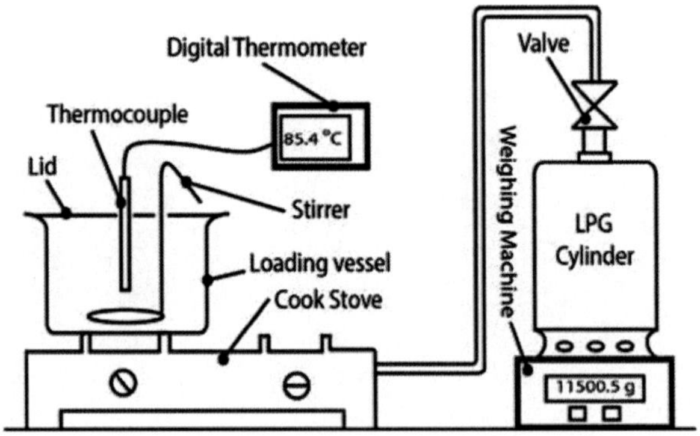

Fig. 3 Schematic of the experimental setup

$$\eta = \frac{100(M_w + M_{we})(T_2 - T_1)}{M_f \times CV_{net}} \tag{1}$$

where M_w and M_{we} are the quantity of water in the vessel and the water equivalent of the vessel together with stirrer and lid in kg, respectively. M_f is total LPG consumption in kg during the test. The net calorific value (CV_{net}) of LPG is taken as 10,900 kcal/kg for efficiency measurement.

3 Results and Discussion

3.1 Grid Independency Test and Model Validation

Grid independence tests are done for different types of mesh domains. The number of elements of the domain is increased with increasing mesh refinement near the combustion zone and vessel wall. A grid independence study has been conducted to find out enough number of mesh elements to achieve satisfactory results. In this context, thermal efficiency is calculated for different types of mesh domains and results are shown in Fig. 4. Both of mesh type $\mathbf{e}_1$ and $\mathbf{e}_2$ give poorer results than other types of mesh domains. The efficiency results of e_3 and other finer types of mesh domains are fairly close to each other and obviously, e_3 will take less computational time than others. Hence, the mesh type 3 ($\mathbf{e}_3$) is considered for the rest computational works. The present computational model is validated against experimental data. In the experimental measurements, loading height, equivalence ratio, and fuel flowrate were maintained at 18 mm, 1.4, and 71 L/h, respectively. The experimentally calculated value of the efficiency was 67.30 ± 0.66 which deviates only 0.4% from the numerically predicted efficiency.

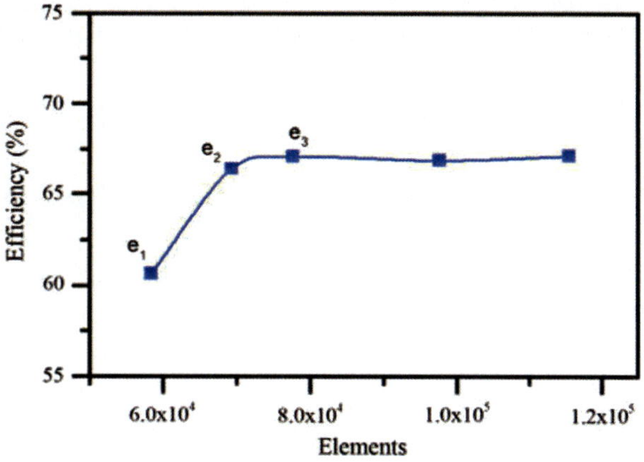

Fig. 4 Grid independency test for different number of elements

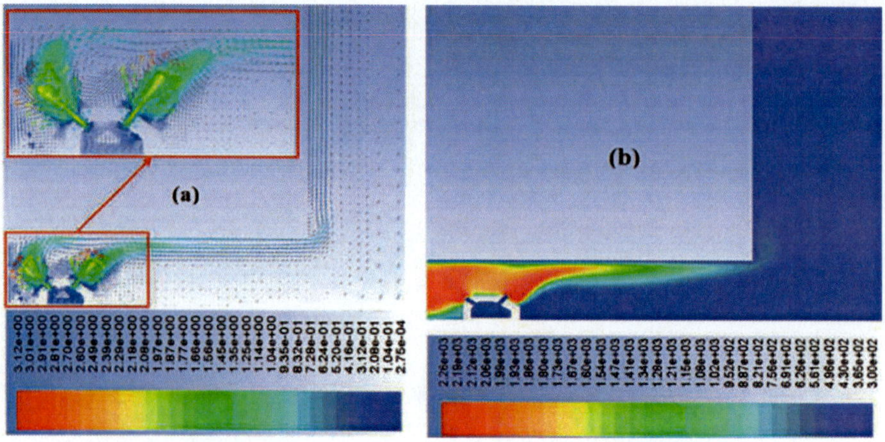

Fig. 5 **a** Velocity (m/s) field and **b** temperature (K) contour in the mid-plane of the domain

3.2 Velocity and Temperature Distributions

The distributions of velocity and temperature along the midplane $(x - y)$ of the 12-degree sector domain are shown in Fig. 5a, b, respectively. In the velocity distribution plot, we observe that the premixed fuel-air mixture is coming out from the burner holes, again mixing with the secondary air and recirculating below the vessel wall before flowing out. The maximum velocity (3.12 m/s) of the hot gases observe in the vicinity of the burner ports. The smaller size of velocity vectors occurs in the pressure outlet regions. The effect of the buoyancy-driven forces is clearly seen in the mixture region of the hot gases and cold air. In Fig. 5b, the flame temperature higher than 2000 K near the burner holes. The hot gases from the burner are mixed with cold air and moved towards the bottom and side wall of the vessel. The temperature of the hot gases is decreased near to the vessel bottom wall, that means heat is transferred through the vessel wall. Some parts of the heat of the hot gases are escaping into the atmosphere.

3.3 Effect of Surface Types of the Loading Vessel Wall

Thermal properties of the vessel wall materials and the effect of the surface types on the thermal efficiency and the total heat transfer rate through the vessel wall are shown in Table 1 and Fig. 6, respectively. In the present work, aluminum alloy (3003 grade) and stainless steel (304/316 grade), which are generally used for cooking utensils, are considered as the loading vessel wall. Thermal properties mainly thermal conductivity and internal emissivity of the vessel wall material are considered that directly effect the heat transfer rate into the loading vessel. Thermal conductivity of

Table 1 Thermal properties of the vessel wall

Material types	Thermal conductivity (W/m-K)	Emissivity	References
Aluminum alloy (3003)			
Polished (p) Oxidized (o) Anodized (a)	162	0.10 0.40 0.55	[12, 13]
Stainless steel			
ss 304 (p) ss 304/316 (o/a) ss 316 (p)	16	0.40 0.80 0.57	[14, 15]

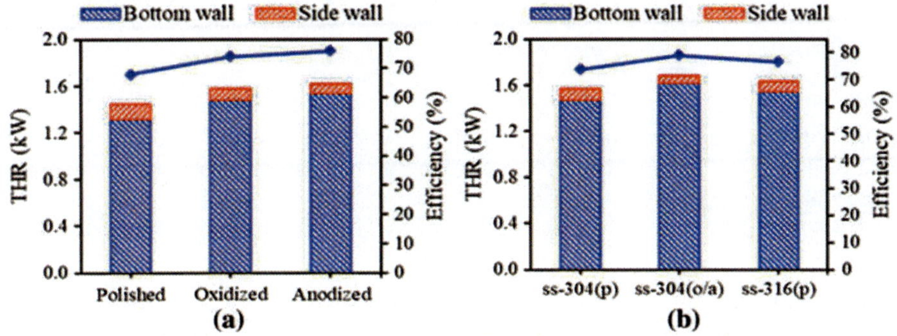

Fig. 6 Effect of material types—**a** aluminum alloy (3003 grade) and **b** stainless steel (304 and 316 grade)—of the vessel wall on total heat transfer rate (THR) through the vessel walls and thermal efficiency

aluminum alloy (3003 grade) and stainless steel (for both 304 and 316 grade) are taken as 162 W/m-K and 16 W/m-K, respectively. The emissivity of the materials is considered based on the surface type, like polished (p), oxidized (o), and anodized (a). The loading height, equivalence ratio and fuel flowrate are kept constant as 18 mm, 1.4, and 71 L/h, respectively, for comparing the other effects.

Maximum parts of fuel energy transfer to the loading vessel bottom and side walls and remaining parts escape through the flue gas. Total heat transfer rate (THR) through the bottom wall of the vessel is about 90% as shown in Fig. 6. The THR for the polished surfaces is lower than oxidized or anodized surfaces as the radiational losses are more for polished surfaces. Radiational loss more means radiation heat flux through the vessel wall is less (see Fig. 7). That's why the overall thermal performance for the polished surface is lower.

Maximum thermal efficiency for the aluminum alloy is ~75% for the anodized surface as the THR is highest. Thermal efficiency for the stainless steel 316 grade is also about 75.6% for the polished surface. It is observed that the thermal conductivity of steel (16 W/m-K) is about 10 times lower than aluminum alloy but, for the same

Fig. 7 Radiation heat flux through the vessel bottom wall for different emissivity (*e*) of aluminum alloy

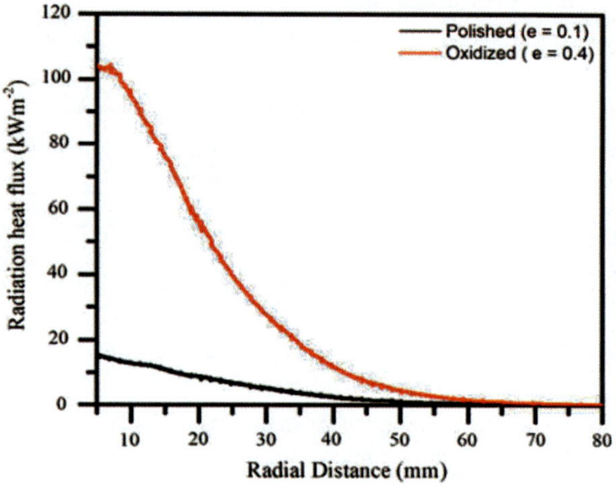

emissivity, the efficiency values are similar. The stainless steel 304/316 grade with oxidized or anodized surface provides maximum thermal efficiency which equals to about 78%.

4 Conclusion

A CFD analysis of a domestic LPG cookstove burner has been studied to analyze the effect of surface types of the vessel wall on the performance of the stove. Aluminum alloy and stainless steel with different surface types are chosen for the vessel wall. The simulation works have been done for a 3-D domain of 12° section of the cooktop burner. The loading height, equivalence ratio, and fuel flowrate are kept constant as 18 mm, 1.4, and 71 L/h, respectively, for comparing the effect of loading wall surface types.

Heat transfer rate through oxidized or anodized surfaces of loading vessel is more compared to the polished surface. This is due to the higher radiative heat flux through the vessel wall. Aluminum alloy with anodized surface and stainless steel with anodized/oxidized surface of the cooking vessel offer about 75% and 78% efficiency, respectively. It can conclude that oxidized/anodized surface of the vessel wall is higher effective and reduces cooking time.

Acknowledgements The author is grateful to Rashtriya Uchchatar Shiksha Abhiyan (RUSA 2.0) and LERC Bangalore, India for partial funding of the present work. The author also would like to acknowledge the Department of Power Engineering, Jadavpur University for using ANSYS CFD software and experimental Setup.

References

1. Silva RL, Sant BV, Patelli JR, Vieira MM (2017) Performance improvements in cooker-top gas burners for small aspect ratio changes
2. Li HB, Wong TT, Leung CW, Probert SD (2006) Thermal performances and CO emissions of gas-fired cooker-top burners. Appl Energy 83(12):1326–1338
3. Hou SS, Lee CY, Lin TH (2007) Efficiency and emissions of a new domestic gas burner with a swirling flame. Energy Convers Manag 48(5):1401–1410
4. Boggavarapu P, Ray B, Ravikrishna RV (2014) Thermal efficiency of LPG and PNG-fired burners: experimental and numerical studies. Fuel 116:709–715
5. Kuntikana P, Prabhu SV (2017) Thermal investigations on methane-air premixed flame jets of multi-port burners. Energy 123:218–228
6. Ashman PJ, Junus R, Stubington JF, Sergeant GD (1994) The effects of load height on the emissions from a natural gas-fired domestic cooktop burner. Combust Sci Technol 103(1–6):283–298
7. Hou SS, Ko YC (2005) Influence of oblique angle and heating height on flame structure, temperature field and efficiency of an impinging laminar jet flame. Energy Convers Manag 46(6):941–958
8. Zhen HS, Leung CW, Wong TT (2014) Improvement of domestic cooking flames by utilizing swirling flows. Fuel 119:153–156
9. Dey S, Das M, Ganguly R, Datta A, Verma MM, Bera AK (2020) Computational fluid dynamic analyses of flow and combustion in a domestic liquified petroleum gas cookstove burner—part I: design optimization of mixing tube-burner assembly. J Therm Sci Eng Appl 12(3):
10. Das M, Ganguly R, Datta A, Verma MM, Bera AK (2020) Computational fluid dynamic analyses of flow and combustion in a domestic liquefied petroleum gas cookstove burner—part II: burning characteristics and overall performance. J Therm Sci Eng Appl 12(3):
11. BIS (2002) IS 4246:2002 Domestic gas stoves for use with liquefied petroleum gases—specification (Fifth Revision)
12. CenturyLife.Org Thermal properties of metals. [Online]. Available: https://www.centurylife.org/thermal-properties-of-metals/
13. L. Technologies Table of emissivity of various surfaces
14. A. Materials Stainless steel—grade 304 (UNS S30400). [Online]. Available: https://www.azom.com/properties.aspx?ArticleID=965
15. Sika.net Table of total emissivity

Low-Velocity Impact Analysis of Pre-twisted Composite Sandwich Shallow Conical Shells in Hygrothermal Environment

Tripuresh Deb Singha, Tanmoy Bandyopadhyay, and Amit Karmakar

1 Introduction

The widespread applications of sandwich structures are found in numerous engineering fields such as aerospace, civil construction, defense, transportation, etc. owing to their low weight-to-stiffness ratio and other preferred properties [1]. An important concern for the trustworthy structural design of any sandwich structure is damage that occurred due to unintended low-velocity impacts (i.e., dropping of tools, striking with birds, hails, etc.). The resulting damages degrade stiffness and residual strength of the sandwich structures considerably [2]. The exposure of these structures in moist and/or hot environments leads to further degradation of stiffness and strength due to moisture absorption and heat spikes. The combined effects of low-velocity impact and environmental conditions may influence the performance of the sandwich structures and cause its premature failure. For this reason, the dynamical behavior of such impacted structures in moist and/or hot environments is needed to study in predicting its service life.

The dynamic behavior of multi-layered orthotropic and sandwich structures subjected to impact loadings was studied extensively over the past decades [3, 4]. One of the remarkable numerical studies on the transient response of impacted orthotropic plate was performed by Sun and Chen [5] employing finite element method (FEM). Maity and Sinha [6] used FEM for investigating the dynamic response of impacted doubly-curved shallow shells with different core proportions. Zhang et al. [7] performed the FE analysis for predicting the dynamical response of sandwich-type

T. D. Singha (✉)
Department of Mechanical Engineering, Government College of Engineering and Textile Technology, Serampore, Hooghly 712201, India

T. Bandyopadhyay · A. Karmakar
Department of Mechanical Engineering, Jadavpur University, Kolkata 700032, India

S. K. Ghosh et al. (eds.), *Advances in Thermal Engineering, Manufacturing, and Production Management*, Lecture Notes in Mechanical Engineering,
https://doi.org/10.1007/978-981-16-2347-9_8

beams having fiber/metal laminate facings and metal foam core subjected to low-velocity impact. Erickson et al. [8] showed the thermal effects on dynamic behavior of impacted sandwich plates through experimental investigations. Yang et al. [9] employed the hybrid method (i.e., combination of FE modeling and experiment) to predict the performance of impacted composite-foam sandwich plates in thermal environments. Reis et al. [10] demonstrated the effects of hygrothermal conditions on impact behavior of multi-layered composites. The influences of humidity and temperature on dynamic response of imperfect multi-layered orthotropic conical shell impacted by multiple impactors with low-velocity have been studied by Bandyopadhyay et al. [11, 12] using FEM. Das et al. [13, 14] carried out the FEM-based numerical study on transient response of impacted functionally graded conical shell. It is found from the literature review that a very little number of works on the dynamic behavior of impacted sandwich panels subjected to moist and/or hot environmental conditions has been reported. In this work, the dynamic behavior of pre-twisted shallow conical shell with composite facings and foam core in hygrothermal environments have been investigated using FEM.

2 Theoretical Formulations

2.1 *Governing Equations*

The schematic diagram of a composite sandwich shallow conical shell (CSSCS) with span length (L), cone length (s), reference breadth (b_0), shell thickness (h), core thickness (h_c), facings thickness (h_f), angle of vertex (ϕ_v), and subtending angle of base (ϕ_0) is shown in Fig. 1. [15–18].

The CSSCS is considered to be impacted perpendicularly with an elastic spherical impactor on its top surface at the central point as shown in Fig. 1b.

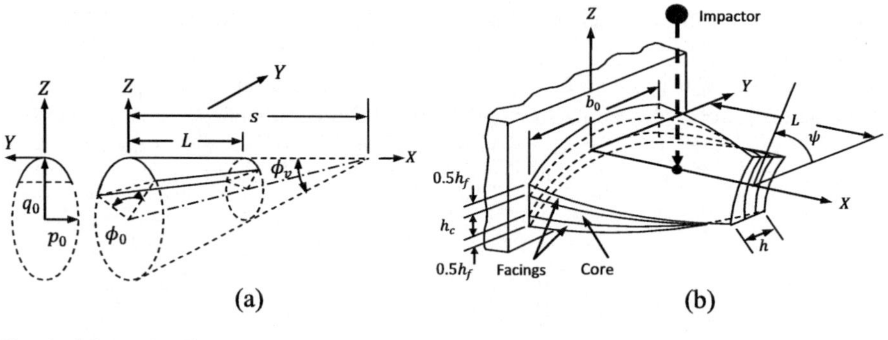

Fig. 1 Schematic diagram of CSSCS (**a**) coordinate system and geometrics (**b**) pre-twisted configuration

The pre-twist radius (r_{XY}) and curvature radius in Y-direction (r_Y) are given by [15–18]

$$r_{XY} = -L/\tan\psi \tag{1}$$

$$|r_Y| = \frac{1}{p^4 q}\left[p^4 + \left(q^2 - p^2\right)Y^2\right]^{1.5} \tag{2}$$

where ψ indicates the pre-twist angle, while $p(X)$ and $q(X)$ represent the minor and major radii of any elliptical cross-section at X, respectively, and are expressed as

$$p(X) = \frac{b_0(s - X)\tan\frac{\phi_v}{2}\tan\frac{\phi_0}{2}}{\sqrt{4s^2\tan^2\frac{\phi_v}{2}\tan^2\frac{\phi_0}{2} - b_0^2}} \tag{3}$$

$$q(X) = (s - Y)\tan\frac{\phi_v}{2} \tag{4}$$

The expression of stress-strain relationship for the CSSCS under hygrothermal loads may be given as

$$\{\sigma\} = \left[\bar{Q}\right][\{\varepsilon\} - \{\beta\}\Delta C - \{\alpha\}\Delta T] \tag{5}$$

in which $\{\sigma\} = \left[\sigma_X\ \sigma_Y\ \tau_{XY}\ \tau_{XZ}\ \tau_{YZ}\right]^{\mathrm{T}}$ represents the stress vector, $\bar{Q}_{ij}$ represents the reduced stiffness coefficient matrix, $\{\beta\} = \left[\beta_X\ \beta_Y\ 0\ 0\ 0\right]^{\mathrm{T}}$ denotes the coefficients of moisture expansion vector, and $\{\alpha\} = \left[\alpha_X\ \alpha_Y\ 0\ 0\ 0\right]^{\mathrm{T}}$ denotes the coefficients of thermal expansion vector. Further, the changes in moisture concentration ΔC and temperature ΔT are determined as

$$\Delta C = C - C_0 \tag{6}$$

$$\Delta T = T - T_0 \tag{7}$$

where C and T signify elevated moisture concentration and temperature, while C_0 (=0.0%) and T_0 (=300 K) indicate their reference values, respectively. The hygrothermal field is assumed to be uniform across the thickness of the CSSCS; thus, there are no differences in moisture concentration and temperature between the outermost layers of top and bottom facings.

The entire plan-form of the CSSCS has been descretized into 8-node serendipity quadrilateral shell element [16] as shown in Fig. 2. The mid-plane displacements ($u_0, v_0, w_0, \theta_X, \theta_Y$) within the element may be expressed in terms of shape function (N_i) of the ith node and nodal displacements ($u_{0i}, v_{0i}, w_{0i}, \theta_{Xi}, \theta_{Yi}$) as

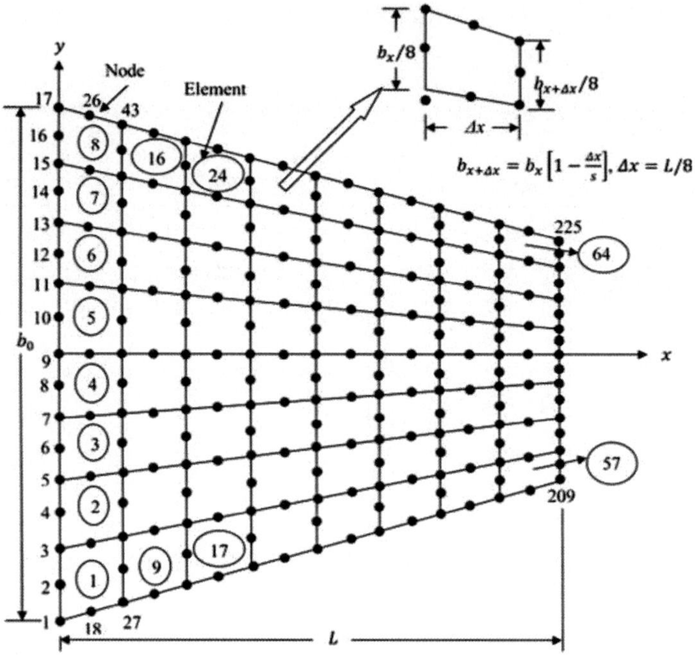

Fig. 2 Descretized plan-form of the CSSCS

$$(u_0, v_0, w_0, \theta_X, \theta_Y) = \sum_{i=1}^{8} [N_i](u_{0i}, v_{0i}, w_{0i}, \theta_{Xi}, \theta_{Yi}) \tag{8}$$

The dynamic equilibrium equation for the CSSCS subjected to hygrothermal loads is deduced using Lagrange's equation as given [11]

$$[M]\{\ddot{\delta}\} + ([K] + [K_\sigma]_{HT})\{\delta\} = \{F_{HT}\} + \{F_C\} \tag{9}$$

where $[M]$ indicates the global mass matrix. $[K]$ and $[K_\sigma]_{HT}$ are the stiffness matrices due to linear strain and non-linear strain, respectively. $\{\delta\}$ denotes the global displacement vector and $\{F_{HT}\}$ represents the global hygrothermal load vector. $\{F_C\}$ is the contact force vector arising out due to the impact load and is expressed as

$$\{F_C\} = \{0\,0\,0 \ldots F_{ci} \ldots 0\,0\,0\} \tag{10}$$

in which F_{ci} is the contact force at ith node where impact occurs.

The dynamic equilibrium equation of the impactor is derived as [5]

$$m_I \ddot{\delta}_I + F_c = 0 \tag{11}$$

in which m_I and $\ddot{\delta}_I$ represent mass and acceleration of the impactor, respectively.

The dynamic equilibrium equations of the CSSCS and impactor as given in Eqs. (9) and (11) can be solved using "Newmark time integration scheme" [19].

2.2 Contact Law

The contact force F_c in CSSCS-impactor system is related with the indentations as follows [5]:

For loading cycle

$$F_C = k_c a^{1.5}, \quad 0 < a \leq a_m \tag{12}$$

where a and a_m are local indentation and maximum local indentation, respectively. The contact stiffness (k_c) is computed as

$$k_c = \frac{4}{3}\sqrt{R_I}\frac{1}{\frac{1-\gamma_I^2}{E_I} + \frac{1}{E_2}} \tag{13}$$

where R_I, E_I, and γ_I indicate the impactor's radius, elasticity modulus, and Poisson's ratio, respectively, while E_2 is the transverse elasticity modulus of the CSSCS.

For unloading and reloading cycles, the contact forces are, respectively

$$F_C = F_m\left[\frac{a - a_0}{a_m - a_0}\right]^{2.5} \tag{14}$$

$$F_C = F_m\left[\frac{a - a_0}{a_m - a_0}\right]^{1.5} \tag{15}$$

where F_m indicates the peak value of contact force. The permanent indentation (a_0) is computed by

$$a_0 = 0 \text{ if } a_m < a_{\text{cr}} \tag{16}$$

$$a_0 = \beta_c(a_m - a_{\text{cr}}) \text{ if } a_m \geq a_{\text{cr}} \tag{17}$$

where the constants, $\beta_c = 0.094$ and $a_{\text{cr}} = 1.667 \times 10^{-2}$ are considered for critical indentation.

The local indentation $a(t)$ of the impactor-CSSCS system is expressed as

$$a(t) = w_I - w_p \cos\psi \tag{18}$$

where w_I is impactor's displacement, w_p is mid-plane displacement of the CSSCS at the impacted node, and ψ is the pre-twist angle.

3 Results and Discussions

Based on the preceding theoretical formulation, the in-house computer codes are developed to analyze the impact response of the pre-twisted CSSCS with a lay-up of [0°/90°/0°/90°/0°/core/0°/90°/0°/90°/0°]. For this investigation, material properties of the foam core (Syntac 350) are considered as [16]: $E_{1c} = E_{2c} = 2.25\,\text{GPa}$, $G_{12c} = G_{13c} = G_{23c} = 1.02\,\text{GPa}$, $\nu_{12c} = 0.31$, $\alpha_{1c} = \alpha_{2c} = 31.6 \times 10^{-6}/°\text{C}$, $\beta_{1c} = \beta_{2c} = 0.4$, $\rho_c = 600\,\text{kg/m}^3$. The material properties of the facings corresponding to different moisture concentrations as well as temperatures [20] are furnished in Table 1. The geometrical parameters of the pre-twisted CSSCS are considered as: $s = 0.4$ m, $L/s = 0.7$, $\phi_v = \phi_0 = 20°$, $\psi = 30°$, and $h_c/h_f = 5$. The following boundary conditions are assumed for the CSSCS

$$\text{At}\, X = 0, \quad u = v = w = \theta_X = \theta_Y = 0 \tag{19}$$

For steel impactor, the following geometrical parameters and elastic properties have been considered: $R_I = 6.35\,\text{mm}$, $E_I = 210\,\text{GPa}$, $\rho_I = 7960\,\text{kg/m}^3$.

3.1 Comparisons

The dimensionless fundamental frequencies of isotropic shallow conical shell with pre-twist are compared against those of Ref. [15] in Table 2. The dimensionless frequency parameters for simply-supported composite panels at different hygrothermal conditions are also compared with those of Ref. [20] as presented in Table 3. Another comparison of the non-dimensional natural frequencies of the composite sandwich plate with those of Ref. [21] is presented in Table 4. Finally, the

Table 1 Material properties of composite facings (graphite/epoxy) corresponding to various hygrothermal conditions [20]; $G_{13f} = G_{12f}$, $G_{23f} = 0.5G_{12f}$, $\nu_{12f} = 0.3$, $\beta_{1f} = 0$, $\beta_{2f} = 0.4$, $\alpha_{1f} = -0.3 \times 10^{-6}/\text{K}$, $\alpha_{2f} = 28.1 \times 10^{-6}/\text{K}$, $\rho_f = 1600\,\text{kg/m}^3$

Elastic modulii (GPa)	Moisture concentration, (C), %			Temperature, (T), K		
	0.0	0.5	1.5	300	350	425
E_{1f}	130	130	130	130	130	130
E_{2f}	9.5	9.0	8.5	9.5	8.0	6.75
G_{12f}	6.0	6.0	6.0	6.0	5.5	4.5

Subscripts 'c' and 'f' indicate the core and facings, respectively

Table 2 Dimensionless fundamental frequency parameters $\left(\lambda = \omega_n b_0^2 \sqrt{\rho h/D}\right), D = Eh^3/12(1 - \nu^2)$ of pre-twisted shallow conical shell having $\nu = 0.3, s/h = 1000, \phi_v = 15°, \phi_0 = 30°$

ψ	L/s	Ref. [15]	Present
0°	0.6	0.35997	0.34241
	0.7	0.30608	0.29412
	0.8	0.27832	0.26915
30°	0.6	0.28828	0.27855
	0.7	0.25752	0.25077
	0.8	0.24179	0.23642

Table 3 Non-dimensional frequencies $\left(\lambda = \omega_n L^2 \sqrt{\rho/E_2 h^2}\right)$, $L/b = 1$, $L/h = 100$, for simply-supported graphite/epoxy multi-layered orthotropic panels (0°/90°/90°/0°)

Mode	$C = 0.1\%$		$T = 325$ K	
	Ref. [20]	Present	Ref. [20]	Present
1	9.429	9.395	8.088	8.049
2	20.679	19.944	19.196	18.409
3	40.068	39.439	39.324	38.686

Table 4 Non-dimensional natural frequency parameters $\left[\varpi = \omega_n \left(L^2/h\right)\sqrt{(\rho/E_2)_c}\right]$ of sandwich plate (0°/90°/0°/core/0°/90°/0°), $[L/b = 1, L/h = 10, h_c/h = 0.88$ and $k_s = 0.19$; material properties for facings: $E_{1f} = 24.51$ GPa, $E_{2f} = 7.77$ GPa, $G_{12f} = G_{13f} = 3.34$ GPa, $G_{13f} = 1.34$ GPa, $\rho_f = 1800\,\text{kg/m}^3$, $\nu_f = 0.078$; core: $E_c = 103.63$ MPa, $G_c = 50$ MPa, $\rho_c = 130\,\text{kg/m}^3$, $\nu_c = 0.32]$

Mode	Ref. [21]		Present
	FOSDT	HOSDT	
1	15.22	15.28	15.216
2	27.75	28.69	27.778
3	30.36	30.01	30.385
4	38.43	38.86	38.465

temporal variation of contact force of impacted simply-supported composite plate is compared with Sun and Chen [5] as shown in Fig. 3. It has been found that the results computed from the present FE formulation are reasonably close to those in the published articles.

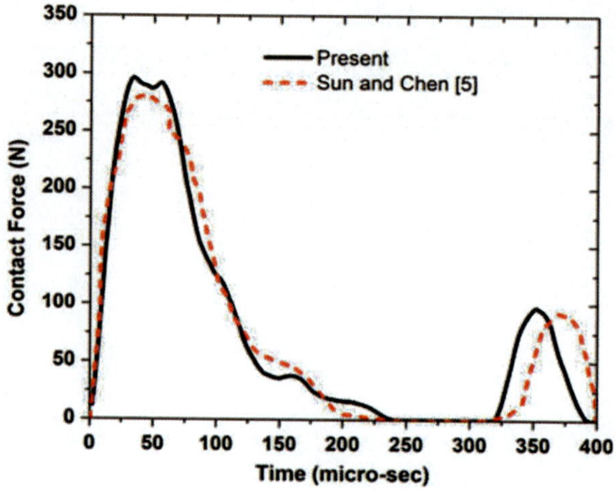

Fig. 3 Temporal variation of contact force of impacted simply-supported composite $[0°/90°/0°/90°/0°]_s$ panel [$L = 20$ cm, $b = 20$ cm, $h = 0.269$ cm, $E_1 = 120$ GPa, $E_2 = 7.9$ GPa, $G_{12} = G_{23} = G_{13} = 5.5$ GPa, $\nu_{12} = 0.30$, $\rho = 1.58 \times 10^{-5}$ N-s^2/cm^4]

3.2 *Influence of Cone Length-to-Thickness Ratio*

The effect of cone length-to-thickness ratio (s/h) on the contact force arising out due to the impact on CSSCS at its central point with a spherical steel impactor having initial velocity $V_i = 3$ m/sec corresponding to reference temperature ($T_0 = 300$ K), elevated temperature ($T = 425$ K) and elevated moisture concentration ($C = 1.5\%$) are depicted in Fig. 4. It is found that the contact duration and maximum value of contact force during loading cycle decreases with increasing the s/h ratio irrespective of temperature as well as moisture concentration although the trends of temporal variation of contact force are similar. This may be attributed due to the fact that an increase in s/h ratio leads to decrease in the structural stiffness and strength of the CSSCS which in turn decreases the contact duration and value of the contact force. Further, the rise in temperature and moisture concentration from their reference values causes a slight decrease in contact force due to the overall stiffness degradation of the CSSCS. The fluctuations in contact force are found corresponding to elevated temperature ($T = 425$ K) and moisture concentration ($C = 1.5\%$) as well as for higher values of s/h ratio (150 and 200).

Figure 5 shows the temporal variation of normalized shell displacement (i.e., shell displacement/thickness) of the CSSCS at its central point corresponding to different cone length-to-thickness ratios ($s/h = 100$, 150, and 200). At any instant of time, the maximum value of normalized shell displacement is found for $s/h = 200$ and its minimum value is observed for $s/h = 100$. The changes in temperature and moisture concentration do not affect the normalized shell displacement, significantly.

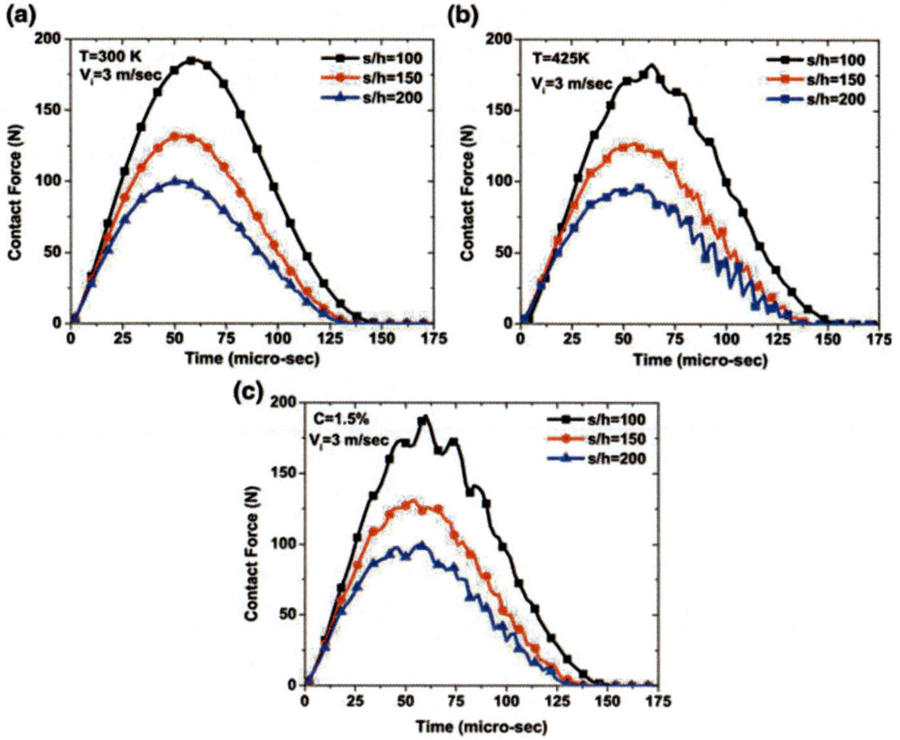

Fig. 4 Influence of cone length-to-thickness ratio on contact force of the impactor-CSSCS system, $V_i = 3\,\text{m/sec}$. **a** $T = 300\,\text{K}$ **b** $T = 425\,\text{K}$ **c** $C = 1.5\%$

3.3 *Influence of Initial Impact Velocity*

Figures 6 and 7 depict the influence of initial impact velocity (V_i) on the contact force and normalized shell displacement (at central point) of the CSSCS under different hygrothermal conditions (T = 300 K, T = 425 K, and C = 1.5%). The trends of temporal variation of contact force and normalized shell displacement corresponding to V_i = 1, 3, 5, and 10 m/sec are found to be similar except for their magnitudes. The value of the contact force is maximum for V_i = 10 m/sec while its minimum value is found for V_i = 1 m/sec. During loading cycle, the shortest duration of contact is observed for V_i = 10 m/sec while longest duration of contact is found for V_i = 1 m/sec. At reference temperature (T_0 = 300 K), small fluctuations in contact forces are observed only corresponding to higher initial impact velocity ($V_i = 10\,\text{m/sec}$) while at elevated temperature (T = 425 K) and moisture concentration (C = 1.5%), fluctuations in contact force are found for all initial impact velocities.

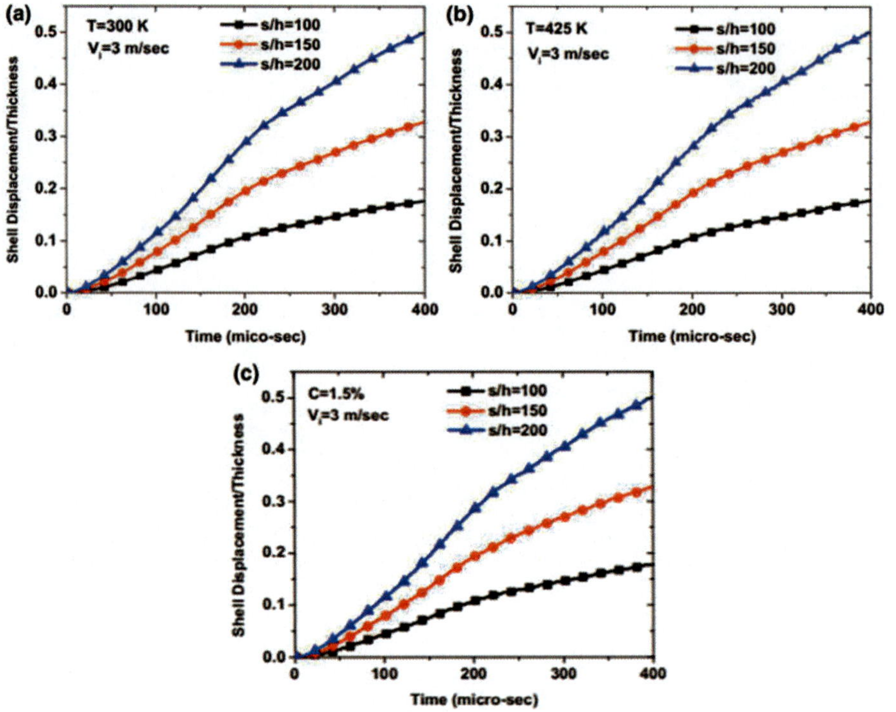

Fig. 5 Influence of cone length-to-thickness ratio on the normalized shell displacement of pre-twisted CSSCS, $V_i = 3\,\text{m/sec}$. **a** $T = 300\,\text{K}$ **b** $T = 425\,\text{K}$ **c** $C = 1.5\%$

4 Conclusions

The low-velocity impact response of pre-twisted CSSCS in uniform hygrothermal environments has been studied numerically employing FEM. The conclusions drawn from the numerical investigation are listed as:

- The contact force and normalized shell displacement of the CSSCS-impactor system are greatly influenced by the cone length-to-thickness ratio under hygrothermal loads. The lower values of contact force with shorter contact duration and higher values of normalized shell displacement are observed corresponding to higher values of cone length-to-thickness ratio.
- The fluctuations in contact force are found at elevated temperature and moisture concentration. The normalized shell displacement of the CSSCS is not much affected by the hygrothermal conditions.
- The initial impact velocity amplifies the magnitude of contact force as well as normalized shell displacement irrespective of the hygrothermal conditions.

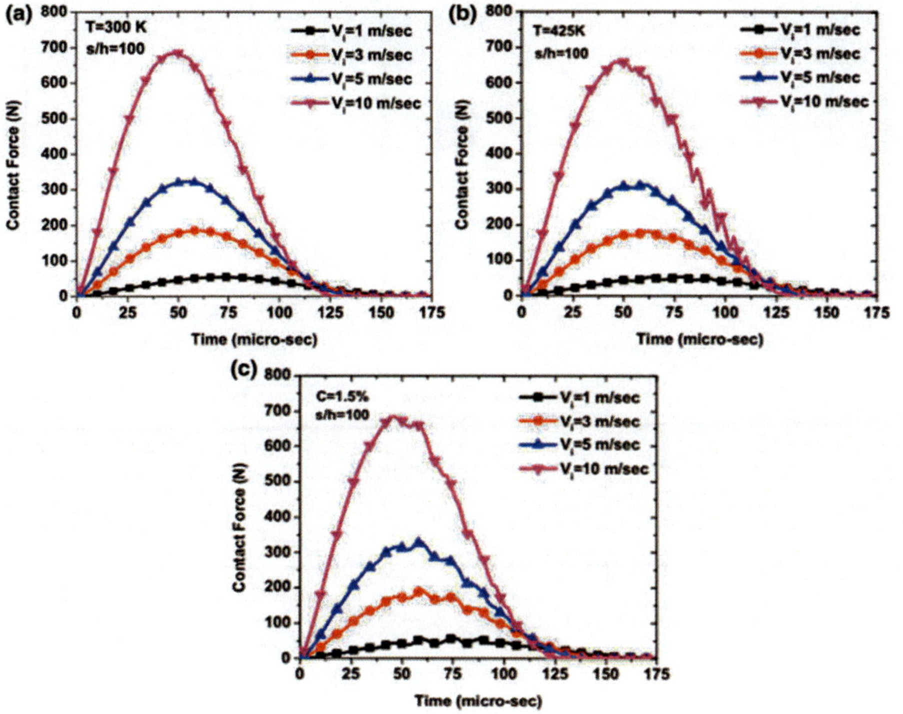

Fig. 6 Influence of initial impact velocity on the contact force of impactor—CSSCS system, $s/h = 100$. **a** $T = 300\,\text{K}$ **b** $T = 425\,\text{K}$ **c** $C = 1.5\%$

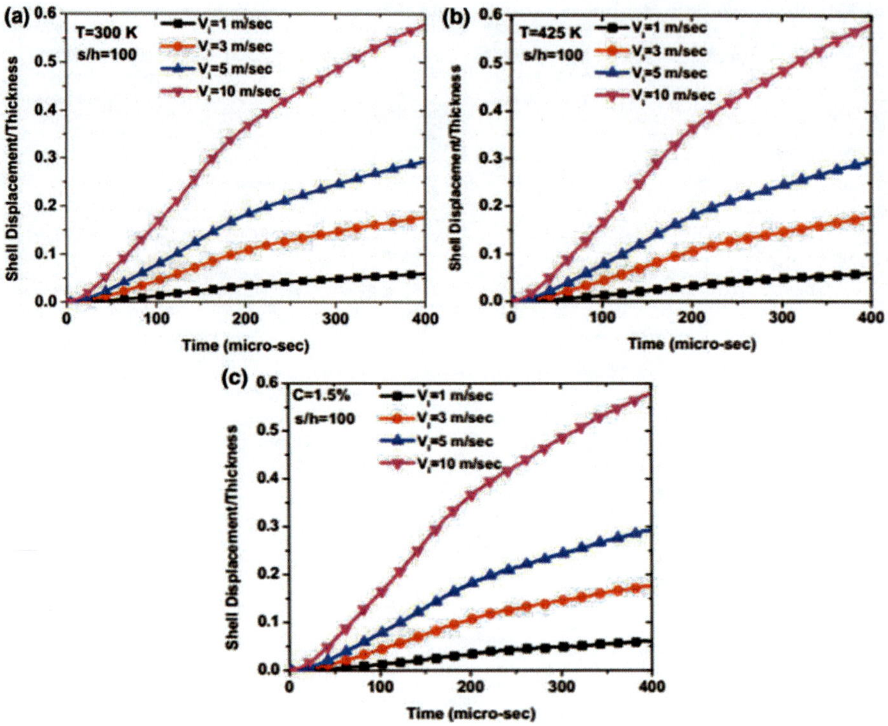

Fig. 7 Influence of initial impact velocity on the normalized shell displacement of the pre-twisted CSSCS, $s/h = 100$. **a** $T = 300\,\text{K}$ **b** $T = 425\,\text{K}$ **c** $C = 1.5\%$

References

1. Birman V, Kardomateas GA (2018) Review of current trends in research and applications of sandwich structures. Compos Part B Eng 142:221–240
2. Caprino G (1984) Residual strength prediction of impacted CFRP laminates. J Compos Mater 18(6):508–518
3. Richardson MOW, Wisheart MJ (1996) Review of low-velocity impact properties of composite materials. Compos Part A-App S 27(12):1123–1131
4. Chai GB, Zhu S (2011) A review of low-velocity impact on sandwich structures. Proc Inst Mech Eng Part L J Mater Des Appl 225(4):207–230
5. Sun CT, Chen JK (1985) On the impact of initially stressed composite laminates. J Compos Mater 19(6):490–504
6. Maity DK, Sinha PK (1996) Low velocity impact analysis of composite sandwich shells using higher-order shear deformation theories. Sadhana 21(5):597–622
7. Zhang J, Ye Y, Qin Q et al (2018) Low-velocity impact of sandwich beams with fiber-metal laminate facesheets. Compos Sci Technol 168:152–159
8. Erickson MD, Kallmeyer AR, Kellogg KG (2005) Effect of temperature on the low-velocity impact behavior of composite sandwich panels. J Sand Struct Mater 7(3):245–264
9. Yang P, Shams SS, Slay A et al (2015) Evaluation of temperature effects on low velocity impact damage in composite sandwich panels with polymeric foam cores. Compos Struct 129:213–223
10. Reis PNB, Silva AP, Santos P et al (2013) Hygrothermal effect on the impact response of carbon composites with epoxy resin enhanced by nanoclays. Mech Compos Mater 49:429–436

11. Bandyopadhyay T, Karmakar A, Kishimoto K (2016) Transient response of delaminated composite conical shells due to multiple low velocity impacts in hygrothermal environment. Compos Struct 143:202–219
12. Bandyopadhyay T, Karmakar A (2017) Low-velocity impact response of delaminated composite conical shells in hygrothermal environment due to time-delay. Procedia Eng 173:463–470
13. Das A, Singha TD, Karmakar A (2019) Low velocity normal impact performance of functionally graded conical shell with simple power law. Mater Today Proc 11:729–739
14. Das A, Banerjee R, Karmakar A (2018) Transient dynamic analysis of pretwisted functionally graded conical shells subject to low velocity impact: a finite element approach. In: Proceedings of the ASME 2017 Gas Turbine India Conference: V002T10A002-V002T10A002
15. Liew KM, Lim CM, Ong LS (1994) Vibration of pretwisted cantilever shallow conical shells. Int J Sol Struct 31:2463–2474
16. Singha TD, Rout M, Bandyopadhyay T et al (2020) Free vibration analysis of rotating pretwisted composite sandwich conical shells with multiple debonding in hygrothermal environment. Eng Struct 204:
17. Singha TD, Das A, Agarwal G et al (2020) Free vibration characteristics of sandwich conical shells with FGM face sheets: a finite element approach. In: Proceedings of the ASME 2019 Gas Turbine India Conference: V001T05A010(1–8)
18. Singha TD, Rout M, Bandyopadhyay T et al (2021) Free vibration of rotating pretwisted FG-GRC sandwich conical shells in thermal environment using HSDT. Compos Struct 257:
19. Bathe KJ (1990) Finite element procedures in engineering analysis. Prentice Hall of India, New Delhi
20. Sai Ram KS, Sinha PK (1992) Hygrothermal effects on the free vibration of laminated composite plates. J Sound 158(1):133–148
21. Meunier M, Shenoi RA (1999) Free vibration analysis of composite sandwich plates. Proc. Instn Mech Eng Part C 213(7):715–727

Mixed Convection Condensation of Vapor with Non-condensable Gas Over a Vertical Plate: ODE-Based Integral Solution

Sayan Banerjee and Koushik Ghosh

Nomenclature

U_∞	Free stream velocity [m/s]
T_∞	Free stream temperature [K]
P_∞	Free stream pressure [Pa]
W_∞	Non-condensable gas mass fraction in the free stream
T_w	Wall temperature [K]
u_i	Interfacial velocity along liquid-mixture layer interface [m/s]
T_i	Interfacial temperature along liquid-mixture layer interface [K]
W_i	Interfacial air mass fraction along liquid-mixture layer interface
δ_L	Liquid hydrodynamic boundary layer thickness [m]
δ_m	Mixture hydrodynamic boundary layer thickness [m]
δ_T	Mixture thermal boundary layer thickness [m]
δ_w	Mixture non-condensable concentration boundary layer thickness [m]
$\dot{m}$	Mass flux at the interface between liquid-mixture layer [kg m^{-2} s^{-1}]
K_L	Thermal conductivity for liquid layer [W/m-K]
u_L	Liquid layer velocity [m/s]
T_L	Liquid layer temperature [m/s]
α_L	Thermal diffusivity for liquid layer [m^2/s]
α_m	Thermal diffusivity for mixture layer [m^2/s]
ρ_L	Density of liquid layer [kg/m^3]
ρ_m	Density of mixture layer [kg/m^3]
ρ_∞	Density of the free stream [kg/m^3]

S. Banerjee (✉)
School of Nuclear Studies and Application, Jadavpur University, Kolkata, West Bengal 700032, India

K. Ghosh
Department of Mechanical Engineering, Jadavpur University, Kolkata, West Bengal 700032, India

S. K. Ghosh et al. (eds.), *Advances in Thermal Engineering, Manufacturing, and Production Management*, Lecture Notes in Mechanical Engineering,
https://doi.org/10.1007/978-981-16-2347-9_9

R_v	Universal gas constant [J/mol-K]
h_{fg}	Latent heat of vaporization [kJ/kg]
D_m	Mass diffusivity of mixture [m^2/s]
μ_L	Viscosity of liquid layer [kg/m-s]
μ_m	Viscosity of mixture layer [kg/m-s]
ν_L	Kinematic viscosity of liquid layer [kg/m-s]
ν_m	Kinematic viscosity of mixture layer [kg/m-s]

1 Introduction

Nusselt (1916) pioneered the experimental work on laminar film condensation [1]. Rohsenow (1956) developed the original model to incorporate thermal convection [2]. Sparrow and Gregg (1959) introduced mathematical techniques of boundary layer theory to incorporate inertia effects [3]. The forced convection condensation and free convection condensation of flowing vapor over a flat cooled plate were analyzed by Koh et al. [4]. A similarity solution for both free and forced convection regimes in the two-phase boundary layer regime was developed by Koh. An analytical closed-form solution on laminar condensation for forced and mixed convection regimes was developed by Fujii et al. [5]. Sparrow and Minkowycz worked on Koh's work by simplifying the model, neglecting inertial and convective effects in the condensate flow [6]. Transient film condensation in laminar flow was analyzed by Sparrow and Siegel [7]. Shu and Wilks developed a detailed asymptotic, exact numerical solutions for the flow of pure vapor stream (saturated) over a vertical plate for pure forced convection laminar film condensation [8]. Similarity solutions does not always give fruitful results on mixed convection problems. Liao et al. developed local non-similarity method for mixed convection regime [9]. A boundary layer formulation was developed and reduced that formulation to two limiting cases: forced convection and free convection condensation identical to Koh's solution. Balasubramaniam and Hasan performed a scaling analysis of both laminar and turbulent condensation of pure vapor flowing over a flat cooled plate [10]. Their goal was to determine the time taken to establish a steady-state condensed liquid film by transient region analysis, liquid film thickness, and heat transfer coefficient. Das et al. developed a numerical model to study mixed convection dominated film boiling over vertical plate [11]. They transformed the integral form of conservation equations of different phases to ordinary differential equation form and solved it. They also incorporate the radiation heat transfer effect into it.

We perform an integral analysis on mixed convection condensation of flowing vapor stream with non-condensable gas over a subcooled vertical plate in this work. This analysis is done for laminar flow. Our goal is to determine the effect of non-condensable gas on liquid and mixture layer thickness and also on average heat transfer coefficient.

2 Mathematical Modeling

2.1 Physical Model Detailing

Figure 1 illustrates physical model where free stream of saturated water vapor with non-condensable gas (air) flows over a subcooled vertical wall of length L maintained at uniform temperature $T_w (T_w < T_\infty)$. The free stream of mixture flows over the wall at a velocity U_∞, temperature T_∞, pressure P_∞, and gas concentration W_∞. Due to heat rejection, a thin liquid layer of thickness δ_L is formed very near to the wall. Beyond the hydrodynamic liquid layer, several mixture boundary layers are formed. They are mixture hydrodynamic boundary layer δ_m, mixture thermal boundary layer δ_T, and mixture concentration boundary layer δ_w. In this model, following assumptions are considered:

- Steady state, laminar flow is assumed.
- Density of free stream $(\rho_\infty) \ll$ density of liquid (ρ_L).
- Boundary layer approximation is valid $[\delta_L, \delta_m, \delta_T, \delta_w \ll L]$.
- Wall temperature is constant.
- Gravity appears as only body force term.
- Boussinesq approximation is valid in mixture layer.

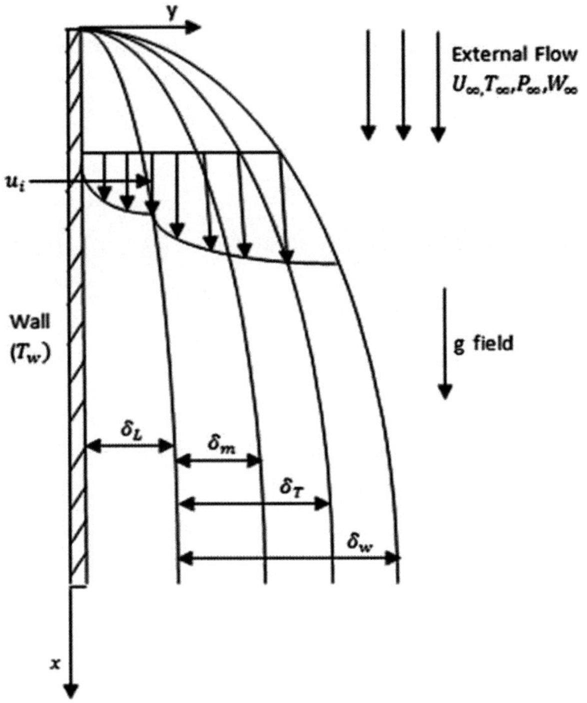

Fig. 1 Physical model and coordinate system

- Viscous dissipation term in energy equation is neglected.

2.2 Governing Equations

Based on the above assumptions, following governing equations are written in Cartesian coordinate systems.

2.2.1 Continuity Equation (in Liquid Layer)

$$\frac{\partial u_L}{\partial x} + \frac{\partial v_L}{\partial y} = 0 \tag{1}$$

2.2.2 Momentum Conservation Equation (in Liquid Layer)

$$\rho_L\left(\frac{\partial\left(u_L^2\right)}{\partial x} + \frac{\partial(u_L v_L)}{\partial y}\right) = -\frac{\mathrm{d}p}{\mathrm{d}x} + \mu_L\left(\frac{\partial^2 u_L}{\partial y^2}\right) + (\rho_L - \rho_\infty)g \tag{2}$$

2.2.3 Energy Conservation Equation (in Liquid Layer)

$$\frac{\partial(u_L T_L)}{\partial x} + \frac{\partial(v_L T_L)}{\partial y} = \alpha_L \frac{\partial^2 T_L}{\partial y^2} \tag{3}$$

2.2.4 Continuity Equation (in Mixture Layer)

$$\frac{\partial u_m}{\partial x} + \frac{\partial v_m}{\partial y} = 0 \tag{4}$$

2.2.5 Momentum Conservation Equation (in Mixture Layer)

$$\frac{\partial\left(u_m^2\right)}{\partial x} + \frac{\partial(u_m v_m)}{\partial y} = v_m\left(\frac{\partial^2 u_m}{\partial y^2}\right) + \beta_m g(T_\infty - T_m) \tag{5}$$

2.2.6 Energy Conservation Equation (in Mixture Layer)

$$\frac{\partial(u_m T_m)}{\partial x} + \frac{\partial(v_m T_m)}{\partial y} = \alpha_m \frac{\partial^2 T_m}{\partial y^2} \tag{6}$$

2.2.7 Species Conservation Equation (in Mixture Layer)

$$\frac{\partial(u_m W_{\mathrm{nc}})}{\partial x} + \frac{\partial(v_m W_{\mathrm{nc}})}{\partial y} = D_m \frac{\partial^2 (W_{\mathrm{nc}})}{\partial y^2} \tag{7}$$

$$W_{\mathrm{nc}} + Wv = 1 \tag{8}$$

2.2.8 Boundary and Interfacial Conditions

$$(u_L)_{y=0} = 0 \tag{9.a}$$

$$(u_L)_{y=\delta_L} = (u_m)_{y=\delta_L} = u_i \tag{9.b}$$

$$(u_m)_{y=\delta_L+\delta_M} = U_\infty \tag{9.c}$$

$$(T_L)_{y=0} = T_w \tag{9.d}$$

$$(T_L)_{y=\delta_L} = (T_m)_{y=\delta_L} = T_i \tag{9.e}$$

$$(T_m)_{y=\delta_L+\delta_T} = T_\infty \tag{9.f}$$

$$(W_{\mathrm{nc}})_{y=\delta_L} = W_i \tag{9.g}$$

$$(W_{\mathrm{nc}})_{y=\delta_L+\delta_W} = W_\infty \tag{9.h}$$

Mass balance equation at interface is written as:

$$\dot{m} = \rho_L \left[u_i \frac{\mathrm{d}\delta_L}{\mathrm{d}x} - v_L|_{y=\delta_L} \right] = \rho_m \left[u_i \frac{\mathrm{d}\delta_m}{\mathrm{d}x} - v_m|_{y=\delta_m} \right] \tag{9.i}$$

Shear balance at interface is described as:

$$\mu_L \left. \frac{\partial u_L}{\partial y} \right|_{y=\delta_L} = \mu_M \left. \frac{\partial u_M}{\partial y} \right|_{y=\delta_L} \tag{9.j}$$

Energy balance equation at interface is written as:

$$\dot{m}h_{\mathrm{fg}} = K_L \left.\frac{\partial T_L}{\partial y}\right|_{y=\delta_L} - K_m \left.\frac{\partial T_m}{\partial y}\right|_{y=\delta_L} \tag{9.k}$$

Gas concentration balance equation at interface is written as:

$$\rho_m D_m \left(\frac{\partial W_{\mathrm{nc}}}{\partial y}\right)_{y=\delta_L} + \dot{m} W_i = 0 \tag{9.l}$$

Interface temperature is calculated using Clausius-Clapreyon equation as follows:

$$\ln\left(\frac{P_{Vi}}{P_\infty}\right) = \frac{h_{\mathrm{fg}}}{R_V}\left[\frac{1}{T_\infty} - \frac{1}{T_i}\right] \tag{9.m}$$

The concentration of air at interface is calculated from an equation which is based on ratio of molecular weight of water vapor and air as follows:

$$P_{Vi} = \frac{1.61 * (1 - W_i) * P_\infty}{1 + 0.61 * (1 - W_i)} \tag{9.n}$$

2.3 ODE Systems with Eight ODEs

Now with the help of governing Eqs. 1–8 and boundary conditions 9.a–9.h and interfacial conditions (9.i)–(9.n), eight ordinary differential equations are developed which are mentioned below in terms of variables δ_L, δ_m, δ_T, δ_W, d_1, c_2, T_i and W_i as follows:

$$\begin{aligned}
&\left[\frac{(2-d_1)}{4}\left\{\frac{2A}{E^2}\left(\frac{(1+2d_1)}{\mathrm{Re}_L\overline{\delta_L}} + \frac{\mathrm{Gr}_L}{2\mathrm{Re}_L^2}\left(\frac{\mu_L}{\mu_M}\overline{\delta_M}(1+2d_1) + \overline{\delta_L}\right)\right)\right\}\right.\\
&\left.\quad + \frac{(2-d_1)}{4E}\left(\frac{2A}{\mathrm{Re}_L\overline{\delta_L}^2} + \frac{\mathrm{Gr}_L}{\mathrm{Re}_L^2}\right) + 0.25\frac{\mathrm{Gr}_L}{\mathrm{Re}_L^2}\right]\frac{\mathrm{d}\overline{\delta_L}}{\mathrm{d}\overline{x}}\\
&\quad - \left[\frac{(2-d_1)}{4}\left\{\frac{2A\overline{\delta_L}}{\overline{\delta_M}E^2}\left(\frac{(1+2d_1)}{\mathrm{Re}_L\overline{\delta_L}^2} - \frac{0.5\mathrm{Gr}_L}{\mathrm{Re}_L^2}\right)\right\}\right]\frac{\mathrm{d}\overline{\delta_M}}{\mathrm{d}\overline{x}}\\
&\quad - \left[\frac{(2-d_1)}{4}\left\{\frac{2A}{\overline{\delta_L}E^2}\left(\frac{2}{\mathrm{Re}_L} + \frac{\mathrm{Gr}_L}{2\mathrm{Re}_L^2}\left(\frac{\mu_L}{\mu_M}\right)\overline{\delta_L\delta_M}\right)\right\}\right.\\
&\left.\quad + \frac{1}{8E\overline{\delta_L}}\left\{\frac{4A}{\mathrm{Re}_L} + \frac{\mathrm{Gr}_L}{\mathrm{Re}_L^2}\right\}\right]\frac{\mathrm{d}d_1}{\mathrm{d}\overline{x}} = \frac{\mathrm{Ja}_L(1-c_2)}{\mathrm{Re}_L\mathrm{Pe}_L\overline{\delta_L}^3}
\end{aligned}$$

$$-\frac{2\mathrm{Ja}_M}{\mathrm{Re}_L\mathrm{Pe}_M\overline{\delta_L}^2\overline{\delta_T}}\left(\frac{\rho_L}{\rho_M}\right) \tag{10}$$

$$\left[\begin{array}{l} 0.667\frac{\mathrm{Gr}_L\overline{u_i}\overline{\delta_L}^2\overline{D}}{\mathrm{Re}_L^2}+\frac{0.33\overline{u_i}^2\overline{D}^2}{\mathrm{Re}_L}+\frac{\overline{u_i}^2d_1^2}{7\mathrm{Re}_L}-0.1\frac{\mathrm{Gr}_L\overline{u_i}\overline{\delta_L}^2d_1}{\mathrm{Re}_L^2}-0.75\frac{\mathrm{Gr}_L\overline{u_i}\overline{\delta_L}^2\overline{D}}{\mathrm{Re}_L^2}+\frac{0.4\overline{u_i}^2d_1\overline{D}}{\mathrm{Re}_L}\\ -\frac{0.5\overline{u_i}^2}{\mathrm{Re}_L}+\frac{0.25\overline{u_i}^2d_1}{\mathrm{Re}_L}-0.25\frac{\mathrm{Gr}_L\overline{u_i}\overline{\delta_L}^2}{\mathrm{Re}_L^2}+\overline{U}\left[\frac{2A}{E^2}\left\{\frac{(1+2d_1)}{\mathrm{Re}_L\overline{\delta_L}}+\frac{\mathrm{Gr}_L}{2\mathrm{Re}_L^2}\left(\frac{\mu_L}{\mu_M}\overline{\delta_M}(1+2d_1)+\overline{\delta_L}\right)\right\}\right]\end{array}\right]\frac{\mathrm{d}\overline{\delta_L}}{\mathrm{d}\overline{x}}$$

$$-\overline{U}\left[\frac{2A}{\overline{\delta_M}E^2}\left\{\frac{(1+2d_1)}{\mathrm{Re}_L}-\frac{\mathrm{Gr}_L\overline{\delta_L}^2}{2\mathrm{Re}_L^2}\right\}\right]\frac{\mathrm{d}\overline{\delta_M}}{\mathrm{d}\overline{x}}+\left[\begin{array}{l}-\frac{4}{15}\frac{\overline{u_i}^2\overline{\delta_L}\overline{D}}{\mathrm{Re}_L}-\frac{4}{35}\frac{\overline{u_i}^2d_1\overline{\delta_L}}{\mathrm{Re}_L}+\frac{1}{12}\frac{\mathrm{Gr}_L\overline{u_i}\overline{\delta_L}^3}{\mathrm{Re}_L^2}\\ +\frac{0.25\overline{u_i}^2\overline{\delta_L}}{\mathrm{Re}_L}-\overline{U}\left\{\frac{2A}{E^2}\left(\frac{2}{\mathrm{Re}_L}+\frac{Gr_L}{2\mathrm{Re}_L^2}\left(\frac{\mu_L}{\mu_M}\right)\overline{\delta_L\delta_M}\right)\right\}\end{array}\right]\frac{\mathrm{d}d_1}{\mathrm{d}\overline{x}}=\frac{3\overline{u_i}d_1}{\mathrm{Re}_L^2\overline{\delta_L}} \tag{11}$$

$$\left[\begin{array}{l}\left[0.5(1-d_1)\overline{T_w}+0.25d_1\overline{T_w}-\frac{\overline{\theta_l}(1+c_2)(1-d_1)}{3}-0.2\overline{\theta_l}(1+c_2)d_1+0.25\overline{\theta_l}(1-d_1)c_2+\frac{\overline{\theta_l}d_1c_2}{6}\right]*\\ \left[\frac{\overline{u_i}}{\mathrm{Re}_L}+\overline{\delta_L}\left[\frac{2A}{E^2}\left\{\frac{(1+2d_1)}{\overline{\delta_L}\mathrm{Re}_L}+\frac{\mathrm{Gr}_L}{2\mathrm{Re}_L^2}\left(\frac{\mu_L}{\mu_M}\overline{\delta_M}(1+2d_1)+\overline{\delta_L}\right)\right\}\right]\right]\end{array}\right]\frac{\mathrm{d}\overline{\delta_L}}{\mathrm{d}\overline{x}}$$

$$-\left[\begin{array}{l}\left[\begin{array}{l}0.5(1-d_1)\overline{\delta_L}\overline{T_w}+0.25d_1\overline{\delta_L}\overline{T_w}-\frac{(1+c_2)(1-d_1)\overline{\delta_L}\overline{\theta_l}}{3}-\\ 0.2(1+c_2)\overline{\delta_L}d_1\overline{\theta_l}+0.25(1-d_1)c_2\overline{\delta_L}\overline{\theta_l}+\frac{d_1c_2\overline{\delta_L}\overline{\theta_l}}{6}\end{array}\right]*\\ \left[\frac{2A}{\overline{\delta_M}E^2}\left\{-\frac{\mathrm{Gr}_L\overline{\delta}_L^2}{2\mathrm{Re}_L^2}+\frac{(1+2d_1)}{\mathrm{Re}_L}\right\}\right]\end{array}\right]\frac{\mathrm{d}\overline{\delta_M}}{\mathrm{d}\overline{x}}$$

$$-\left[0.5\overline{\delta_L}\overline{T_w}\overline{U_1}-0.25\overline{\delta_L}\overline{T_w}\overline{U_2}-0.33\overline{\theta_l}(1+c_2)\overline{\delta_L}\overline{U_1}+0.2\overline{\theta_l}(1+c_2)\overline{\delta_L}\overline{U_2}+0.25\overline{\theta_l}c_2\overline{\delta_L}\overline{U_1}-\frac{c_2\overline{\theta_l}\overline{\delta_L}\overline{U_2}}{6}\right]\frac{1}{\mathrm{Re}_L}\frac{\mathrm{d}d_1}{\mathrm{d}\overline{x}}$$

$$-\left[\begin{array}{l}\frac{0.33(1-d_1)\overline{\theta_l}\overline{\delta_L}\overline{u_i}}{\mathrm{Re}_L}+\frac{\mathrm{Gr}_L\overline{\delta}_L^3\overline{\theta_l}}{24\mathrm{Re}_L^2}+\frac{0.2\overline{\delta_L}d_1\overline{\theta_l}\overline{u_i}}{\mathrm{Re}_L}-\frac{\mathrm{Gr}_L\overline{\delta}_L^3\overline{\theta_l}}{40\mathrm{Re}_L^2}\\ -\frac{0.25(1-d_1)\overline{\delta_L}\overline{\theta_l}\overline{u_i}}{\mathrm{Re}_L}-\frac{\overline{\delta_L}d_1\overline{\theta_l}}{6\mathrm{Re}_L}\overline{u_i}\end{array}\right]\frac{\mathrm{d}c_2}{\mathrm{d}\overline{x}}$$

$$-\left[\begin{array}{l}-\frac{\mathrm{Gr}_L\overline{\delta}_L^3(1+c_2)}{24\mathrm{Re}_L^2}-\frac{0.33(1-d_1)(1+c_2)\overline{\delta_L}\overline{u_i}}{\mathrm{Re}_L}-\frac{0.2d_1(1+c_2)\overline{\delta_L}\overline{u_i}}{\mathrm{Re}_L}+\\ \frac{0.25(1-d_1)c_2\overline{\delta_L}\overline{u_i}}{\mathrm{Re}_L}+\frac{\mathrm{Gr}_L\overline{\delta}_L^3c_2}{40\mathrm{Re}_L^2}+\frac{d_1c_2\overline{\delta_L}}{6\mathrm{Re}_L}\overline{u_i}\end{array}\right]\frac{\mathrm{d}\overline{\theta_l}}{\mathrm{d}\overline{x}}$$

$$=\frac{2c_2\overline{\theta_l}}{\overline{\delta_L}\mathrm{Pe}_L\mathrm{Re}_L}+\frac{\mathrm{Ja}_L(1-c_2)\overline{\theta_i}}{\overline{\delta_L}\mathrm{Pe}_L\mathrm{Re}_L}+\frac{2\mathrm{Ja}_M\overline{\theta_i}}{\overline{\delta_T}\mathrm{Pe}_M\mathrm{Re}_L} \tag{12}$$

$$\left[\left(\frac{0.4A\overline{\delta_M}}{\mathrm{Re}_LE}+\frac{\mathrm{Gr}_L}{10E\mathrm{Re}_L^2}\overline{\delta_L}^2\overline{\delta_M}-\frac{\overline{\delta_M}}{\mathrm{Re}_L15}\right)\left\{\frac{2A}{E^2}\left(\frac{(1+2d_1)}{\mathrm{Re}_L\overline{\delta_L}}+\frac{\mathrm{Gr}_L}{2\mathrm{Re}_L^2}\left(\frac{\mu_L}{\mu_M}\overline{\delta_M}(1+2d_1)+\overline{\delta_L}\right)\right)\right\}\right]\frac{\mathrm{d}\overline{\delta_L}}{\mathrm{d}\overline{x}}$$

$$+\left[\begin{array}{l}\left(\frac{2A}{\mathrm{Re}_LE}+\frac{\mathrm{Gr}_L\overline{\delta_L}^2}{2E\mathrm{Re}_L^2}-\frac{1}{\mathrm{Re}_L}\right)\left(\frac{0.4A}{\mathrm{Re}_LE}+\frac{0.1\mathrm{Gr}_L\overline{\delta_L}^2}{E\mathrm{Re}_L^2}+\frac{2}{15\mathrm{Re}_L}\right)-\\ \left(\frac{0.8A}{\mathrm{Re}_LE}+\frac{0.2\mathrm{Gr}_L\overline{\delta_L}^2}{E\mathrm{Re}_L^2}-\frac{1}{15\mathrm{Re}_L}\right)\left\{\frac{2A}{E^2}\left(\frac{-\mathrm{Gr}_L\overline{\delta_L}^2}{2\mathrm{Re}_L^2}+\frac{(1+2d_1)}{\mathrm{Re}_L}\right)\right\}\end{array}\right]$$

$$*\frac{\mathrm{d}\overline{\delta_M}}{\mathrm{d}\overline{x}}-\left[\left(\frac{0.8A\overline{\delta_M}}{\mathrm{Re}_LE}+\frac{0.2\mathrm{Gr}_L\overline{\delta_L}^2\overline{\delta_M}}{E\mathrm{Re}_L^2}-\frac{\overline{\delta_M}}{15\mathrm{Re}_L}\right)\left\{\frac{2A}{E^2}\left(\frac{2}{\mathrm{Re}_L}+\frac{0.5\mathrm{Gr}_L\mu_L\overline{\delta_M}\overline{\delta_L}}{\mathrm{Re}_L^2\mu_M}\right)\right\}\right]\frac{\mathrm{d}d_1}{\mathrm{d}\overline{x}}$$

$$=\left[\left(\frac{2A}{\mathrm{Re}_LE}+\frac{\mathrm{Gr}_L\overline{\delta_L}^2}{2E\mathrm{Re}_L^2}-\frac{1}{\mathrm{Re}_L}\right)\left\{\frac{2}{\mathrm{Re}_L\mathrm{Re}_M\overline{\delta_M}}+\frac{2\mathrm{Ja}_M}{\mathrm{Re}_L\mathrm{Pe}_M\overline{\delta_T}}+\frac{\mathrm{Ja}_L(c_2-1)\rho_L}{\overline{\delta_L}\mathrm{Re}_L\mathrm{Pe}_L\rho_M}\right\}\right]+\frac{\mathrm{Gr}_M}{6\mathrm{Re}_L^2\mathrm{Re}_M^2} \tag{13}$$

$$\left[\overline{\delta_T}F\left[\frac{2A}{E^2}\left\{\frac{(1+2d_1)}{\overline{\delta_L}\mathrm{Re}_L}+\frac{\mathrm{Gr}_L}{2\mathrm{Re}_L^2}\left(\frac{\mu_L}{\mu_M}\overline{\delta_M}(1+2d_1)+\overline{\delta_L}\right)\right\}\right]\right]\frac{\mathrm{d}\overline{\delta_L}}{\mathrm{d}\overline{x}}$$

$$
\begin{aligned}
&+\Bigg[\overline{\delta_T F}\Bigg[\frac{2A}{\overline{\delta_M}E^2}\left(\frac{-\mathrm{Gr}_L\overline{\delta_L}^2}{2\mathrm{Re}_L^2}+\frac{(1+2d_1)}{\mathrm{Re}_L}\right)\Bigg]\\
&+\left(\frac{2A}{E\mathrm{Re}_L}+\frac{\mathrm{Gr}_L\overline{\delta_L}^2}{2E\mathrm{Re}_L^2}-\frac{1}{\mathrm{Re}_L}\right)\left(\frac{\overline{\phi}}{15}-\frac{1}{6}\right)\overline{\phi}^2\Bigg]\frac{\mathrm{d}\overline{\delta_M}}{\mathrm{d}\overline{x}}\\
&-\Bigg[\Bigg[\overline{\delta_T F}\Bigg\{\frac{2A}{E^2}\left(\frac{2}{\mathrm{Re}_L}+\frac{0.5\mathrm{Gr}_L\mu_L\overline{\delta_M\delta_L}}{\mathrm{Re}_L^2\mu_M}\right)\Bigg\}\Bigg]\Bigg]\frac{\mathrm{d}d_1}{\mathrm{d}\overline{x}}\\
&+\frac{\overline{\delta_T}}{3\mathrm{Re}_L}+\left(\frac{2A}{E\mathrm{Re}_L}+\frac{\mathrm{Gr}_L\overline{\delta_L}^2}{2E\mathrm{Re}_L^2}-\frac{1}{\mathrm{Re}_L}\right)\overline{\delta_T F}\Bigg]\frac{1}{\overline{\theta_M}}\frac{\mathrm{d}\overline{\theta_M}}{\mathrm{d}\overline{x}}\\
&+\Bigg[\left(\frac{2A}{E\mathrm{Re}_L}+\frac{\mathrm{Gr}_L\overline{\delta_L}^2}{2E\mathrm{Re}_L^2}-\frac{1}{\mathrm{Re}_L}\right)\overline{F}+\frac{1}{3\mathrm{Re}_L}\\
&+\left(\frac{2A}{E\mathrm{Re}_L}+\frac{\mathrm{Gr}_L\overline{\delta_L}^2}{2E\mathrm{Re}_L^2}-\frac{1}{\mathrm{Re}_L}\right)\left(\frac{\overline{\phi}}{15}-\frac{1}{6}\right)\overline{\phi}\Bigg]\frac{\mathrm{d}\overline{\delta_T}}{\mathrm{d}\overline{x}}\\
&=\frac{2(1+\mathrm{Ja}_M)}{\overline{\delta_T}\mathrm{Pe}_M\mathrm{Re}_L}+\frac{\mathrm{Ja}_L(c_2-1)}{\mathrm{Pe}_L\mathrm{Re}_L\overline{\delta_L}}
\end{aligned}
\tag{14}
$$

$$
\begin{aligned}
&\Bigg[\overline{\delta_W S_1}\Bigg[\frac{2A}{E^2}\Bigg\{\frac{(1+2d_1)}{\overline{\delta_L}\mathrm{Re}_L}+\frac{\mathrm{Gr}_L}{2\mathrm{Re}_L^2}\left(\frac{\mu_L}{\mu_M}\overline{\delta_M}(1+2d_1)+\overline{\delta_L}\right)\Bigg\}\Bigg]\Bigg]\frac{\mathrm{d}\overline{\delta_L}}{\mathrm{d}\overline{x}}\\
&-\Bigg[\overline{S_1}\Bigg[\frac{2A\overline{\phi}}{E^2}\left(\frac{-\mathrm{Gr}_L\overline{\delta_L}^2}{2\mathrm{Re}_L^2}+\frac{(1+2d_1)}{\mathrm{Re}_L}\right)\Bigg]\\
&+\left(\frac{2A}{E\mathrm{Re}_L}+\frac{\mathrm{Gr}_L\overline{\delta_L}^2}{2E\mathrm{Re}_L^2}-\frac{1}{\mathrm{Re}_L}\right)\left(\frac{1}{3\overline{\phi}}-\frac{1}{3}-\frac{1}{10\overline{\phi}^2}\right)\Bigg]\frac{\mathrm{d}\overline{\delta_M}}{\mathrm{d}\overline{x}}\\
&-\Bigg[\overline{\delta_W S_1}\Bigg\{\frac{2A}{E^2}\left(\frac{2}{\mathrm{Re}_L}+\frac{0.5\mathrm{Gr}_L\mu_L\overline{\delta_M\delta_L}}{\mathrm{Re}_L^2\mu_M}\right)\Bigg\}\Bigg]\frac{\mathrm{d}d_1}{\mathrm{d}\overline{x}}\\
&+\Bigg[\left(\frac{2A}{E\mathrm{Re}_L}+\frac{\mathrm{Gr}_L\overline{\delta_L}^2}{2E\mathrm{Re}_L^2}-\frac{1}{\mathrm{Re}_L}\right)\overline{\delta_W S_1}\\
&+\frac{\overline{\delta_W S_2}}{3\mathrm{Re}_L}+\frac{\overline{\delta_W}}{3\mathrm{Re}_L}\left(1-\frac{1}{\overline{\phi}}\right)^3\Bigg]\frac{\mathrm{d}W_i/(W_i-W_\infty)}{\mathrm{d}\overline{x}}\\
&\Bigg[\left(\frac{2A}{E\mathrm{Re}_L}+\frac{\mathrm{Gr}_L\overline{\delta_L}^2}{2E\mathrm{Re}_L^2}-\frac{1}{\mathrm{Re}_L}\right)\overline{S_1}+\frac{\overline{S_2}}{3\mathrm{Re}_L}+\frac{1}{3\mathrm{Re}_L}\left(1-\frac{1}{\overline{\phi}}\right)^3\\
&+\left(\frac{2A}{E\mathrm{Re}_L}+\frac{\mathrm{Gr}_L\overline{\delta_L}^2}{2E\mathrm{Re}_L^2}-\frac{1}{\mathrm{Re}_L}\right)\left(\frac{1}{3\overline{\phi}^2}-\frac{1}{3\overline{\phi}}-\frac{1}{10\overline{\phi}^3}\right)\Bigg]\frac{\mathrm{d}\overline{\delta_W}}{\mathrm{d}\overline{x}}\\
&=\frac{2}{\mathrm{Sc}_M\mathrm{Re}_M\mathrm{Re}_L\overline{\delta_W}}+\frac{2\mathrm{Ja}_M}{\mathrm{Re}_L\mathrm{Pe}_M\overline{\delta_T}}+\frac{\mathrm{Ja}_L(c_2-1)}{\mathrm{Re}_L\mathrm{Pe}_L\overline{\delta_L}}
\end{aligned}
\tag{15}
$$

$$\frac{(W_i-1)(1.61-0.61W_i)C_{\mathrm{pM}}}{R_V\mathrm{Ja}_{\mathrm{in}}\left[\overline{\theta_M}+\overline{T_w}\right]^2}\frac{\mathrm{d}\overline{\theta_M}}{\mathrm{d}\overline{x}}-\frac{\mathrm{d}W_i}{\mathrm{d}\overline{x}}=0 \tag{16}$$

$$\begin{aligned}
&\frac{\mathrm{Ja}_L(c_2-1)}{\mathrm{Pe}_L}\frac{\mathrm{d}\overline{\delta_L}}{\mathrm{d}\overline{x}}+\frac{2\mathrm{Ja}_M\rho_M}{\mathrm{Pe}_M\rho_L}\frac{\mathrm{d}\overline{\delta_T}}{\mathrm{d}\overline{x}}-\frac{\mathrm{Ja}_L}{\mathrm{Pe}_L}\frac{\mathrm{d}c_2}{\mathrm{d}\overline{x}}\\
&+\left(1-\frac{W_\infty}{W_i}\right)\left(\frac{\rho_M}{\rho_L}\right)\frac{2}{\overline{\delta_W}\mathrm{Sc}_M\mathrm{Re}_M}\frac{\mathrm{d}\overline{\delta_W}}{\mathrm{d}\overline{x}}\\
&+\left[\frac{\mathrm{Ja}_L(1-c_2)}{\overline{\delta_L}\mathrm{Pe}_L}\frac{\mathrm{d}\overline{\theta_l}/\overline{\theta_l}}{\mathrm{d}\overline{x}}+\frac{2\mathrm{Ja}_M\rho_M}{\overline{\delta_T}\mathrm{Pe}_M\rho_L}\frac{\mathrm{d}\overline{\theta_M}/\theta_M}{\mathrm{d}\overline{x}}\right]\\
&+\left[\frac{\mathrm{Ja}_L(1-c_2)}{\overline{\delta_L}\mathrm{Pe}_L}-\frac{2\mathrm{Ja}_M\rho_M}{\overline{\delta_T}\mathrm{Pe}_M\rho_L}-\frac{2\rho_M}{\mathrm{Sc}_M\mathrm{Re}_M\rho_L}\right]\frac{\mathrm{d}W_i/W_i}{\mathrm{d}\overline{x}}=0
\end{aligned} \tag{17}$$

where

$$A=\frac{\mu_M\overline{\delta_L}}{\mu_L\overline{\delta_M}},\quad E=\{(1+2d_1)+2A\},$$

$$\overline{U_1}=\overline{u_i}+(1-d_1)\left[\frac{2A}{E^2}\left\{2+\frac{\mathrm{Gr}_L}{2\mathrm{Re}_L}\left(\frac{\mu_L}{\mu_M}\right)\overline{\delta_L\delta_M}\right\}\right],$$

$$\overline{U_2}=\overline{u_i}+d_1\left[\frac{2A}{E^2}\left\{2+\frac{\mathrm{Gr}_L}{2\mathrm{Re}_L}\left(\frac{\mu_L}{\mu_M}\right)\overline{\delta_L\delta_M}\right\}\right],$$

$$\overline{\theta_l}=\frac{T_w-T_i}{T_w-T_\infty},\quad \overline{\theta_M}=\frac{T_i-T_\infty}{T_w-T_\infty},\quad \overline{T_w}=\frac{T_w}{T_w-T_\infty}$$

$$\overline{\theta_i}=\frac{T_i}{T_w-T_\infty},\quad \overline{u_i}=\frac{2A}{E}+\frac{\mathrm{Gr}_L\overline{\delta_L^2}}{2\mathrm{Re}_LE},\quad \overline{D}=\left[1+\frac{\frac{\mathrm{Gr}_L\overline{\delta_L}^2}{2\mathrm{Re}_L}}{\overline{u_i}}-d_1\right],$$

$$\overline{S_1}=\frac{1}{3\overline{\overline{\phi}}}+\frac{1}{30\overline{\phi}^3}-\frac{1}{6\overline{\phi}^2},\quad \overline{\phi}=\frac{\overline{\delta_W}}{\overline{\overline{\delta_M}}}$$

$$\overline{S_2}=1-\left(1-\frac{1}{\overline{\overline{\phi}}}\right)^3,\quad \overline{F}=\frac{\overline{\phi}^2}{30}-\frac{\overline{\phi}}{6}+\frac{1}{3},\quad \overline{\phi}=\frac{\overline{\delta_T}}{\overline{\overline{\delta_M}}}$$

$$\begin{aligned}
\overline{U}=&-\frac{7}{12}\frac{\mathrm{Gr}_L\overline{\delta_L}^3\overline{D}}{\mathrm{Re}_L}+0.667\overline{u_i\delta_LD}^2+\frac{2}{7}\overline{u_i\delta_L}d_1^2-\frac{11}{30}\frac{\mathrm{Gr}_L\overline{\delta_L}^3d_1}{\mathrm{Re}_L}\\
&+0.125\frac{\mathrm{Gr}_L^2\overline{\delta_L}^5}{\mathrm{Re}_L^2\overline{u_i}}+0.8\overline{u_i\delta_LD}d_1-\overline{u_i\delta_L}\\
&+0.5\overline{u_i\delta_L}d_1-\frac{1}{12}\frac{\mathrm{Gr}_L\overline{\delta_L}^3}{\mathrm{Re}_L}
\end{aligned}$$

This system of eight ordinary differential equations is solved using MATLAB ODE solver, e.g., "ODE15s." The results, which are gotten from that numerical solution, are processed to evaluate the following results.

3 Results and Discussion

Local heat transfer coefficient can be calculated from following equation:

$$h_x(T_w - T_\infty) = q_x \tag{18}$$

where q_x = total heat flux $= K_L\left(\frac{\partial T_L}{\partial y}\right)_{y=0}$

So,

$$h_x = \frac{K_L(T_w - T_i)(1 + c_2)}{\delta_L(T_w - T_\infty)} \tag{19}$$

Average heat transfer coefficient can be calculated from following equation:

$$h_{\text{avg}} = \frac{\int_0^L h_x \text{d}x}{L} \tag{20}$$

3.1 Validation of Model

Developed model is validated with those of Jacobs [12] and mathematical model of Liao et al. [9] for pure vapor stream for a wide range of Richardson number.

We calculate three non-dimensional numbers as Nusselt number $\left[\text{Nu}_L = \frac{h_x L}{K_L}\right]$, Grashof number $\left[\text{Gr}_L = \frac{(\rho_L - \rho_\infty) g L^3}{\rho_L \nu_L^2}\right]$, Reynolds number $\left[\text{Re}_L = \frac{U_\infty L}{\nu_L}\right]$ for pure vapor.

We plot $\frac{\text{Nu}_L}{\text{Re}_L^{0.5}}$ versus Richardson Number $\left[\text{Ri}_L = \frac{\text{Gr}_L}{\text{Re}_L^2}\right]$. We compared these values with those of experimental results of Jacobs [12] and mathematical model of Liao et al. [9] in Figs. 2 and 3. The entire range of mixed convection condensation predicted by present model is validated well against experimental and non-similarity solution for pure vapor condensation.

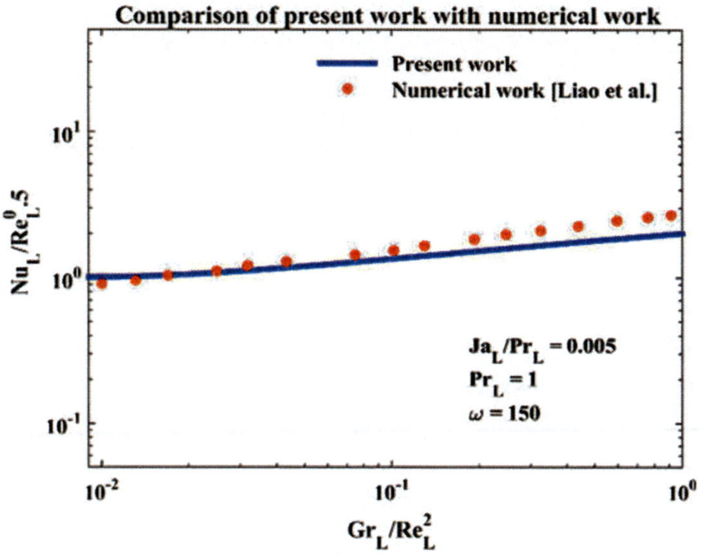

Fig. 2 Comparison of present work with non-similarity solution of Liao et al

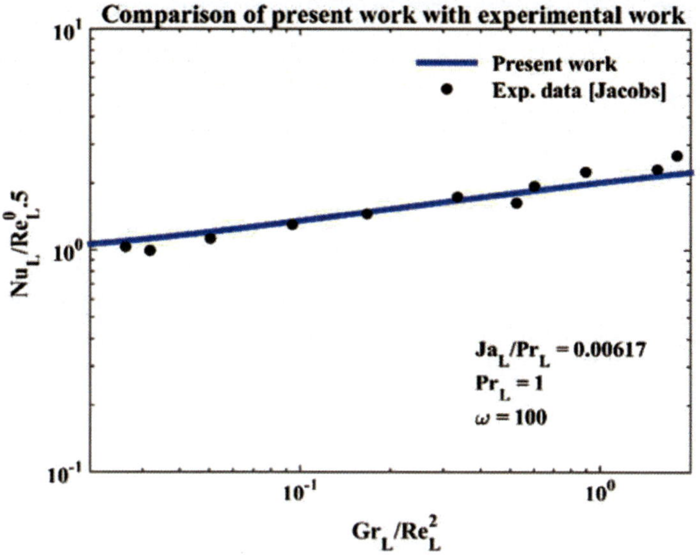

Fig. 3 Comparison of present work with experimental work by Jacobs

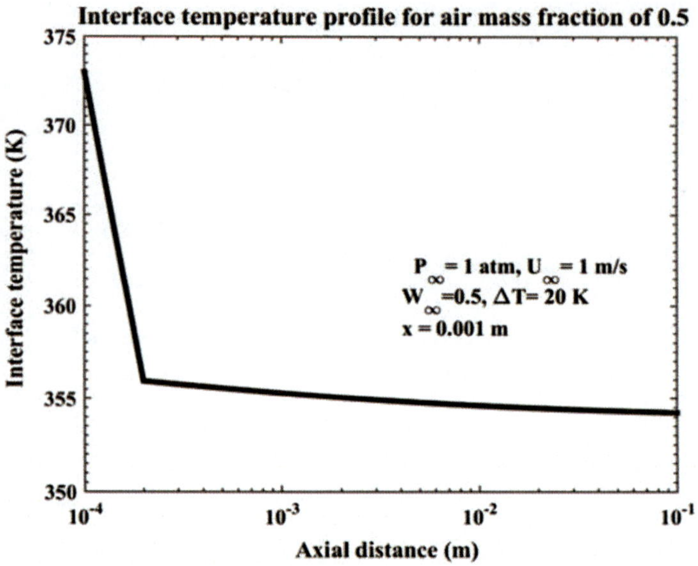

Fig. 4 Interface temperature profile for air mass fraction = 0.5

3.2 Interface Temperature and Interface Concentration Profile

Figures 4 and 5 show variation of interface temperature and concentration profile along the isothermal wall. Even small amount of non-condensable gas has great influence on heat and mass transfer process due to extra resistance offered by it. Accumulation of gases increases partial pressure of non-condensable gases and decreases partial pressure of vapor along the interface region as total pressure remains constant at each and every point of mixture. This increase in partial pressure of gases cause decease in saturation pressure of vapor and leads to lowering in saturation temperature along the axial direction of wall. Due to accumulation of gases, interface concentration along the axial direction of wall also increases continuously.

3.3 Effect of Free Stream Non-condensable Gas Concentration

The effect of non-condensable gas concentration on condensation process has been analyzed for three different mass fractions of 1, 20, and 50% in Fig. 6. The analysis has been done with $P_\infty = 1$ atm, $U_\infty = 3$ m/s, $\Delta T = 20$ K. The increase in free stream gas concentration creates additional resistance near liquid-gas interface region. As a result, vapor mass fraction at the interface decreases to maintain total mass fraction

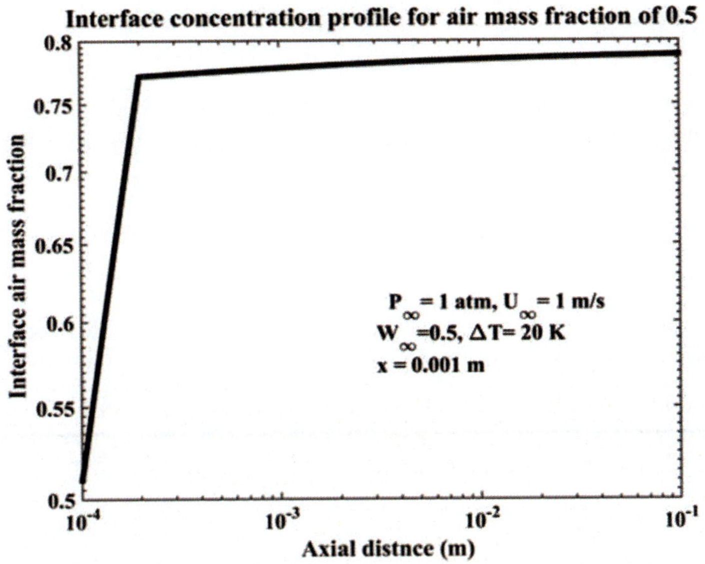

Fig. 5 Interface concentration profile for air mass fraction = 0.5

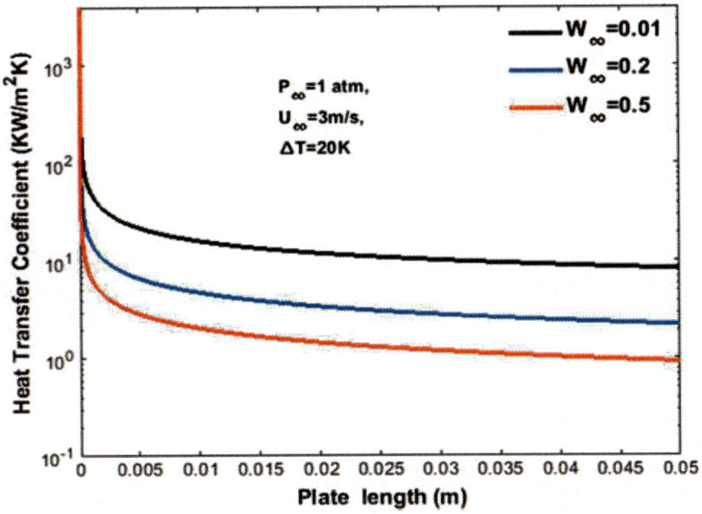

Fig. 6 Variation of heat transfer coefficient with different air mass fraction

to unity at the interface. This leads to decreasing in saturation vapor temperature along the axial direction through the wall. Due to this, condensation process as well as heat transfer coefficient decreases significantly along the wall. Figure 6 shows the variation of heat transfer coefficient with non-condensable gas concentration.

4 Conclusion

In this report, a generalized heat transfer model is developed for mixed convection regime. A detailed analytical study is performed to study the effects of non-condensable gases on the condensation process. This report mainly deals with forced convection and natural convection condensation over a subcooled vertical isothermal surface. The results obtained from this model contain various parameters such as film thickness of each layer, interface velocity, interface temperature, and air mass fraction variation along the liquid-vapor interface. The validation with the current work shows the assumptions to be reasonably acceptable. The presented results support the following conclusions.

- The presented analytical results for laminar film condensation agree closely with the numerical results of Liao et al. [9] at $W_\infty \sim 0\%$, i.e., pure saturated vapor. The presented results also match closely with the experimental results of Jacobs [12].
- Interface temperature of saturated vapor decreases continuously along the axial direction of the wall. Similarly, the interface concentration of non-condensable gases increases along the axial direction. This phenomenon has been shown in the presented figures.
- Due to the non-diffusive nature, non-condensable gas accumulates at the liquid-vapor interface and creates resistance to heat and mass transfer processes. It reduces saturation vapor temperature in interface region and therefore rate of condensation and heat transfer coefficient decreases.

References

1. Nusselt W (1916) Die oberflachenkondensation des wasserdamphes. VDI-Zs 60:541
2. Rohsenow WM (1956) Heat transfer and temperature distribution in laminar film condensation. Trans Asme 78:1645–1648
3. Sparrow EM, Gregg JL (1959) A boundary-layer treatment of laminar-film condensation. J Heat Transfer 81(1):13–18
4. Koh JCY, Sparrow EM, Harnett JP (1961) The two-phase boundary layer in laminar film condensation. Int J Heat Mass Transf 2(1–2):69–82
5. Tetsu F, Haruo U, Chikatoshi K (1972) Laminar filmwise condensation of flowing vapor on a horizontal cylinder. Int J Heat Mass Transf 15(2):235–246
6. Sparrow EM, Minkowycz WJ, Saddy M (1967) Forced convection condensation in the presence of non-condensable and interfacial resistance. Int J Heat Mass Transf 10(12):1829–1845
7. Sparrow EM, Siegel R (1959) Transient film condensation. J Appl Mech 26:120
8. Shu JJ, Wilks G (2014) Mixed convection laminar film condensation on a semi-infinite vertical plate. arXiv preprint arXiv:1403.3240
9. Liao Y, Guentay S, Vierow K (2010) Local non-similarity method for the two-phase boundary layer in mixed convection laminar film condensation. Heat Mass Transf 46(4):447–455
10. Balasubramaniam R, Hasan MM (2015) Transient condensation of flowing vapor on a flat-plate: a scaling analysis. Int J Heat Mass Transf 91:739–799

11. Das DC, Ghosh K, Sanyal D, Meigen R (2014) A novel approach for modelling mixed convection film boiling for a vertical flat plate. Numer Heat Transf Part A Appl 66(10):1112–1130
12. Jacobs HR (1966) An integral treatment of combined body force and forced convection in laminar film condensation. Int J Heat Mass Transf 9(7):637–648

Magneto-Convective Heat Transfer in a Cavity Under Partial Magnetic Fields

Aparesh Datta, Nirmalendu Biswas, and Nirmal K. Manna

Nomenclature

•

B	Magnetic field, Tesla, N/A·m^2
Ha	Hartmann number
L	Length of the cavity/length scale, m
L_B	Length of the partially applied magnetic-field, m
Nu	Average Nusselt number
P	Pressure, Pa
Pr	Prandtl number
Ra	Rayleigh number
T	Temperature, K
u, v	Velocity components, m/s
U, V	Dimensionless velocity components
x, y	Cartesian coordinates, m
X, Y	Dimensionless coordinates

A. Datta
Department of Mechanical Engineering, NIT Durgapur, Durgapur 713209, India

N. Biswas (✉)
Department of Power Engineering, Jadavpur University, Salt Lake, Kolkata 700106, India

N. K. Manna
Department of Mechanical Engineering, Jadavpur University, Kolkata 700032, India

S. K. Ghosh et al. (eds.), *Advances in Thermal Engineering, Manufacturing, and Production Management*, Lecture Notes in Mechanical Engineering,
https://doi.org/10.1007/978-981-16-2347-9_10

Greek symbols

α	Thermal diffusivity, m^2/s
β	Thermal expansion coefficient of fluid, K^{-1}
θ	Dimensionless temperature
υ	Kinematic viscosity, m^2/s
Π	Dimensionless heat function
ρ	Density, kg/m^3
κ	Electrical conductivity ($\mu S\ cm^{-1}$)
ψ	Dimensionless stream function
a, h	Ambient, heating

1 Introduction

The fastest growth in technological advancement and demand for miniaturization of the object size researchers are paying their attention to improving the performance of any device. In this regard, the application of natural convective heat transfer is widely used in the electronics industry. Apart from electronic cooling system, there are several processes, which demand precious control on its thermal-hydrodynamic behavior. The imposing magnetic field is of good means of controlling fluid-flow and thermal pattern (which is known as magneto-hydrodynamics—MHD). For example, thermo-magnetic mixing, targeted drug delivery in the human body, cancer treatment, cell separation, magnetic endoscopy, cancer and tumor treatments, antivibration devices, etc. A detailed account of the review on the application of magneto-hydrodynamics is well documented [1–4].

Natural convective heat transfer in a confined space has been analyzed extensively by Biswas et al. [5, 6] under different thermal boundary conditions. Matt et al. [7] have examined natural convective heat transfer under the influence of magnetohydrodynamics. Thermomagnetic buoyancy-driven convection in a left-hot and right-cold cavity in presence of line dipole magnetic field has been analyzed by Ganguly et al. [8] and reported about augmented heat transfer. Applying inclined magnetic field, Al-Balushi and Rahman [9] have examined the convective heat transfer under different thermal boundary conditions.

Introducing a non-uniformly active upright partial magnetic field, Jalil et al. [10] have investigated natural convective heat transfer in a differentially heated cavity and observed oscillation behavior disappear with the increasing magnetic field. Thereafter a three-dimensional natural convective heat transfer of nanoliquid filled differentially heated cavity along with the influence of horizontally applied partially active magnetic field (in vertical direction) has been investigated by Al-Rashed [11]. They found that the heat transfer process modifies notably with the change in length and location of the imposed magnetic field. Very recently, Geridonmez and Oztop [12] have examined the natural convective phenomena in a left-hot right-cold porous cavity in presence of horizontally imposed partial magnetic field. It is noted that by adjusting location and its strength of partial magnetism, flow and heat transfer process could be controlled. In other class of works, convective heat transfer has been studied imposing variable magnetic field in a lid-driven cavity [13], periodic magnetic field in a differentially heated cavity [14] and along a vertical surface [15], non-uniform magnetic field in a differentially heated cavity [16].

Although few researchers have studied the effect of partial magnetism on thermal convection, still fundamental research is needed to explore more underneath physics under the effect of partial magnetism. The effect of partial magnetic fields on different working fluids is not reported till now. The aim of this research is to examine thermal convection in a bottom hot, sides cold cavity under the influence of vertically applied partial magnetic field considering different working fluid. The study is carried out varying Rayleigh number (Ra), Hartmann number (Ha), length of the partial magnetic field (L_B) and different types of fluid (Pr). The outcome of this study can be applied for the design of any device subjected to the heat transfer process along with its control in several industrial applications.

2 Problem Formulation

2.1 Problem Descriptions

The configuration of the studied problem including boundary conditions is depicted in Fig. 1. The square cavity (with side/length L) is heated isothermally (at temperature

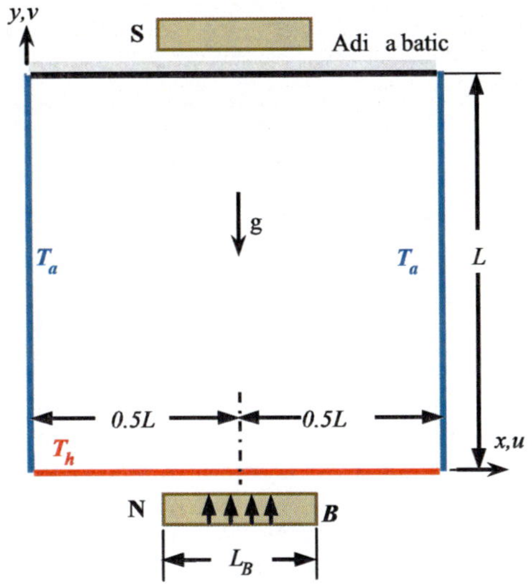

Fig. 1 Schematic diagram of the problem geometry, boundary conditions

T_h) from the bottom and is allowed to exchange heat with the surroundings (at temperature T_a) through the sidewalls; whereas the top wall is thermally insulated. Externally generated a magnetic field of intensity B is uniformly applied partially over length L_B in the vertical direction (bottom to top). The cavity is filled with fluid, three different types of fluid are considered having Prandtl number Pr = 0.054, 0.71, and 6.93.

The evolved flow physics is amazed assuming: flow is two-dimensional, steady, laminar, incompressible, Newtonian within the limit of Boussinesq approximation. Viscous dissipation, Joule heating, Hall effect, and radiation are ignored. Resulting dimensionless governing equations for mass, momentum, and energy based on conservation principles are given as-

$$\frac{\partial U}{\partial X}+\frac{\partial V}{\partial Y}=0 \tag{1}$$

$$U\frac{\partial U}{\partial X}+V\frac{\partial U}{\partial Y}=-\frac{\partial P}{\partial X}+\mathrm{Pr}\left(\frac{\partial^2 U}{\partial X^2}+\frac{\partial^2 U}{\partial Y^2}\right)-\mathrm{Pr}\,\lambda_B \mathrm{Ha}^2 U \tag{2}$$

$$U\frac{\partial V}{\partial X}+V\frac{\partial V}{\partial Y}=-\frac{\partial P}{\partial Y}+\mathrm{Pr}\left(\frac{\partial^2 V}{\partial X^2}+\frac{\partial^2 V}{\partial Y^2}\right)+\mathrm{Ra}\,\mathrm{Pr}\,\theta \tag{3}$$

$$U\frac{\partial \theta}{\partial X}+V\frac{\partial \theta}{\partial Y}=\left(\frac{\partial^2 \theta}{\partial X^2}+\frac{\partial^2 \theta}{\partial Y^2}\right) \tag{4}$$

where X and Y, U and V, P, and are the dimensionless coordinates, velocity components, pressure, and temperature as detailed below.

$$(X, Y)=(x, y)/L,\ (U, V)=(u, v)/(\alpha/L),$$
$$P=p/\rho(\alpha/L)^2, \theta=(T-T_a)/(T_h-T_a) \tag{5}$$

Partial magnetic field is imposed through the factor λ_B and it is defined as $\lambda_B = 0$ ($0 \le X < 0.5L\text{-}0.5L_B$) or 1 ($0.5L\text{+}0.5L_B \le X \le 1$). Derived dimensionless parameters are Prandtl, Rayleigh, and Hartmann number (Pr, Ra, and Ha respectively) obtained as

$$\mathrm{Ra}=g\beta(T_h-T_a)L^3/\alpha\upsilon,\ \mathrm{Pr}=\upsilon/\alpha,\ \mathrm{Ha}=BL\sqrt{\kappa/\mu} \tag{6}$$

For the numerical computation, following boundary conditions are specified:

- at the bottom hot wall, $U = 0,\ V = 0,\ \theta = 1$
- at the side cold walls, $U = 0,\ V = 0,\ \theta = 0$
- at the adiabatic walls, $U = V = 0,\ \partial\theta/\partial Y = 0$

The heat transport pathway from the bottom wall to the sidewalls are visualized using Bejan's heatlines [5, 17]. Heatfunction equations in the first-order differential form can be written as

$$-\frac{\partial \Pi}{\partial X} = V\theta - \frac{\partial \theta}{\partial Y} \text{ and } \frac{\partial \Pi}{\partial Y} = U\theta - \frac{\partial \theta}{\partial X} \tag{7}$$

Using the solved velocity and temperature fields, heatlines (which are constant heatfunctions) are generated by solving Eq. (7) iteratively. The heat transfer characteristics (local and average) from the bottom wall is calculated by the Nusselt number (Nu) as

$$\mathrm{Nu}(X) = -\left.\frac{\partial \theta}{\partial Y}\right|_{Y=0} \text{ and } \mathrm{Nu} = \int_0^1 \mathrm{Nu}(X)\, \mathrm{d}X \tag{8}$$

2.2 *Numerical Technique*

The dimensionless Eqs. (1–4), with appropriate conditions are computationally solved using FVM based well-validated written algorithm [18] in a staggered grid scheme, utilizing the SIMPLE algorithm [19], TDMA algorithm and ADI sweep. The converged solution is obtained in an iterative process by setting the maximum residuals and mass-defect < 10^{-6} and 10^{-8}. The same code has been validated extensively by simulating the problems available in the open literature [5, 6, 18, 20, 21]. Before conducting extensive simulations, a mesh sensitivity test is also carried out and finally 200 × 200 grid size is chosen, which is not incorporated here for brevity.

3 Result and Discussion

Thermo-magnetic convection of the studied geometry is investigated for wide ranges of parameters: Rayleigh numbers (Ra = 10^3, 10^4, 10^5, and 10^6), Hartmann number (Ha = 0, 10, 30, 50, 70, and 100), length of the partial magnetic field (L_B = 0, 0.1, 0.3, 0.5, 0.7, and 1) and different type of fluid (Pr = 0.054, 0.71, and 6.93). Obtained results are visualized using streamlines, isotherms, heatlines, average Nu and discussed below systematically.

3.1 *Impact of Rayleigh Number (Ra)*

The convection mechanism, as well as strength of the buoyancy force within the simulated domain, is governed by the Rayleigh number (Ra). The impact of varying Ra on the thermo-fluid flow structure is illustrated in Fig. 2, in terms of combined streamlines (firm line—purple color) and isotherms (dashed line—red color) in the

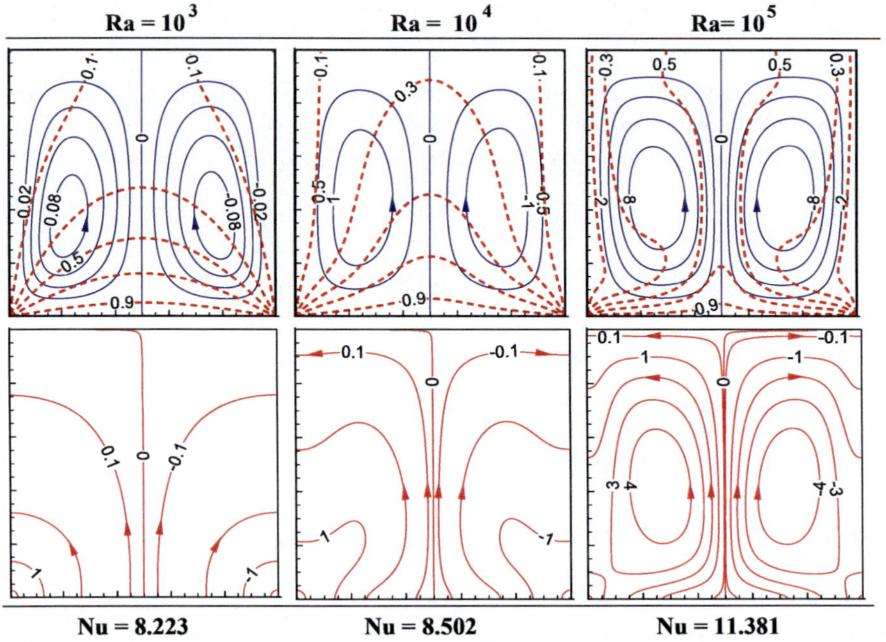

Fig. 2 Impact of varying Rayleigh number (Ra) on the flow fields, and thermal behavior for Ha = 30, $L_B = 0.5$, Pr = 0.71

top row and heatlines in the bottom row Ha = 30, $L_B = 0.5$, Pr = 0.71. At lower Ra = 10^3 value, the strength of the buoyancy force is weak, resulting in weaker flow circulation velocity. Due to the symmetrical heating at the bottom wall and cooling at the sidewalls of the cavity, two symmetrical circulating cells (rotating in a counter-clockwise direction at the left and clockwise direction at the right) forms inside the cavity. Thus, the cavity fluid being heated near the bottom wall and moves upward thereafter obstructed from the top insulated wall and releases heat to the ambient through side cold walls. From the heated wall to the cold walls, the heat energy transportation is mainly dictated by conduction mode of heat transfer. Thus, the isotherms are distributed inside the cavity with gradual decrement towards the cold sidewalls (as contour value varies in between 0 and 1). Isotherms of higher contour values are clustered with the heated bottom wall. Corresponding, heatlines (in the second row) originates from the heated bottom wall and directly terminated at the cold sidewalls symmetrically.

Now, with the increasing Ra value, to Ra = 10^4 (second column), and 10^5 (third column), the strength of the fluid circulation increases significantly as reflected by the streamline contours values. At Ra = 10^4, both the conduction and convection modes of heat transfer assist in the heat transfer process. As the heat energy added into the cavity fluid increases with increasing Ra, temperature distribution pattern, as well as heatlines contours, modifies significantly compared to Ra = 10^3. Further, increase in

Ra to 10^5 heat transfer process is dictated by the convection mode. Although the fluid flow pattern remains similar, the circulation strength increases notably. The isotherms are uniformly and densely distributed near the heated wall and it indicates a sharp temperature gradient. Of course, the rest of the isotherm contours are connected with the top adiabatic wall. It is interesting note that, as the input heat energy to the cavity fluid increases with increasing Ra, the heat energy from the bottom heated surface to the side cold walls transport through a long passageway. As a result, energy circulation cells appear in both parts of the cavity. As noticeable from the value of heatfunction, maximum heatfunction is obviously increased with the increasing Ra. Further to this, an increasing trend of heat transfer from the hot bottom wall is clearly noted from the indicated average Nu—indicated below the heatlines contours.

3.2 Impact of Magnetic Field Strength (Ha)

The impacts of an externally generated uniform magnetic fields (of strength Ha) on the thermo-fluid behavior are analyzed and presented in Fig. 3 for Ha = 0, 10, and 100 at a fixed value of Ra = 10^6, $L_B = 0.5$, Pr = 0.71. First, the without magnetic field (Ha = 0) case is analyzed. In consistent with Fig. 2 (as in the first column), two

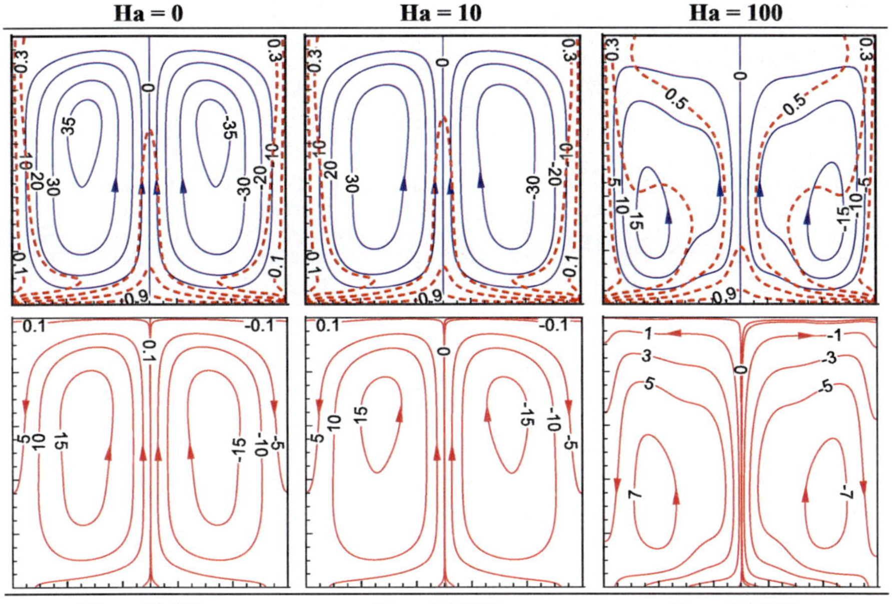

Fig. 3 Impact of varying Hartmann number (Ha) on the flow fields and thermal behavior for Ra = 10^6, $L_B = 0.5$, Pr = 0.71

circulation cells rotating in anticlockwise (at left portion) and clockwise (at right portion) directions are formed symmetrically. The core of these circulating cells positioned at the top portion of the enclosure. The isotherms of higher contour values are densely distributed near the heated bottom wall and show temperature gradient; whereas the isotherms with lower magnitude are clustered near the side cold walls and connected with the top adiabatic wall, resulting in higher fluid velocity and stronger thermal convection. Corresponding heatlines contours show a symmetrical distribution along with the formation of stronger energy circulation cells.

Now imposing the magnetic field with Hartmann number Ha = 10, the force due to the magnetic field increases, which partially counteracts the buoyancy force. Resulting decrease in the fluid velocity—which is reflected by the streamline contour value. The reason for such decrement can be realized by noting a negative source term in the X-momentum equation as in Eq. (2). The dampening effect of the imposed magnetic field increases substantially as the Ha value increases from 10 to 100. This implies, partially applied middle-centered magnetic field effect impacts flow structure as well as thermal behavior. Fluid circulation strength becomes weaker (as noted from the streamline contour value). Although circulation cells remain symmetrical about the mid-vertical plane, it deforms in its shape and core of the circulation cells shifts from upper to lower part of the cavity. Distribution of isotherms also distorted inside the cavity (compared to Ha = 10). As fluid circulation strength reduces, at the same time heat energy transportation from heated wall to the cold sidewalls also reduces, and distribution of the heatlines contours distorted significantly. As a result, convective heat transport reduces with increasing Ha, which is reflected in reduced Nu value—indicated below the heatlines contours.

3.3 *Impact of Length of the Applied Magnetic Field* (L_B)

Effect of change in the length, over which magnetic field is applied, on the hydrothermal behavior, is illustrated in Fig. 4 considering three different lengths $L_B = 0.1$, 0.3, and 1 keeping the fixed value of Ra = 10^6, Ha = 50, Pr = 0.71. Inconsistent with earlier findings on flow structure and thermal behavior (as in Fig. 3 for Ha = 0), flow circulation pattern reveals similar at $L_B = 0.1$ as in Fig. 4. As the length L_B increases from 0.1 to 0.3, fluid circulation strength reduces, and at the same time, the isotherms are stretched towards the upper part of the cavity. It results in decrement of buoyancy force. Thus, convective heat transport reduces which lowers average Nu. Further, the increase in L_B from 0.3 to 1, symmetrical circulation cells deformed substantially with a reduction in fluid flow velocity (indicated by the streamlines contours). Of course, the temperature gradient inside the cavity decreases. Heat energy-transportation also reduced significantly. As a result, heat transfer retarded and Nu value decreases considerably.

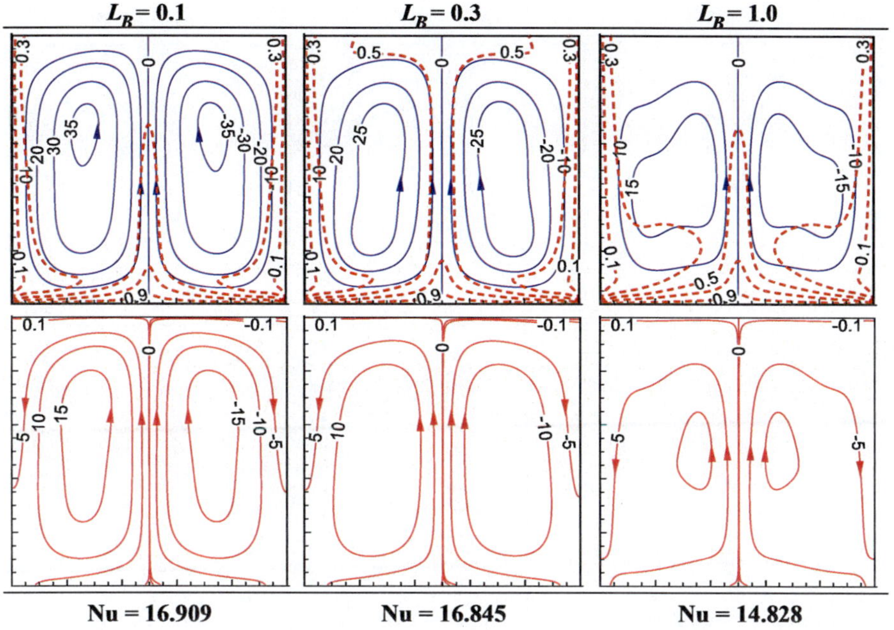

Fig. 4 Effect of length of applied magnetic field (L_B) on flow fields and thermal behavior for Ra = 10^6, Ha = 50, Pr = 0.71

3.4 *Effect of Different Prandtl Number (Pr)*

The effect of partial magnetism on the hydrothermal behavior under different Prandtl number (Pr = 0.054, 0.71, and 6.93) are also investigated and presented in Fig. 5 at Ra = 10^6, Ha = 50, L_B = 0.5. As found earlier, the fluid flow structures, isotherm distribution, and heatlines pattern exhibit comparable to that with air. Of course, the shapes of circulation cells are different from each other. For lower Pr value, circulation cells stretched vertically; whereas for Pr = 0.71 it stretched in horizontally and vertically; at Pr = 6.93 it stretches horizontally. At Pr = 0.71, the core of the flow circulation cells remains in the lower part of the cavity; whereas for the other two Pr, the core remains in the middle part of the cavity. Pattern-wise heatlines contours show similar distribution following the fluid circulation cells. Here the distortion of the isotherms is more distinct, leading to different convection strength. Of course for high Prandtl-fluid (Pr = 6.93), the heat transfer rate is higher due to better thermal conductivity. Heatfunctions contours also show a higher value than that of the other two fluids, which implies a higher heat transfer rate. However, due to the presence of partial magnetism, the heat transfer rate is influenced considerably and Nu value shows maximum with Pr = 0.71. This happens so, with lower Pr value buoyancy force which dampens the convection strength retarded substantially than that of Pr = 0.71.

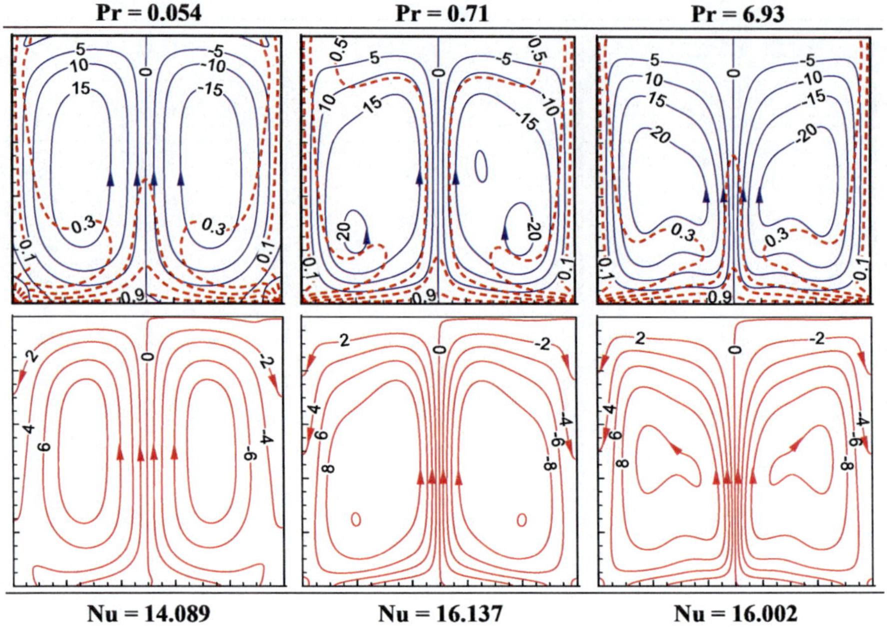

Fig. 5 Effect different Prandtl number (Pr) on the flow fields and thermal behavior for Ra $= 10^6$, Ha $= 50$, $L_B = 0.5$

3.5 Heat Transfer Analysis

The overall trends of heat transfer characteristics (expressed by average Nu) with varying Ra are presented in Fig. 6a for three different Pr $= 0.054$, 0.71, and 6.93 keeping fixed value of Ha $= 50$, $L_B = 0.5$. Figure 6a, shows no variation in average Nu until Ra $= 10^4$, after which it shows monotonous increment in Nu as Ra increases until Ra $= 10^6$. Beyond Ra $= 10^5$ as convection mode of heat transfer dominates, thus buoyancy force increases, leading to an increase in heat transfer significantly. However, with Pr $= 0.71$ and 6.93, the heat transfer rate is found to be superior compared to Pr $= 0.054$. Now, the change in Nu with different Ha is illustrated in Fig. 6b considering Ra $= 10^6$, $L_B = 0.5$, Pr $= 0.054$, 0.71, and 6.93. It shows that as Ha increases, Nu decreases monotonously indicating to the dampening effect of magnetic force (as reflected by negative term associated with Ha in X momentum equation). Of course, this effect is not so significant with the fluid of Pr $= 0.054$ compared to the fluid of Pr $= 0.71$, and 6.93. On the other hand, when the length of the partially imposed magnetic field increases considering Ra $= 10^6$, Ha $= 50$, Pr $=$ 0.054, 0.71 and 6.93, the trend of overall heat transfer shows a decreasing trend (as in Fig. 6c). For the same strength of the magnetic field (Ha $= 50$), as L_B increases from no-magnetic field to $L_B = 1$, the impact of magnetic force is modified within the entire area of the enclosure. This effect on the flow and thermal field is already

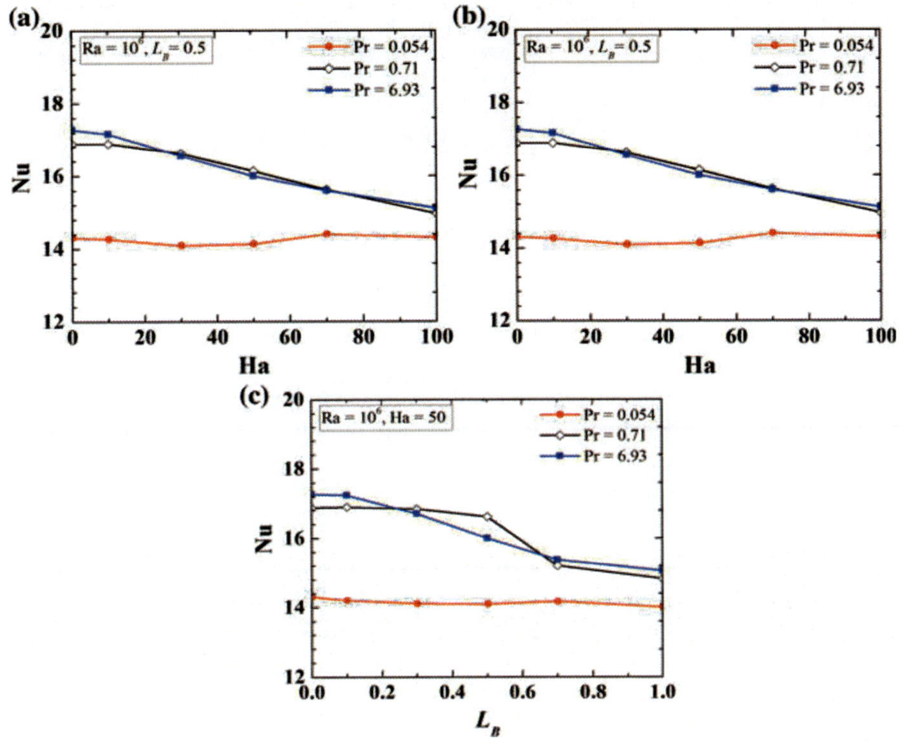

Fig. 6 Trend of overall heat transfer characteristics with varying **a** Ra, **b** Ha and **c** L_B for Pr = 0.054, 0.71, 6.93

illustrated in Fig. 4. Similar to Fig. 6b, a rapid decrement in the heat transfer process is recognized with Pr = 0.71 and 6.93 as L_B increases at a fixed Ha value. This indicates increase in suppression of the buoyancy force due to increase in magnetic field affected zone. With a lower value of Pr = 0.054, small changes in Nu is noted as L_B increases. It implies no significant effect of magnetic force on hydrothermal behavior with Pr = 0.054.

For a better understanding of the global trend of heat transfer characteristics, the analysis of local thermo-fluid behavior is presented in Fig. 7. Thus, vertical profiles of V-velocity and temperature (θ) with the respect to the mid-horizontal plane (at Y = 0.5) at different Ra (for Ha = 50, L_B = 0.5) and Ha (for L_B = 0.5, Ra = 10^6) are processed and demonstrated in Fig. 7 for three Prandtl numbers Pr = 0.054, 0.71, and 6.93, respectively.

The effect of varying Rayleigh number (Ra = 10^3, 10^5 and 10^6) indicates stronger convection at higher Ra (Ra $\geq 10^5$) as V-velocity is higher; whereas for lower Ra, V-velocity is negligible. Of course, positive peak V-velocity (about mid vertical plane) is observed with Pr = 0.71. Higher flow velocity within the cavity leads to transport of more amount of heat energy from the heated bottom wall to the cold sidewalls.

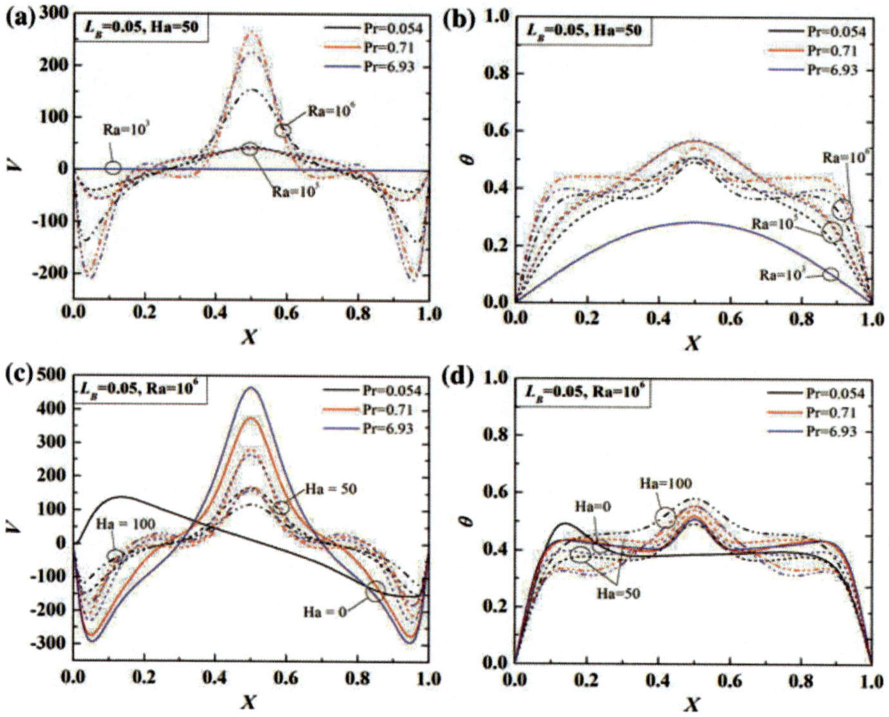

Fig. 7 Horizontal mid-plane V-velocity and temperature (θ) profile for varying Ra **a** and **b**, Ha **c** and **d** for L_B= 0.5 for Pr = 0.054, 0.71, and 6.93 respectively

This is consistent with the findings of the higher Nu as in Fig. 6a. It is also noticed that in the V-velocity profile—there are two negative peaks towards the sides of the X-axis.

This happens so, as thermal convection dominates counter-rotating symmetrical fluid circulations. As the V-velocity component is very weaker at $\text{Ra} = 10^3$, it leads to lower thermal energy transportation and lower Nu value. Of course, the fluid velocity is comparatively lower with Pr = 0.054 than that of Pr = 0.71 and 6.93; as a result, Nu is significantly lower. The temperature (θ) peaks are found to be higher for higher Rayleigh values ($\text{Ra} = 10^5$ and 10^6) for Pr = 0.71 and 6.93. The effect of increasing Hartman number (Ha) on the V-velocity and temperature (θ) are presented in Fig. 7c and d respectively. In the case of no magnetic field (Ha = 0), velocity peaks are maximum, but it reduces substantially at Ha = 100 (as in Fig. 7c). Similarly, the temperature profiles higher values at Ha = 100. As the applied magnetic field strength increases from 0 to 100, the flow velocity decreases due to the dampening effect of the magnetic field; which reduces the convection strength.

4 Conclusions

In this research work, an attempt has been taken to explore the fundamentals of thermo-magnetic convection under the influence of partially applied magnetic field on a classical bottom heated, sides cold cavity. The uniform magnetic field is applied partially from the bottom in a vertical direction. Evolved flow physics is analyzed and visualized numerically for a range of parametric variations of Rayleigh number, Hartmann number, length of the applied magnetic field, and different Prandtl number. The heat energy transportation from heat source to heat sink is visualized using Bejan's heatlines. The major findings are summarized as follows:

- As Rayleigh number increases ($Ra \geq 10^4$), convective heat transfer characteristics increase monotonously due to an increase in buoyancy force.
- The heatlines contours illustrate the energy flow path from the heated surface to the cold walls. It shows the formation of energy recirculation cells when $Ra \geq 10^5$. Else heatlines are connected directly between the heated surfaces and the cold walls.
- At $Pr = 0.71$ and 6.93, the convective heat transfer process improves markedly compared to $Pr = 0.054$. With lower Pr value buoyancy force retarded substantially; which dampens the convection strength.
- As the employed magnetic field intensity in terms of Hartmann number (Ha), the heat transfer rate decreases substantially as the magnetic force dominates the buoyancy force. Fluid circulation strength becomes weaker and distortion of fluid circulation cells as well as energy circulation cells due to the increasing effect of partially applied middle centered magnetic field. Such an effect is noticeable with the fluid of $Pr = 0.71$ and 6.93 compared to $Pr = 0.054$.
- Overall heat transfer characteristics show a decreasing trend with the increasing active length L_B of partial magnetic field. The effect of magnetic force is dominating at a higher Rayleigh number and higher length (L_B) of the applied magnetic field (Ha). Lower Prandtl number fluid shows less response to the partial applied magnetic field. Such an effect significant with the fluid of $Pr = 0.71$ and 6.93 compared to $Pr = 0.054$.

The present study on buoyancy-driven thermomagnetic convection has practical relevance in numerous industrial as well as medical science applications like targeted drug delivery, cancer treatment, antivibration devices, etc. Both the strength length of the externally generated magnetic field could be used as a controlling parameter of the heat transfer process.

Acknowledgements None.

References

1. Turcotte D, Lyons J (1962) A periodic boundary-layer flow in magnetohydrodynamics. J Fluid Mech 13(04):519–528
2. Davidson PA (2001) An introduction to magnetohydrodynamics, Cambridge texts in applied mathematics. Cambridge University Press, Cambridge, UK
3. Ozoe H (2005) Magnetic convection. Imperial College Press, Singapore
4. Rashidi S, Esfahani JA, Maskaniyan M (2017) Applications of magnetohydrodynamics in biological systems-a review on the numerical studies. J Magn Magn Mater 439:358–372
5. Biswas N, Mahapatra PS, Manna NK (2016) Buoyancy-driven fluid and energy flow in protruded heater enclosure. Meccanica 51:2159–2184
6. Biswas N, Mahapatra PS, Manna NK, Roy PC (2015) Influence of heater aspect ratio on natural convection in a rectangular enclosure. Heat Transf Eng 37(2):1–15
7. Matt CFT, Quaresma JNN, Cotta RM (2017) Analysis of magnetohydrodynamic natural convection in closed cavities through integral transforms. Int J Heat Mass Transf 113:502–513
8. Ganguly R, Sen S, Puri IK (2004) Thermomagnetic convection in a square enclosure using a line dipole. Phys Fluids 16(7):2228–2236
9. Al-Balushi LM, Rahman MM (2019) Convective heat transfer utilizing magnetic nanoparticles in the presence of a sloping magnetic field inside a square enclosure. J Thermal Sci Eng Appl 11:041013-1-19
10. Jalil JM, Al-Tae'y KA, Ismail SJ (2013) Natural convection in an enclosure with a partially active magnetic field. Numer Heat Transf A 64(1):72–91
11. Al-Rashed AAAA, Kolsi L, Oztop HF, Aydi A, Malekshah EH, Abu-Hamdeh N, Borjini MN (2018) 3D magneto-convective heat transfer in CNT-nanofluid filled cavity under partially active magnetic field. Physica E 99:294–303
12. Geridonmez BP, Oztop HF (2019) Natural convection in a cavity filled with porous medium under the effect of a partial magnetic field. Int J Mech Sci 161–162:
13. Sheikholeslami M, Chamkha AJ (2016) Flow and convective heat transfer of a ferro-nanofluid in a double-sided lid-driven cavity with a wavy wall in the presence of a variable magnetic field. Numer Heat Transf A 69(10):1186–1200
14. Mehryan SAM, Izadi M, Chamkha AJ, Sheremet M (2018) Natural convection and entropy generation of a ferrofluid in a square enclosure under the effect of a horizontal periodic magnetic field. J Mol Liquids 263:510–525
15. Siddiqa S, Hossain MA, Gorla RSR (2012) Conduction-radiation effects on periodic magnetohydrodynamic natural convection boundary layer flow along a vertical surface. Int J Therm Sci 53:119–129
16. Szabo PSB, Früh W-G (2018) The transition from natural convection to thermomagnetic convection of a magnetic fluid in a non-uniform magnetic field. J Magn Magn Mater 447:116–123
17. Kimura S, Bejan A (1983) The heatline visualization of convective heat transfer. J Heat Transf Trans 105(4):916–919
18. Biswas N, Manna NK (2018) Magneto-hydrodynamic Marangoni flow in bottom-heated lid-driven cavity. J Mol Liquids 251:249–266
19. Patankar SV (1980) In: Numerical heat transfer and fluid flow. Taylor and Francis
20. Biswas N, Sarkar UK, Chamkha AJ, Manna NK (2020) Magneto-hydrodynamic thermal convection of Cu–Al_2O_3/water hybrid nanofluid saturated with porous media subjected to half-sinusoidal nonuniform heating. J Thermal Anal Calorim. https://doi.org/10.1007/s10973-020-10123-0
21. Biswas N, Chamkha AJ, Manna Manna NK (2020) Energy-saving method of heat transfer enhancement during magneto–thermal convection in typical thermal cavities adopting aspiration. SN Appl Sci 2–1911:1–25

MHD Convection in Cavity Under Partially Applied Magnetic Fields

Nirmalendu Biswas, Ritesh Samanta, Aparesh Datta, and Nirmal K. Manna

Nomenclature

•

B	Magnetic field, Tesla, N/A·m^2
Ha	Hartmann number
L	Length of the cavity/length scale, m
L_B	Length of the partially applied magnetic-field, m
Nu	Average Nusselt number
P	Pressure, Pa
Ra	Rayleigh number
T	Temperature, K
U, V	Dimensionless velocity components
X, Y	Dimensionless coordinates

Greek symbols

α	Thermal diffusivity, m^2/s
β	Thermal expansion coefficient of fluid, K^{-1}

N. Biswas
Department of Power Engineering, Jadavpur University, Salt Lake, Kolkata 700106, India

R. Samanta · N. K. Manna
Department of Mechanical Engineering, Jadavpur University, Kolkata 700032, India

A. Datta (✉)
Department of Mechanical Engineering, NIT Durgapur, Durgapur 713209, India

S. K. Ghosh et al. (eds.), *Advances in Thermal Engineering, Manufacturing, and Production Management*, Lecture Notes in Mechanical Engineering,
https://doi.org/10.1007/978-981-16-2347-9_11

θ	Dimensionless temperature
λ_B	Dimensionless length of the partially applied magnetic-field
υ	Kinematic viscosity, m^2/s
ρ	Density, kg/m^3
κ	Electrical conductivity ($\mu S\ cm^{-1}$)
ψ	Dimensionless stream function
γ	Magnetic field inclination angle
c,h	Cooling, heating

1 Introduction

Due to its multifarious nature, buoyancy-driven natural convection heat transfer is encountered in numerous natural processes as well as industrial appliances like cooling of electronic devices, biomedical applications, metal casting, heat exchanger, solar energy storage, and many more. In many such processes, precious control of its thermo-fluid behavior is of utmost importance from a process operation point of view. In this regards, externally generated uniformly applied magnetic field (also known as magneto-hydrodynamics - MHD) is a good means for controlling (reduction/enhancing) thermo-fluid pattern as well as heat transfer process; such as thermo-magnetic mixing, magnetic endoscopy, cell separation, targeted drug delivery, cancer, and tumor treatments, etc. The application of magneto-hydrodynamics in numerous thermal processes is well reported in [1–4].

Analysis of thermal performance in a classical flow geometries involving natural convective heat transfer in a closed cavity is well reported by Biswas et al. [5, 6] under different thermal conditions. The influence of applied magnetic in an enclosure undergoing natural convection was looked into by Matt et al. [7], Al-Balushi and Rahman [8] considering various thermal boundary conditions. Very recently Jelodari and Nikseresht [9] have investigated the thermal performance of applied magnetic nanofluid in a cubical cavity; Acharya and Dash [10] examined the consequence of externally imposed magnetic field on non-Newtonian power-law CuO–water nanofluid undergoing natural convection. Applying inclined magnetic field Gangawane [11] have studied the buoyancy-driven convection due to partially active walls of an open-ended cavity. In other areas of work based on MHD, Jalil et al. [12] have examined the natural convective heat transfer in a cavity introducing a non-uniform partially active magnetic field in a vertical direction. They observed that oscillation behavior disappears with the increasing magnetic field. Later, the influence of a partially active magnetic field applied horizontally on a three-dimensional cavity heated differentially undergoing natural convection heat transfer has been investigated by Al-Rashed [13]. In their study, they noticed that the location and length of the imposed magnetic field influence the heat transfer process notably. Recently, the influence of horizontally imposed partial magnetic filed in a differentially heated porous cavity undergoing natural convection has been studied by

Geridonmez and Oztop [14]. In their study, it was observed that by adjusting the location and intensity of partial magnetic field can alter the heat transfer process. The MHD convective heat transfer has been also been studied in a lid-driven cavity considering uniform magnetic field [15], and variable magnetic field [16], applying horizontal periodic magnetic field in a differentially heated cavity [17] and along a vertical surface [18], non-uniform magnetic field in a differentially heated cavity [19].

From the thorough literature survey, it is observed that although the effect of partial magnetism on thermal convection has been introduced, still it requires extensive investigation to explore more underneath physics. The main purpose of this research is to analyze the consequence of a partially applied external magnetism on the thermo-fluid behavior in an enclosure heated at the left and cooled at the right. The investigation is conducted for the range of controlling parameters like Rayleigh number (Ra), Hartmann number (Ha), length of the partially active magnetic field (L_B), direction of applied magnetic field (γ), length over which magnetic field is imposed (L_B) and different position of the applied magnetic field.

2 Problem Formulation

2.1 Problem Descriptions

The physical model under the consideration is illustrated in Fig. 1. The flow domain (having equal length and height *L*) is filled with electrically conducting fluid (of $\Pr = 0.71$). The cavity is a differentially heated cavity, whose left wall is heated isothermally (at temperature T_h) and cooled isothermally (at temperature T_c) the right wall. An adiabatic condition is imposed at the horizontal walls.

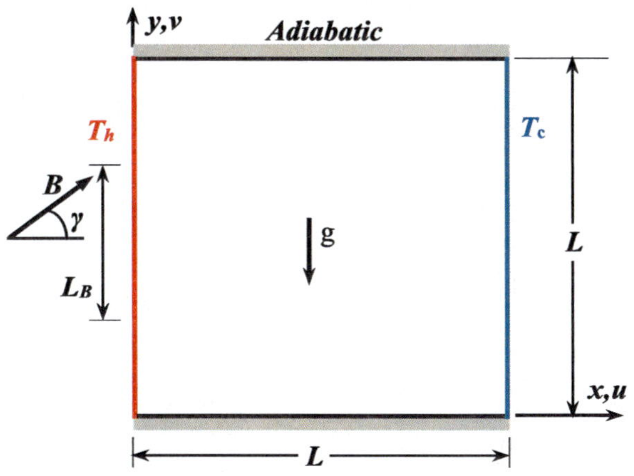

Fig. 1 Schematic diagram of the problem geometry, boundary conditions

For analyzing the MHD effect, magnetic field of intensity B is used externally and partially over the length of L_B in an angle γ with respect to the horizontal axis (as in Fig. 1). The magnetic field can act either horizontally along the left vertical wall ($\gamma = 0^o$) or vertically along the bottom wall ($\gamma = 90^o$). The length L_B may vary and position either in the left vertical wall or in the bottom wall; the magnetic field can act over the entire length of the cavity wall(s) ($L_B = 1$) or over the small parts (in single band or four bands). In order to analyze the studied problem, it is assumed that flow is laminar, steady, incompressible, Newtonian, constant fluid properties, neglecting Hall effect, Joule heating, and viscous dissipation. Boussinesq's approximation is considered to model the buoyancy term. With the above consideration, the non-dimensional governing equations yield as-

$$\frac{\partial U}{\partial X} + \frac{\partial V}{\partial Y} = 0 \tag{1}$$

$$U\frac{\partial U}{\partial X} + V\frac{\partial U}{\partial Y} = -\frac{\partial P}{\partial X} + \Pr\left(\frac{\partial^2 U}{\partial X^2} + \frac{\partial^2 U}{\partial Y^2}\right) + \Pr \lambda_B \mathrm{Ha}^2 (V \sin\gamma \cos\gamma - U \sin^2\gamma) \tag{2}$$

$$U\frac{\partial V}{\partial X} + V\frac{\partial V}{\partial Y} = -\frac{\partial P}{\partial Y} + \Pr\left(\frac{\partial^2 V}{\partial X^2} + \frac{\partial^2 V}{\partial Y^2}\right) + \Pr \lambda_B \mathrm{Ha}^2 (U \sin\gamma \cos\gamma - V \cos^2\gamma) + \mathrm{Ra} \Pr \theta \tag{3}$$

$$U\frac{\partial \theta}{\partial X} + V\frac{\partial \theta}{\partial Y} = \left(\frac{\partial^2 \theta}{\partial X^2} + \frac{\partial^2 \theta}{\partial Y^2}\right) \tag{4}$$

where, (X, Y) and (U, V) represent dimensionless coordinates and components of velocity. The dimensionless pressure and temperature are scaled as P and θ. The scaling parameters were chosen as

$$(X, Y) = (x, y)/L,\ (U, V) = (u, v)/(\alpha/L),$$
$$P = p/\rho(\alpha/L)^2, \theta = (T - T_c)/(T_h - T_c) \tag{5}$$

The dimensionless numbers are defined as- Prandtl, Rayleigh, and Hartmann number (Pr, Ra, and Ha respectively) are defined as

$$\mathrm{Ra} = g\beta(T_h - T_c)L^3/\alpha\upsilon, \Pr = \upsilon/\alpha,\ \mathrm{Ha} = BL/\sqrt{\rho\upsilon/\kappa} \tag{6}$$

For the implementation of partial magnetic field in the computational domain, the factor λ_B is used and it is defined as $\lambda_B = 0$ for inactive zones or 1 for active zones. For the computation, of the problem configuration, following boundary conditions are utilized

(a) $U = V = 0, \theta = 1$ for the left heated wall ($X = 0$),

(b) $U = V = \theta = 0$ for the right cold wall ($X = 1$),
(c) $U = 0,\ V = 0,\ \partial\theta/\partial Y = 0$ for the adiabatic walls.

The average Nusselt number (Nu) – as a heat transfer characteristics for the right cold wall is given by

$$\mathrm{Nu} = \int_0^1 \left(-\left.\frac{\partial\theta}{\partial X}\right|_{X=1} \right) \mathrm{d}Y \tag{7}$$

2.2 *Numerical Technique*

The dimensionless governing Eqs. (1–4) are simulated numerically in an iterative manner using finite volume approach (FVM) and SIMPLE algorithm [20] based well-validated written algorithm [5, 6, 15, 21, 22]. The central differencing scheme (second-order) for the diffusion terms and QUICK (a third-order upwind) scheme for the advection terms are chosen. The converged solution is considered when the maximum residuals is $< 10^{-7}$ and the mass-defect reaches $< 10^{-9}$ respectively. A mesh sensitivity test is also executed for obtaining the correct mesh size. Finally, a 100×100 grid size with uniform distribution is chosen for the entire simulation. Uniform grid distribution is chosen for capturing accurate length and positioning of the length L_B. However, these results are not incorporated here for brevity.

3 Result and Discussion

The thermo-fluid behavior of the studied problem geometry under the influence of partial magnetism is investigated for the range of Rayleigh number ($\mathrm{Ra} = 10^5$ and 10^6), Hartmann number ($\mathrm{Ha} = 0, 10, 30, 50, 100$), direction of applied magnetic field ($\gamma = 0, 90^{\circ}$), length over which magnetic field is applied ($L_B = 0, 0.1, 0.3, 0.4$, and 1) and different positions of the applied magnetic field. Furthermore, imposed magnetic field acts over the length L_B and positioned either in the left vertical wall or in the bottom wall; the magnetic field can act over the entire length of the cavity wall(s) ($L_B = 1$) or over the small parts (in single band or four bands). The magnetic field acts partially to the square cavity within a band with an upper limit and lower limit in different positions. The term "Top band" and "Middle band" will denote a single band in which the magnetic field is employed. The term "4 bands" will denote, that the magnetic field is employed in four single bands in equal intervals. The processing and analysis of the results are carried out using streamlines, isotherms, and average Nusselt number (Nu).

3.1 Effect of the Partially Applied Magnetic Field

The Rayleigh number (Ra) plays a crucial role in the dominance of the convection mechanism. The effect of two different Ra = 10^5 and 10^6 on the flowing fluid and temperature pattern are displayed in Fig. 2, using streamlines and isotherms at Ha = 50, $\gamma = 0^0$, $L_B = 0.3$. Figure 2a and b illustrate 'Top band' and 'Middle band' position of imposed magnetic field respectively. As the cavity is differentially heated, there exists one large circulating cell (rotating in a clockwise direction) appears within the cavity. As a result, fluid adjacent to the left heated wall being heated and moves in upward and obstructed by the top adiabatic wall and finally heat is released to the side cold wall. The isotherms show the temperature distribution of the fluid inside the cavity. At Ra = 10^5 and top band position, the isotherms contours are stretched in-between two active walls; contours of higher values are clustered with the heated wall. When Rayleigh number increases from Ra = 10^5–10^6, it is observed that fluid circulation strength increases significantly (which is indicated by the magnitude of streamlines contour). Additionally, a tiny circulation cell appears within the main circulating vortex and it is located upper and lower portion of the cavity with respect to the horizontal midplane. Corresponding isotherm contours shows horizontally stretched isotherms within the cavity and clustered of isotherms nearly active walls. As Ra increases, the buoyancy force becomes larger, resulting in higher fluid circulation velocity and stronger convection mechanism, and a higher heat transfer.

Now, with the 'Middle band' position of the imposed magnetic field, flow structures, as well as temperature distribution, modifies significantly, as shown in Fig. 2b. At Ra = 10^5 although there exists one large CW circulation, there also appears two more circulating vortices within the large circulation cell—one in the upper part and other in the lower part of the cavity. Corresponding isotherms stretches diagonally in between the bottom-left and top-right corner of the cavity following a zigzag way. The reason behind such behavior is that as the magnetic field is imposed partially (over the length $L_B = 0.3$), about the middle portion of the cavity. As a result of dampening effect of the Lorentz force, fluid velocity reduces in this portion of the cavity. Thus, larger flow vortex split into two cells. Consequently, the isotherm contours modified significantly. On the other hand, at Ra = 10^6 flow structure modifies significantly compared to the 'top band' position of the applied magnetic field. The shape and position of the inner smaller circulating cells changes. Outer circulating cells become compressed towards each of the cavity corner. Of course, there are no significant changes in the isotherms.

3.2 Effect of Different Positions in the Cavity

In this section, four different cases of imposed magnetic field four band ($L_B = 0.1$), top band ($L_B = 0.4$), middle band ($L_B = 0.4$), and bottom band ($L_B = 0.4$) are studied

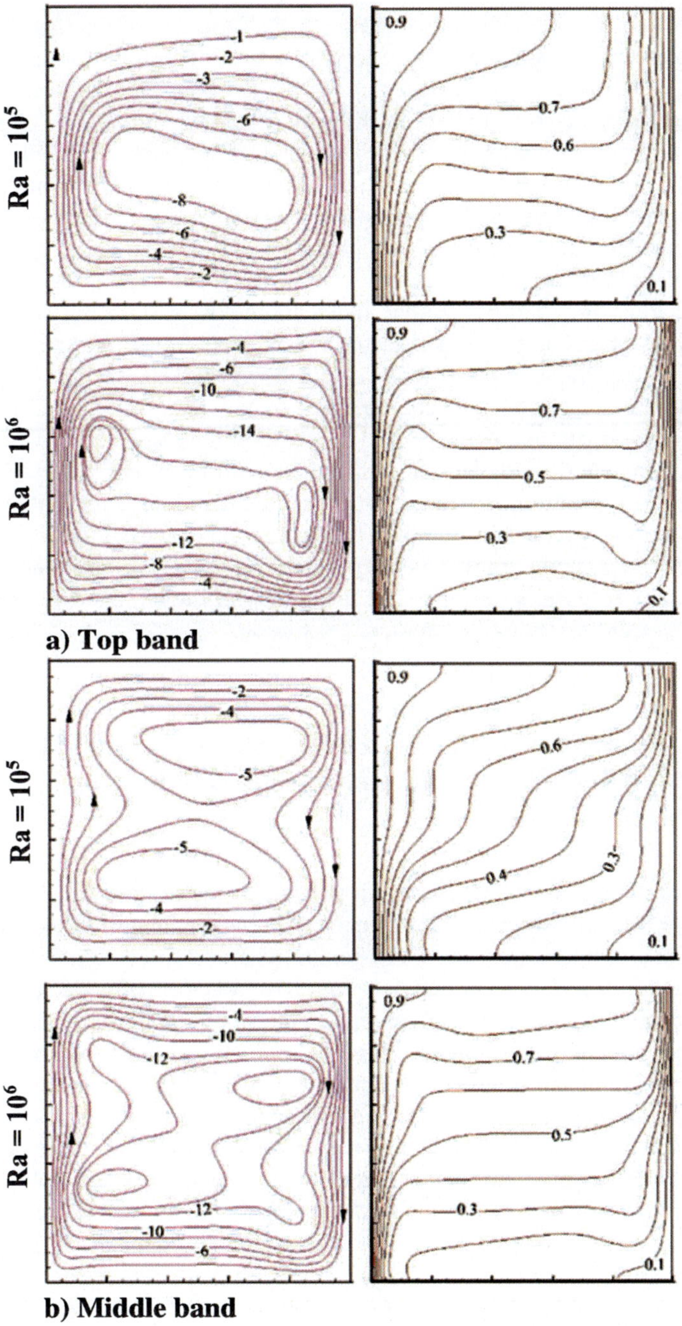

Fig. 2 Effect of changing the band position of employed magnetic field on flowing fluid and thermal behavior at Ra $= 10^5$, 10^6, Ha $= 50$, $\gamma = 0^0$, $L_B = 0.3$

and presented in Fig. 3 for Ha = 30 and Ra = 10^6. In the case of four bands (as shown in the first row) the magnetic field is applied over four different positions (having length $L_B = 0.1$ with a gap of 0.1) symmetrically about the mid-point of vertical walls and at a distance of $0.15L$ from the endpoint of the vertical wall. For this case, it is being observed from the streamlines of the outer and larger circulating cell (CW rotation) is parallel to the walls; inner tiny vortices appear along the mid horizontal plane. Due to the symmetrically applied magnetic field, there are no major changes to streamline. Now for the same Ha = 30 and L_B= 0.4, when there is a single band magnetic filed in the top position (as shown in the second row), the flow structure changes as inner tiny vortices shifted above and below with respect hid horizontal plane. This happens as the magnetic force acts partially in the upper part of the cavity, where magnetic force dampens fluid flow; whereas the lower part of the cavity is unaffected by the magnetic field. Of course, there is no significant change in the isotherm contours. With the middle position (as shown in the third row) of the imposed magnetic field, the shape and position of the inner fluid circulating cells modify. In this case, fluid flow faces resistance in the middle portion of the cavity as the magnetic field acts partially. When the same band is shifted to the bottom-most position (as shown in the fourth row), it is being observed that the streamline pattern for this case is forming a mirror image diagonally of the top band's case. It can be observed that all the isotherms contours are similar.

3.3 Comparison of Length of Partial Magnetic Field (L_B*) Over Whole Length Magnetic Field with Varying Hartmann Number (Ha)*

In this section, the effect of four band partial magnetic field (having each band length $L_B = 0.1$) and whole length magnetic field (having length $L_B = 1$) on the flow structures and temperature pattern are compared and displayed in Fig. 4 for Ra = 10^6, $\gamma = 0$ varying Hartmann number (Ha = 10, 30, and 100). Ha, value is varied to regulate the magnitude of magnetic field intensity. At a lower value of Ha = 10, it signifies that pattern-wise local contours of streamlines and isotherms are almost similar. However, in the case of four bands magnetic field fluid velocity is higher (compared to whole length magnetic field)—as reflected by a higher magnitude of streamline contours. The reason behind this fact is that the magnetic field is imposed in a split manner instead of the whole of the left cavity wall. As a result, flow resistance is comparatively lower with four band case compared to the whole length case. This leads to a smaller reduction of heat transfer (due to imposed magnetic field) with four band cases compared to the whole length case (in Fig. 4a). On increasing Hartmann number from Ha = 10 to 30 (second row), it is observed that for whole length magnetic field circulating cell is symmetric about the mid horizontal plane of the enclosure; whereas with four band case symmetry pattern is lost slightly. Of course, there are no significant changes in the isotherms contours. Further increase in Ha to 100, fluid

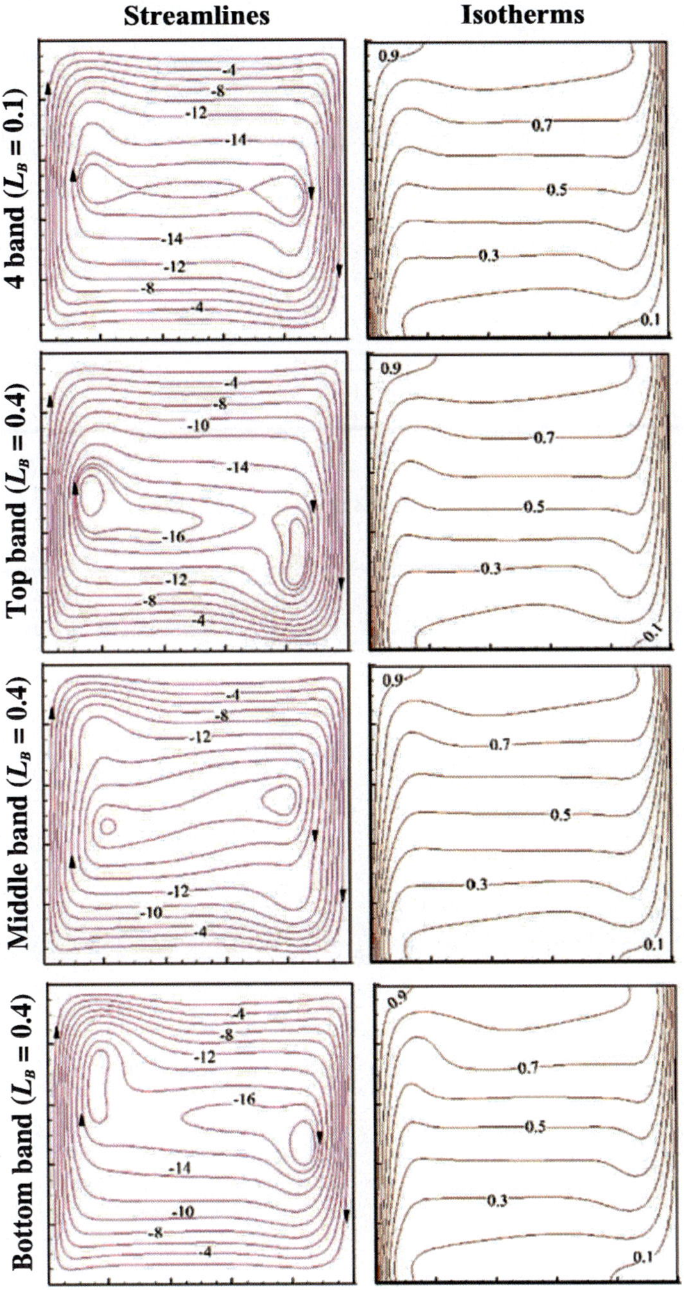

Fig. 3 Effect of different band positions along with 4 band cases on flowing fluid and thermal behavior at Ra $= 10^5$, Ha $= 30$, $\gamma = 0^0$

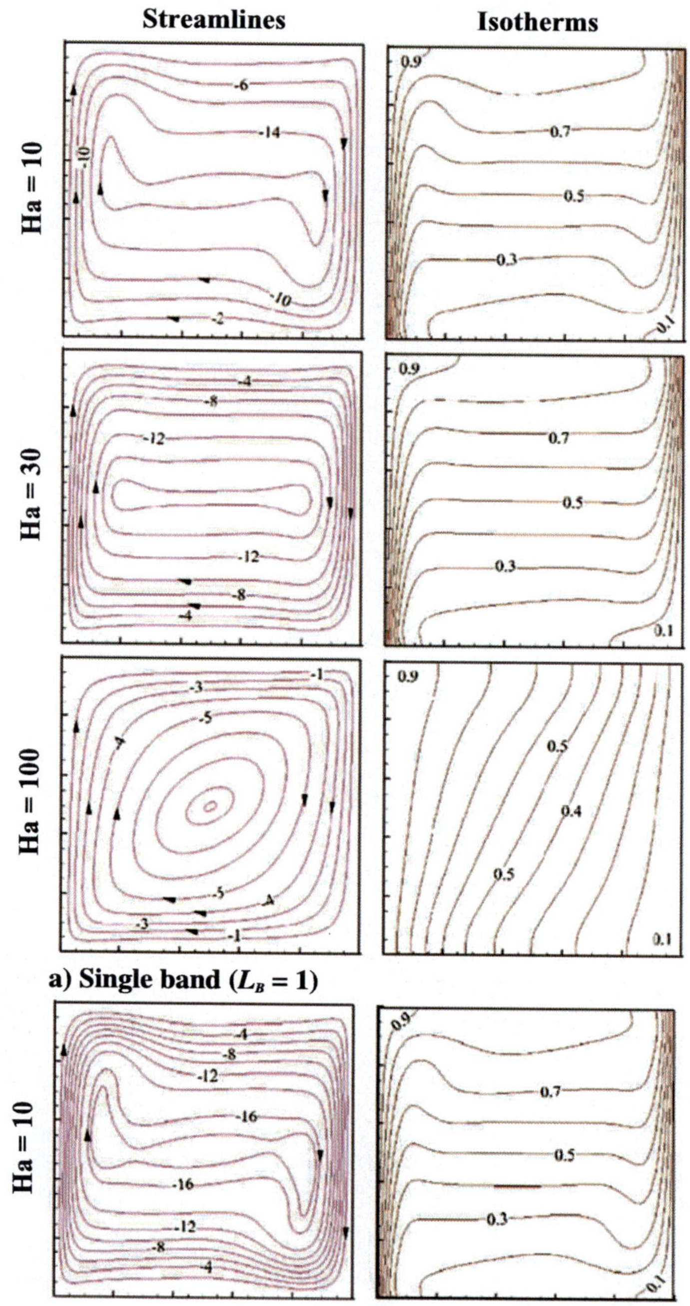

Fig. 4 Impact on DHC cavity and 4 bands partially applied magnetic field cavity on increasing the strength of the magnetic field by varying Hartman number on fluid flow and thermal behavior at Ra $= 10^6$, $\gamma = 0^0$

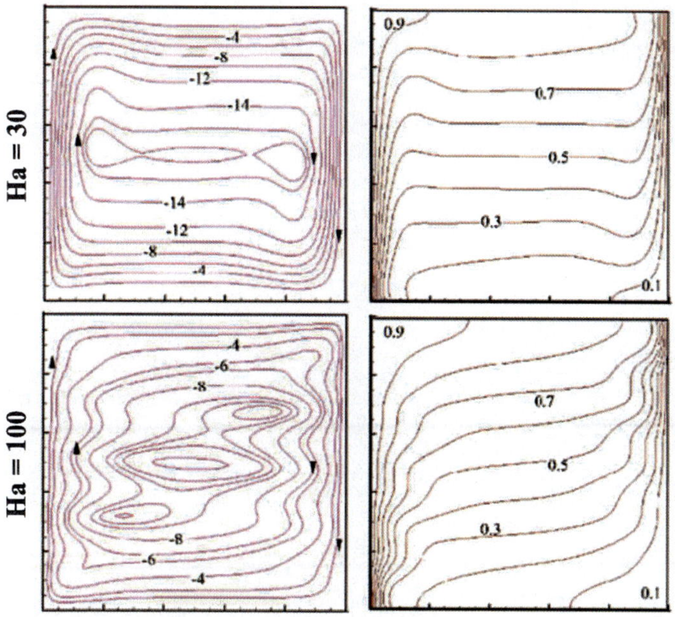

b) Four bands (L_B = 0.1)

Fig. 4 (continued)

flow as well as temperature distribution pattern modifies drastically as presented in the third row of Fig. 4b. For the whole length case circulating cell stretches between left-bottom and right-top corners and circulation velocity reduces substantially as noticed from contours of streamline. Corresponding isotherm lines connected directly between top and bottom adiabatic walls maintaining higher temperature lines at the left and lower temperature lines at the right. This is due to the counteraction or negative impact of magnetic force; the advection in the cavity reduces, which causes decreased circulation strength. In the case of four bands magnetic field, both streamline and isotherm modifies substantially. Lines are not smooth, follow a wavy pattern. Circulating vortex stretches along the left-bottom and right-top corners along with the formation of two numbers additional tiny circulating vortices. Isotherms are distributed diagonally.

3.4 Heat Transfer Analysis

To visualize the overall heat transfer characteristics, the average Nu of the cold wall is estimated and confronted in Fig. 5 varying flow controlling parameters. The first plot (Fig. 5a) shows the variation of Nu for different Ha values considering the whole length of the applied magnetic field ($L_B = 1$) for Ra $= 10^5$ and 10^6.

Fig. 5 Variation of overall heat transfer characteristics (Nu) with varying Ha for different parameters: **a** $L_B = 1$ and $\gamma = 0^0$, **b** $Ra = 10^6$, $L_B = 0.3$, $\gamma = 0^0$ and 90^0, **c** $Ra = 10^6$, $L_B = 0.1$ and 0.4, $\gamma = 0^0$

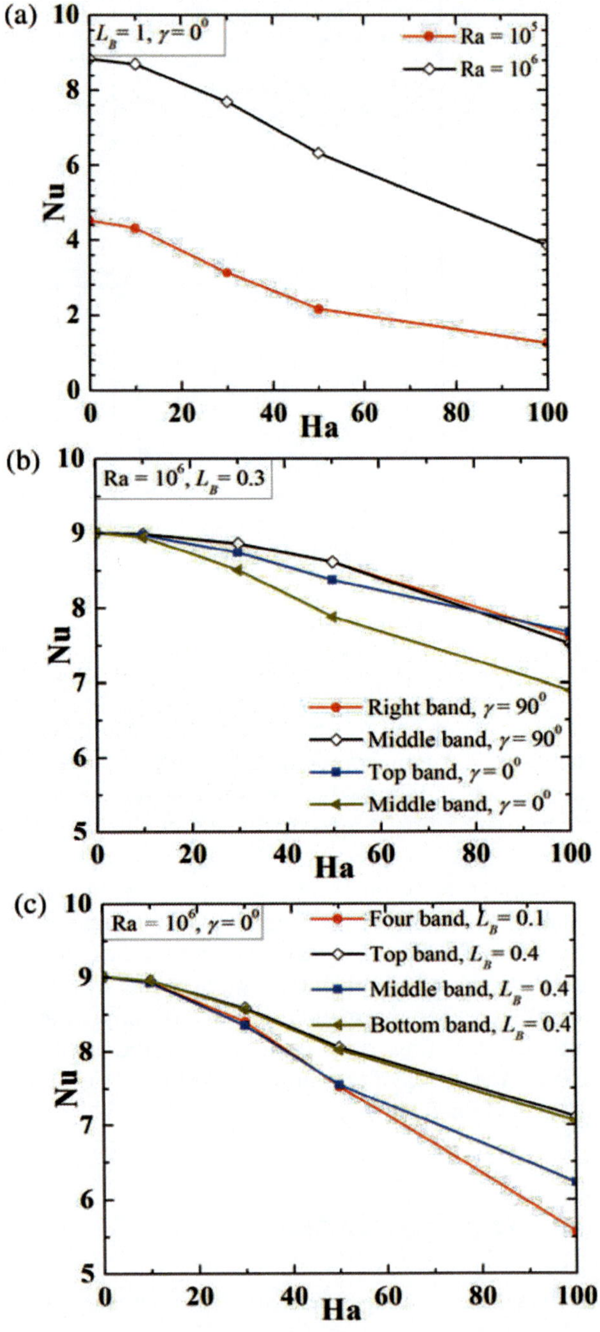

From the figure, it is found that the heat transfer is higher with higher Rayleigh number due to higher convection strength. This results in higher fluid velocity in the cavity (as observed from contour plots in earlier Figs.), which transfers more amount of heat from heat source to the heat sink. From this figure, it is evident that the magnetic force has a significant impact on the heat transfer characteristics and it shows a monotonically decreasing trend of Nu with increasing Ha. In general, it is obvious that the heat energy transport is mainly ruled by the natural convection mechanism, which is again dominated by the external magnetic field. Figure 5b illustrates the effect of different positions of imposed the magnetic field with varying magnetic field strength (Ha) considering a single band ($L_B = 0.3$) at Ra = 10^6. Four positions are considered which are Top and Middle band on the left vertical wall when the magnetic field was employed horizontally ($\gamma = 0^o$) and the rightmost and middle band on the bottom wall when the magnetic field was employed vertically upwards ($\gamma = 90^o$). From the figure, it is apparent that the trend of Nu is decreasing with increasing magnetic field intensity. However, the slop of decrement in the Nu curve is more for the horizontally applied magnetic field compared to the vertically employed magnetic field. The reason is that as the magnetic force acts horizontally buoyancy force is more dominated by the magnetic force (which is obvious due to presence of 'negative' term of Ha in the Eqs. 2 and 3), and this results in dampening of flowing fluid and lowers the heat transfer rate. Furthermore, it is observed that, when the magnetic field applied vertically, both curves corresponding to the rightmost band and the middle band are overlapping each other, this shows that there is no significant effect in the band position in the bottom wall when the magnetic field employed vertically.

In Fig. 5c the effect of different positions of the imposed magnetic field along with increasing Ha on the left vertical wall is analyzed considering four bands ($L_B = 0.1$) and single band ($L_B = 0.4$) at Ra = 10^6. It is observed that the variation of Nu with increasing Ha for 4 band and single band magnetic field shows a monotonically decreasing trend. Moreover, the variation of Nu for the top and middle bands are overlapping each other, which shows that whether the band is in the top or bottom position the rate of heat transfer, will remain the same. Also, it is noted that the rate of heat transfer decrement is faster with four bands compared to the top and bottom band although the overall length of the imposed magnetic field is the same. The reason behind this fact is that due to the splitting of the length of the magnetic field (four bands), fluid flow distortion is more compared to the single band. This results in lowering the heat transfer.

4 Conclusions

In this work, buoyancy driven-convection in an enclosure heated at the left and cooled at the right undergoing partially applied external magnetic filled is analyzed numerically. The study is carried out under a range of pertinent parameters. The major observations are:

- The fluid flow and temperature patterns markedly, changes with the Rayleigh number. Higher Ra results in higher Nu value due to stronger convection.
- Partially imposed magnetic field modifies flow structure and temperature distribution significantly over the whole length magnetic field. The usage of a partial magnetic field in a single band or four bands is always beneficial compared to the whole length magnetic field. The vertical magnetic field has a lesser dampening effect on Nu compared to the horizontal magnetic field.
- By increasing Ha, the heat transfer rate is found to be decreasing substantially. An increase in the magnetic force intensity reduces the buoyancy effect, leading to decrement in heat transfer.

The above-mentioned observations can be implemented in any device for controlling the thermo-fluid flow behavior.

Acknowledgements None.

References

1. Turcotte D, Lyons J (1962) A periodic boundary-layer flow in magnetohydrodynamics. J Fluid Mech 13(04):519–528
2. Davidson PA (2001) An introduction to magnetohydrodynamics, Cambridge Texts in Applied Mathematics. Cambridge University Press, Cambridge, UK
3. Ozoe H (2005) Magnetic convection. Imperial College Press, Singapore
4. Rashidi S, Esfahani JA, Maskaniyan M (2017) Applications of magnetohydrodynamics in biological systems-a review on the numerical studies. J Magn Magn Mater 439:358–372
5. Biswas N, Mahapatra PS, Manna NK, Roy PC (2015) Influence of heater aspect ratio on natural convection in a rectangular enclosure. Heat Transf Eng 37(2):1–15
6. Manna NK, Biswas NN, Mahapatra PS (2019) Convective heat transfer enhancement: effect of multi-frequency heating. Int J Numer Method Heat Fluid 29(10):3822–3856
7. Matt CFT, Quaresma JNN, Cotta RM (2017) Analysis of magnetohydrodynamic natural convection in closed cavities through integral transforms. Int J Heat Mass Transf 113:502–513
8. Al-Balushi LM, Rahman MM (2019) Convective heat transfer utilizing magnetic nanoparticles in the presence of a sloping magnetic field inside a square enclosure. J Therm Sci Eng Appl 11:041013-1–19
9. Jelodari I, Nikseresht AH (2018) Effects of Lorentz force and induced electrical field on the thermal performance of a magnetic nanofluid-filled cubic cavity. J Mole Liquids 252:296–310
10. Acharya S, Dash SK (2019) Natural convection in a cavity with undulated walls filled with water-based non-Newtonian power-law CuO–water nanofluid under the influence of the external magnetic field. Numer Heat Transfer A 76(7):552–575
11. Gangawane KM (2017) Effect of angle of applied magnetic field on natural convection in an open ended cavity with partially active walls. Chem Eng Res Design 127:22–34
12. Jalil JM, Al-Tae'y KA, Ismail SJ (2013) Natural convection in an enclosure with a partially active magnetic field. Numer Heat Transfer A 64(1):72–91
13. Al-Rashed AAAA, Kolsi L, Oztop HF, Aydi A, Malekshah EH, Abu-Hamdeh N, Borjini MN (2018) 3D magneto-convective heat transfer in CNT-nanofluid filled cavity under partially active magnetic field. Physica E 99:294–303
14. Geridonmez BP, Oztop HF (2019) Natural convection in a cavity filled with porous medium under the effect of a partial magnetic field. Int J Mech Sci 161–162:

15. Biswas N, Manna NK (2018) Magneto-hydrodynamic Marangoni flow in bottom-heated lid-driven cavity. J Mol Liquids 251:249–266
16. Sheikholeslami M, Chamkha AJ (2016) Flow and convective heat transfer of a ferro-nanofluid in a double-sided lid-driven cavity with a wavy wall in the presence of a variable magnetic field. Numer Heat Transf A 69(10):1186–1200
17. Mehryan SAM, Izadi M, Chamkha AJ, Sheremet M (2018) Natural convection and entropy generation of a ferrofluid in a square enclosure under the effect of a horizontal periodic magnetic field. J Mol Liquids 263:510–525
18. Siddiqa S, Hossain MA, Gorla RSR (2012) Conduction-radiation effects on periodic magnetohydrodynamic natural convection boundary layer flow along a vertical surface. Int J Therm Sci 53:119–129
19. Szabo PSB, Früh W-G (2018) The transition from natural convection to thermomagnetic convection of a magnetic fluid in a non-uniform magnetic field. J Magn Magn Mater 447:116–123
20. Patankar SV (1980) In: Numerical heat transfer and fluid flow. Taylor and Francis
21. Biswas N, Sarkar UK, Chamkha AJ, Manna NK (2020) Magneto-hydrodynamic thermal convection of Cu–Al_2O_3/water hybrid nanofluid saturated with porous media subjected to half-sinusoidal nonuniform heating. J Thermal Anal Calorim. https://doi.org/10.1007/s10973-020-10123-0
22. Biswas N, Chamkha AJ, Manna Manna NK (2020) Energy-saving method of heat transfer enhancement during magneto–thermal convection in typical thermal cavities adopting aspiration. SN Appl Sci 2–1911:1–25

Comparison of Thermo-Hydraulic Performance Enhancement of Liquid He-Based Cryogenic Nanofluid Flow in Turbulent Region Through Rectangular Plate Fin Heat Exchangers

Anirban Bose and Subhadeep Chakraborty

1 Introduction

Liquefaction of gases, refrigeration, cryogenic energy storage systems, rocket and space applications are few examples where cryogenic fluids are used to exchange heat in compact heat exchangers. Plate fin heat exchangers (PFHE) are very common in exchanging heat at cryogenic temperature. It is possible to save energy if performance of the PFHE can be increased or in other words waste heat can be reduced. Performance of the PFHEs depend on geometry of the fins, properties of the fluid and flow parameters (like mass flow rate). A lot of research work has been done on geometry of the fins to improve thermal performance. But in almost all the cases it has been observed that heat transfer enhancement occurs at the cost of high flow resistant passage which ultimately increases pumping power. Same conclusion can be made to increase the convective heat transfer coefficient by maintaining high mass flow rate, pumping power would be increased. The idea of nanofluid is improving the thermo-hydraulic performance of the PFHE by increasing the thermal conductivity of the fluid. Choi [1] introduced the term "nanofluids", a reference to dispersed nano-particles smaller than 100 nm into base fluid. Commonly used base fluids are namely water (W), ethylene glycol (EG), EG/W base mixture and oil [2]. Studies by Lee et al. [3] proved that nanofluids have higher thermal conductivity compared to their base fluids. Lee et al. [3] found that the 13 nm Al_2O_3 nanoparticles dispersed in water increased by 30% of its thermal conductivity compared to water at 4.3% volume concentration.

Plate fin heat exchangers are significant in recent times and widely used due to high heat transfer rate. Most extensive experimental studies have been published by Kays and London [4]. It was investigated that in compact heat exchangers such as

A. Bose (✉) · S. Chakraborty
Department of Mechanical Engineering, Meghnad Saha Institute of Technology, Kolkata, India
e-mail: a.bose@rediffmail.com

S. K. Ghosh et al. (eds.), *Advances in Thermal Engineering, Manufacturing, and Production Management*, Lecture Notes in Mechanical Engineering,
https://doi.org/10.1007/978-981-16-2347-9_12

plain fin strip, wavy fin, offset fin, and perforated fin, the pressure-drop decreases with respect to increasing the turbulence in working fluid. Onwards 1942 by Norris and Spofford [5] provide the first experimental report they draw out the effect of heat transfer coefficient on the basis of length, thickness and pitch of fins and also reduced the friction factor and Colburn j factor. Experimental studies have been made by London and Shah [6] in 1967 and they concluded that small offset spacing fin thickness and a large number of fins per inch gives better heat transfer. In 1975 Wieting [7] setup a relationship between the variables from earlier experimental results of heat transfer and fluid flow friction data for a plate-fin heat exchanger of offset fin to design accurately within the parameter range of the correlations.

PFHEs performance also has been studied with cryogenic applications. Robertson [8] studied offset PFHE with subcooled liquid nitrogen at around 77 K. Cao et al. [9] worked on PFHE with mixed refrigerant including CH_4, C_2H_4, C_4H_{10}, C_3H_8 and N_2 at about 113 K. They found a large deviation in the correlation of heat transfer coefficient from the existing correlations and experiments. A CFD simulation study done by Yang et al. [10] compared thermal performance of different cryogenic medium at gaseous and liquid state in OSF channels revealed that fin materials with low thermal conductivity deteriorates the performance of the heat exchanger. Thermal performance of aluminum PFHEs were investigated with helium gas cooled to around 77 K as working fluid by Doohan et al. [11] and Goyal et al. [12]. Moreover, many cryogenic experiments are carried out to study Thermo-hydraulic performance of PFHE with helium gas at low temperature by Wieting [13].

It has been noticed after literature survey that no work has been published comparing nanofluids performance applied in PFHE at cryogenic temperature with liquid He as base fluid. Liquid He and nanofluids with nanoparticles Al_2O_3, CuO, Fe_3O_4 and SWCNT of 5% volume fraction in each case are compared in terms of j-factor and f factor and thermo-hydraulic performance parameter $\frac{j}{f^{1/3}}$. In each nanofluids base fluid is considered as liquid He. A numerical simulation using Finite Volume Method (FVM) is used to analyze the thermo-hydraulic performance of the PFHE nanofluid at cryogenic temperature. Reynolds number range of the analysis has been set from 4000 to 8000.

2 Numerical Analysis

2.1 Physical Description of the Model

The PFHE of rectangular cross-section as shown in Fig. 1 is of our interest of study for different ranges as mentioned below with liquid helium-nanofluid as working fluid (Table 1).

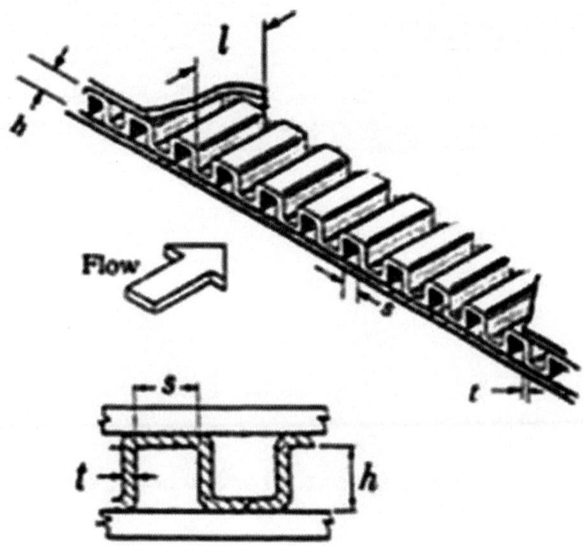

Fig. 1 Geometry of plain rectangular plate fin heat exchanger

Table 1 Working ranges of model parameters

Parameters	Working range
Reynolds number	$4000 \leq \text{Re} \leq 8000$
Volume fraction of nanoparticles	$\emptyset = 5\%$
Height of the fin	$h = 9.75\,\text{mm}$
Fin spacing	$s = 4.85\,\text{mm}$
Thickness of the fin	$t = 0.25\,\text{mm}$
Length of the fin	100 mm

2.2 *Governing Equations*

The governing equations to describe the fluid flow and heat transfer phenomena are continuity, momentum and energy equation. The mentioned equations are solved to get the velocity, pressure and temperature field. We have considered 3-D, turbulent and steady flow model, so the governing equations are as following

Continuity equation:

$$\nabla.(\rho V) = 0 \tag{1}$$

Momentum equation:

$$\nabla.(\rho V V) = -\nabla P + \nabla.(\mu \nabla V) \tag{2}$$

Energy equation:

$$\nabla.(\rho c_p VT) = \nabla.(k\nabla T) \tag{3}$$

Standard K-ε model equations:

$$\bar{u}_j\frac{\partial k}{\partial x_j} = \frac{\partial}{\partial x_j}\left[\frac{\upsilon_t}{\sigma_k}\frac{\partial K}{\partial x_j}\right] + 2\upsilon_t E_{ij}E_{ij} - \varepsilon \tag{4a}$$

$$\bar{u}_j\frac{\partial \varepsilon}{\partial x_j} = \frac{\partial}{\partial x_j}\left[\frac{\upsilon_t}{\sigma_\varepsilon}\frac{\partial \varepsilon}{\partial x_j}\right] + 2\upsilon_t E_{ij}E_{ij}\varepsilon\frac{C_{\varepsilon 1}}{k} - \frac{C_{\varepsilon 2}}{k}\varepsilon^2 \tag{4b}$$

2.3 *Thermophysical Properties of Nanofluid*

By applying the principle of mass conservation to the two species in finite control volume of the nanofluids, the nanofluid density was obtained from the relation:

$$\rho_{nf} = \varphi\rho_{np} + (1 - \varphi) \tag{5}$$

By applying the principle of calorimetry in the mixture the overall specific heat of the nanofluid was calculated from the relation:

$$c_{nf} = \frac{\varphi(\rho c)_{np} + (1 - \varphi)(\rho c)_b}{\varphi(\rho)_{np} + (1 - \varphi)(\rho)_b} \tag{6}$$

Maxwell model of equivalent thermal conductivity as studied by Yu et al. [14] was used for the estimation of thermal conductivity and Einstein model of viscosity as mentioned by Mishra et al. [15] were used to estimate nanofluid properties.

$$\mu_{nf} = \mu_b(1 + 2.5\phi) \tag{7}$$

$$k_{nf} = k_b\frac{k_{np} + 2k_b + 2\varphi(k_{np} - k_b)}{k_{np} + 2k_b - \varphi(k_{np} - k_b)} \tag{8}$$

2.4 *Boundary Conditions*

The inlet, outlet and wall boundary conditions are as shown in Fig. 2. Top, bottom and periodic side walls are assumed no slip boundary condition. Sidewall is kept at

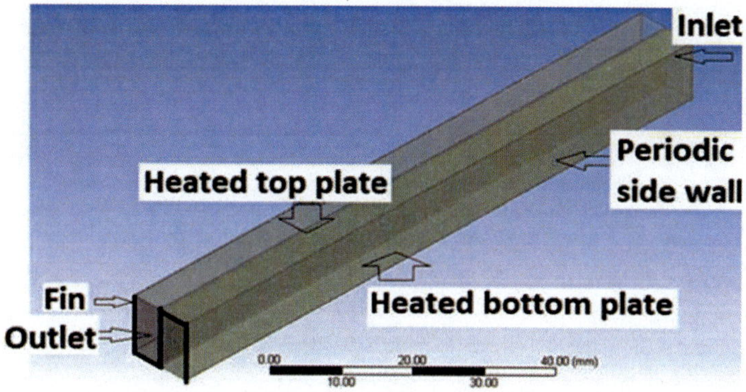

Fig. 2 Geometry of the model and boundary conditions

periodic boundary condition and bottom and top walls are kept at constant temperature boundary condition. Inlet set as velocity inlet boundary condition corresponding to the Re range from 4000 to 8000 to assure the flow is turbulent.

2.5 Meshing and Model Setup

The mesh of the computational domain was generated using a triangular patch conforming method. This mesh contains hexahedral cells having rectangular faces at the boundaries. The generated mesh consisted of 233,460 elements and 245,063 nodes.

A 3-D, incompressible and pressure-based solver was chosen for the computational domain. A viscous turbulent k-€ model with standard wall function is used for numerical simulation. A 2nd order upwind interpolation formula was used for discretization of momentum and energy equation. The conventional SIMPLE (Semi-Implicit Method for Pressure-Linked Equations) algorithm was used to solve the pressure velocity coupled equations, where several iterations were performed to ensure convergence of the numerical solution was assured by monitoring the scaled residuals to a constant level of 10–6 for each variable.

2.6 Grid-Independency

The solution obtained in this work is mesh independent as we have studied the mesh independency test. The result is shown in Fig. 3.

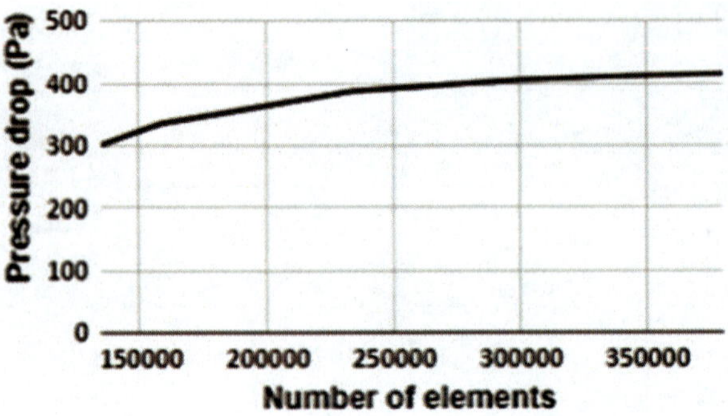

Fig. 3 Grid independence test

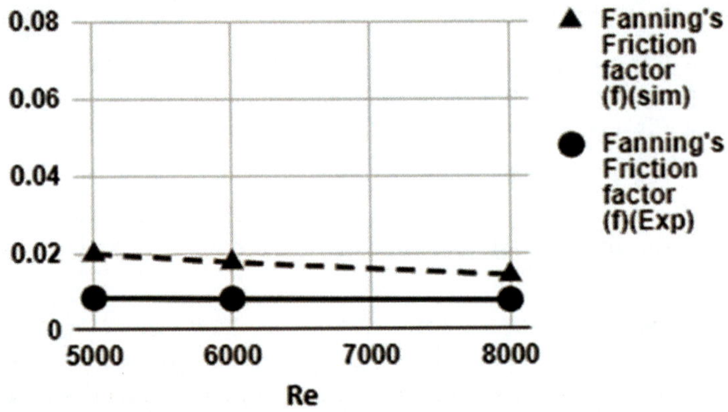

Fig. 4 Validation of simulation result with the experiment of Kays and London [4]

2.7 *Experimental Validation of the Model*

Validation is done to check the authenticity of the CFD tool. We have validated the model by comparing the results of the experiment done by Kays and London [4] as shown in Fig. 4.

3 Results and Discussion

This is clear from the solution of the numerical analysis as shown in Figs. 5 and 6 that with the increase of Re in the turbulent region from 4000 to 8000 friction factor and *j* factor decreases. These variations of *f* and *j* with Re is similar in both cryogenic nanofluids and pure liquid He. Friction factor is a non-dimensional term used to calculate the pressure drop in the PFHE. On the other hand, Colburn *j* factor

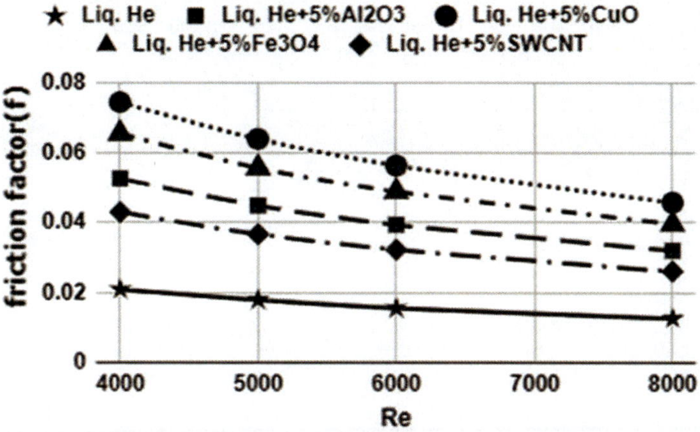

Fig. 5 Comparison of friction factor among different nano-particles with liq. He

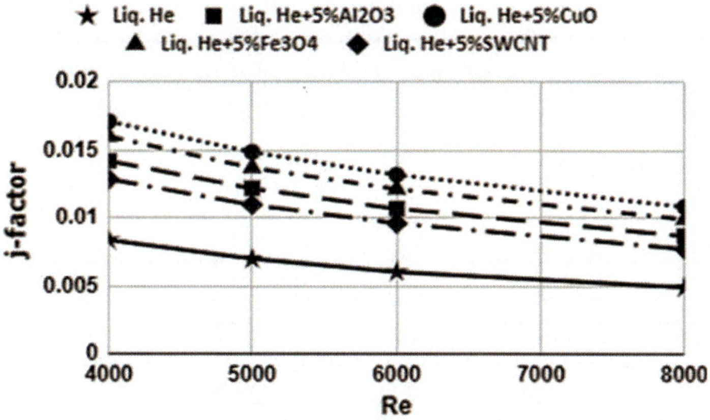

Fig. 6 Comparison of Colburn j factor among different nano-particles with liq. He

is a non-dimensional term used to estimate the heat transfer coefficient and hence the effectiveness of the PFHE. It is observed in Fig. 6 that CuO nanoparticles are very effective in increasing the convective heat transfer coefficient in terms of Colburn j-factor in comparison with Al_2O_3, Fe_3O_4 and SWCNT. But pressure drop in terms of Fanning's friction factor is also high in case of CuO-liquid He cryo-nanofluid in comparison with other cryo-nanofluids as shown in Fig. 5. But this is quite clear from Figs. 5 and 6 that in comparison with liquid He only, He-based cryo-nanofluids are more effective in terms of convective heat transfer.

This is very important to conclude regarding the performance of the PFHE keeping in mind both the heat exchange and pressure drop in the device. That is why volume goodness factor, defined as $\frac{j}{\sqrt[3]{f}}$, is compared among the different cryo-nanofluids as shown in Fig. 7. It is observed that CuO nanoparticles are giving better overall

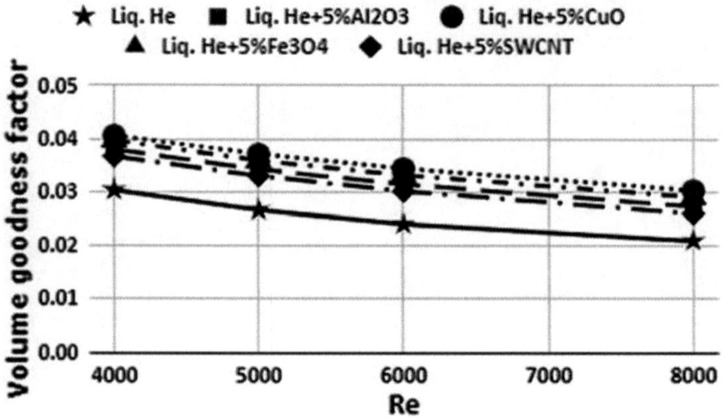

Fig. 7 Comparison of volume goodness factor among different nano-particles with liq. He

performance in terms of volume goodness factor in comparison with other cryo-nanofluids. Overall performance enhancement in He-based cryo-nanofluids is around 25 to 30% in comparison with only liquid He, is clearly observed in Fig. 7.

4 Conclusions

In this research work performance of PFHE with one fluid as He-based cryo-nanofluid is compared with respect to liquid He at turbulent range. It can be concluded from the results that CuO-liq. He-based cryo-nanofluid having best performance among the other cryo-nanofluids studied based on volume goodness factor. Around 25–30% enhancement in the overall performance considering both heat exchange and pressure drop has been found when cryo-nanofluids are used in comparison with liquid He only. But in this research certain aspects like cost of nano-particles, preparation complexity and cost of cry-nanofluid, phase change stability of nanofluids have not been considered in evaluation of overall performance of the PFHE.

Acknowledgements We are thankful to college authority of Meghnad Saha Institute of Technology (M.S.I.T.) to carry out this research work in CFD Lab, Department of Mechanical Engineering.

References

1. Choi SU, Eastman JA (1995) Enhancing thermal conductivity of fluids with nanoparticles (No. ANL/MSD/CP-84938; CONF-951135-29). Argonne National Lab., IL (United States)

2. Redhwan AAM, Azmi WH, Sharif MZ, Mamat R (2016) Development of nano refrigerants for various types of refrigerant based: a comprehensive review on performance. Int Commun Heat Mass Transf 76:285–293
3. Lee S, Choi SS, Li SA, Eastman JA (1999) Measuring thermal conductivity of fluids containing oxide nanoparticles. J Heat Transf 121(2):280–289
4. Kays WM, London AL (1964) In: Compact heat exchangers. 2nd edn
5. Norris R, Spofford WA (1942) High-performance fins for heat transfer. Trans ASME 64(7):489–497
6. London AL, Shah RK (1968) Offset rectangular plate-fin surfaces—heat transfer and flow friction characteristics. J Eng Power 90(3):218–228
7. Wieting AR (1975) Empirical correlations for heat transfer and flow friction characteristics of rectangular offset-fin heat exchangers. Trans ASME Ser C 97:488–490
8. Robertson JM, JM R (1979) Boiling heat transfer with liquid nitrogen in brazed-aluminium plate-fin heat exchangers
9. Cao L, Liu J, Li R, Huang S, Zhang F, Xu X (2016) Experimental study on the mixed refrigerant heat transfer performance in a plate-fin heat exchanger during a single-stage cryogenic cycle. Appl Therm Eng 93:1074–1090
10. Yang Y, Li Y, Si B, Zheng J (2017) Heat transfer performances of cryogenic fluids in offset strip fin-channels considering the effect of fin efficiency. Int J Heat Mass Transf 114:1114–1125
11. Doohan RS, Kush PK, Maheshwari G (2016) Exergy based optimization and experimental evaluation of plate fin heat exchanger. Appl Therm Eng 102:80–90
12. Goyal M, Chakravarty A, Atrey MD (2017) Experimental investigations and validation of two-dimensional model for multi stream plate fin heat exchangers. Cryogenics 82:83–91
13. Jiang Q, Zhuang M, Zhang Q, Zhu Z, Geng M, Sheng L, Zhu P (2018) Experimental study on the thermal hydraulic performance of plate-fin heat exchangers for cryogenic applications. Cryogenics 91:58–67
14. Yu W, Choi SUS (2003) The role of interfacial layers in the enhanced thermal conductivity of nanofluids: a renovated Maxwell model. J Nanoparticle Res 5(1–2):167–171
15. Mishra PC, Mukherjee S, Nayak SK, Panda A (2014) A brief review on viscosity of nanofluids. Int Nano Lett 4(4):109–120

Performance Study of a Small Capacity Thermoacoustic Refrigerator Using DELTA EC Software

Anirban Jana, Sourav Sarkar, and Achintya Mukhopadhyay

1 Introduction

The present concerns on the detrimental effects of conventional refrigeration methods using CFCs have recently brought limelight on Thermoacoustic refrigerators. Pressure-induced temperature oscillations have been observed for centuries. An approximate model of the thermoacoustic oscillations was first developed by Sondhauss in 1850, who experimentally observed the dependency of sound frequency and intensity on length and volume of the glass blowers. Tijani et al. [1] investigated the effect of spacing between the plates on the temperature gradient across the stack. They established that spacing of about four times of the thermal penetration depth (δ_k) between the plates produces the lowest temperature. Kim et al. [2] focused on the stack position in the resonator for a thermoacoustic heat pump for different operating conditions by varying the frequencies to generate the best performances. Setiawan et al. [3] experimentally determined the effects of the stack dimensions and position in the resonator of a thermoacoustic cooler. Wantha et al. [4] proved that in case of a standing wave resonator, the refrigeration effect depends on an optimum sound frequency, 'f' expressed as $f = a/2L$, where 'L' is the resonator length and 'a' is the speed of sound. The poor efficiency of the thermoacoustic devices is a major disadvantage in its implementation on a commercial scale. Thus, many attempts have been mad, mostly experimental in nature to investigate the feasible enhancements in design of the thermoacoustic devices. In this study, a combined study of the parameters using DELTAEC is done which simplifies the design of the experimental model.

A. Jana
Chimie ParisTech, PSL University, 75005 Paris, France
e-mail: anirban.jana@etu.chimieparistech.psl.eu

S. Sarkar (✉) · A. Mukhopadhyay
Department of Mechanical Engineering, Jadavpur University, Kolkata 700032, India
e-mail: souravsarkar.mech@jadavpuruniversity.in

S. K. Ghosh et al. (eds.), *Advances in Thermal Engineering, Manufacturing, and Production Management*, Lecture Notes in Mechanical Engineering,
https://doi.org/10.1007/978-981-16-2347-9_13

The important parameters to ensure achieve performance depends on many factors, the stack design being the most important of all, besides the resonator length, the medium of the tube, operating frequency, size of the tube, and speaker characteristics.

2 Working Principle

In a heat pump or refrigerator, work in different forms is utilized to transport heat from the lower temperature reservoir to the higher temperature reservoir. In a thermoacoustic refrigerator, the standing sound provides the external work, through pressure oscillations. The source of acoustic energy can be a loudspeaker, which generates sound waves in a long hollow tube, called the 'resonance column' or the 'resonator'. This tube is closed at the other end and filled with a gas. A standing pressure wave is introduced in the column and resonance is achieved by adjusting the length of the resonator, by positioning the plug at the closed end. A stack of closely spaced solid plates is kept in the path of sound waves in the resonator, which increases the solid–gas interface area, thereby enhancing heat transfer. The compression and rarefaction of the gas in the stack follow the pressure oscillations, consequently generating a variation in temperature. As the parcel gets compressed it is simultaneously displaced with the wave. Since a standing wave is produced, the gas parcel does not exit the resonator but moves along the length of the stack closely following the displacement wave, which oscillates in the resonator. In Fig. 1, the gas parcel moves to the left, following the sound wave. Here, the pressure increases, causing compression of the packet, which makes it hotter than the nearby stack wall and heat flows from the parcel to the cooler stack. In doing so, the parcel itself cools, expands and moves back to the right, with the displacement wave, where the pressure is lower, and consequently undergoes a rarefaction making it cooler. This results in heat transfer from the hotter stack wall into this packet, resulting in its expansion. This periodic motion continues and with the passage of time, this temperature difference increases as the packets

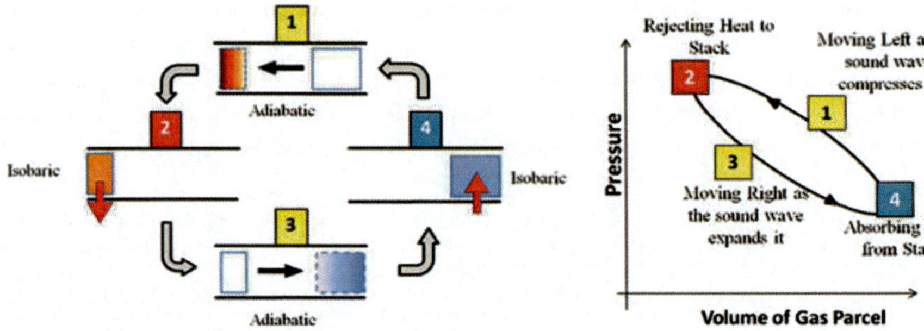

Fig. 1 Parcel oscillation and *P–V*-location diagram for the four stages of a thermoacoustic refrigeration cycle

transfer heat across the ends of the stack, effectively developing a refrigerating effect on one side of the stack [5].

3 Mathematical Formulation

In acoustics, the time-dependent variables are 'small'. The products of these variables are neglected in comparison with the variables themselves. This assumption is called 'Rott's approximations'. The fundamental properties (*p*, *T*, *s*, etc.) are subjected to steady-state sinusoidal oscillations over the mean value are represented in the complex form [68].

$$\rho(x, y, z, t) = \rho_m(x) + \mathrm{Re}\left[\rho_1(x, y, z)e^{i\omega t}\right]$$
$$T(x, y, z, t) = T_m(x) + \mathrm{Re}\left[T_1(x, y, z)e^{i\omega t}\right]$$

The transient behavior is already included in the above equations. Spatial gradients in the *y* and *z* directions are neglected. Similarly, the perpendicular velocity components (*y*, *z*) are neglected in comparison with the velocity, *u* along the propagation direction. The mean velocity is zero since it is a closed column and there is motion only due to the acoustic oscillation, hence u_m is zero. In this notation, the pressure of the acoustic standing wave and the *x*-component of the velocity considering the single plate in Fig. 2 can be written as

$$u(x, y, z, t) = \mathrm{Re}\left[u_1(x, y, z)e^{i\omega t}\right]$$
$$p(x, y, z, t) = p_m + \mathrm{Re}\left[p_1(x)e^{i\omega t}\right]$$

$$p_1 = P_A \sin(2\pi x/\lambda) = p_1(x) \tag{1}$$

$$u_1 = i(P_A/\rho_m a)\cos(2\pi x/\lambda) = iu_1(x) \tag{2}$$

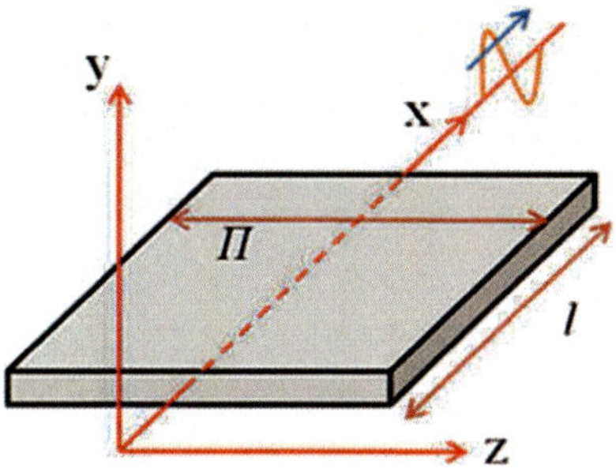

Fig. 2 Geometry of the single plate with negligible thickness

where $0 < x < \lambda/4$.

Under ideal conditions, without any plate, the sound wave is adiabatic and the oscillating temperature in terms of pressure is given as (3), where s is specific entropy and β is the ordinary thermal expansion co-efficient.

$$T_1 = \left(T_m\beta / \rho_m c_p\right) p_1 \tag{3}$$

The presence of the stack plates modifies this undisturbed temperature oscillation, for a fluid about a thermal penetration depth, δ_k away from the plate. The thermal penetration depth is the depth through which heat can diffuse through the fluid over a time $1/\omega$ ($\omega =$ oscillation frequency). A simplified analysis reveals the basic thermoacoustic effects imposed by the plates. In the following analysis, the effect of viscosity has been neglected. It is assumed that a mean-temperature difference exists in the x-direction, and its length is much smaller than the wavelength so that p_1 and u_1 can be considered uniform over the plate. Also, the fluid's thermal conductivity in the x-direction has been neglected. The mean gas temperature and plate temperature is considered to be the same (T_m). The oscillating fluid temperature can be obtained from the general heat transfer equation, combining the temporal entropy change at a point due to convective entropy flow, conduction of heat and generation of entropy (e.g., viscosity).

$$\rho T\left[\frac{\partial s}{\partial t} + \mathbf{v}.\nabla s\right] = \nabla.k\nabla \mathrm{T} + \left(\nabla.\sigma'\right).\mathbf{v} \tag{4}$$

Simplifying by assuming only the first-order term and considering no conduction in the x-direction, (2) becomes

$$\rho_m T_m\left(i\omega s_1 + u_1\frac{\mathrm{d}s_m}{\mathrm{d}x}\right) = k\left(\frac{\partial^2 T_1}{\partial y^2}\right) \tag{5}$$

Expressing s in terms of p and T,

$$\mathrm{d}s = \left(\frac{\partial s}{\partial T}\right)_p \mathrm{d}T + \left(\frac{\partial s}{\partial p}\right)_T \mathrm{d}p;\quad \mathrm{d}s = \frac{c_p}{T}\mathrm{d}T - \frac{\beta}{\rho}\mathrm{d}p \tag{6}$$

$$s_1 = \left(c_p/T_m\right)T_1 - \left(\beta/\rho_m\right)p_1 \tag{7}$$

Substituting (6) and (7) in (5) yields an equation for the unknown $T_1(y)$, subjected to boundary conditions; T_1 (0) = 0, due to the plate and T_1 (∞) is finite.

$$i\omega\rho_m c_p T_1 - k\left(\frac{\partial^2 T_1}{\partial y^2}\right) = i\omega T_m\beta p_1 - \rho_m c_p \nabla T_m u_1$$

$$T_1 = \left(\frac{T_m\beta}{\rho_m c_p}p_1 - \frac{\nabla T_m}{\omega}u_1\right)\left(1 - \mathrm{e}^{-(1+i)y/\delta_k}\right) \tag{8}$$

At $y \gg \delta_k$, the effect of the plate on heat transfer is negligible and $T_1 \rightarrow \left(\frac{\beta T_m}{\rho_m c_p} p_1 - \frac{\nabla T_m}{\omega} u_1\right)$.

- The 1st term arises due to simple acoustics and will exist even in the absence of the stack.
- The mean-temperature gradient in the fluid contributes to the 2nd term. As the fluid oscillates along x-direction with displacement amplitude, u_1/ω, the temperature oscillates by an amount $\nabla T_m u_1 / \omega$ at a given point in space. The linear superposition of the adiabatic motion-induced and adiabatic pressure-induced temperature oscillations produces the resultant distribution, and the sign and phase of the term $\left(\frac{\beta T_m}{\rho_m c_p} p_1 - \frac{\nabla T_m}{\omega} u_1\right)$ depend on the relative values of p_1 and u_1.

Following from Eq. (8), the temperature gradient can become zero if

$$(\nabla T_m)_{\text{crit}} = \frac{T_m \beta \omega p_1(x)}{\rho_m c_p u_1(x)} \tag{9}$$

For this above case, the fluid properties and standing wave position are such that the changes in temperature due to pressure variations are perfectly opposed by the temperature changes arising from the displacement wave. The sound wave-induced temperature variation in the gas must be greater than that across the cold and hot ends of the stack for a net heat transfer across the stack, which will produce the refrigeration effect. The time-averaged heat flux in the x-direction highlights the important parameters on which the temperature distribution depends. Following from the assumption which neglects ordinary thermal conduction in the x-direction, heat flux exists only due to hydrodynamic transport of entropy, carried by the oscillatory velocity, u_1:

$$\begin{aligned}
\dot{q} &= \rho_m T_m \overline{s_1 u_1} \\
\dot{q} &= \rho_m c_p \overline{T_1 u_1} - T_m \beta \overline{p_1 u_1} \\
s_1 &= (c_p / T_m) T_1 - (\beta / \rho_m) p_1
\end{aligned} \tag{10}$$

$$\dot{q} = \frac{1}{2} \rho_m c_p \text{Re}\left[T_1 \tilde{u}\right] - \frac{1}{2} T_m \beta \, \text{Re}\left[p_1 \tilde{u}\right] \tag{11}$$

For a standing wave, the second term has no contribution since p_1 and u_1 have a phase difference of $\pi/2$ and hence is purely imaginary. In the 1st term, Im$[T_1]$ contributes since u_1 is purely imaginary.

$$\dot{q} = \frac{1}{2} \rho_m c_p \text{Im}[T_1] u_1(x) \tag{12}$$

Integration over the y–z plane yields the heat flux along the plate, in the x-direction, as shown in (Fig. 3).

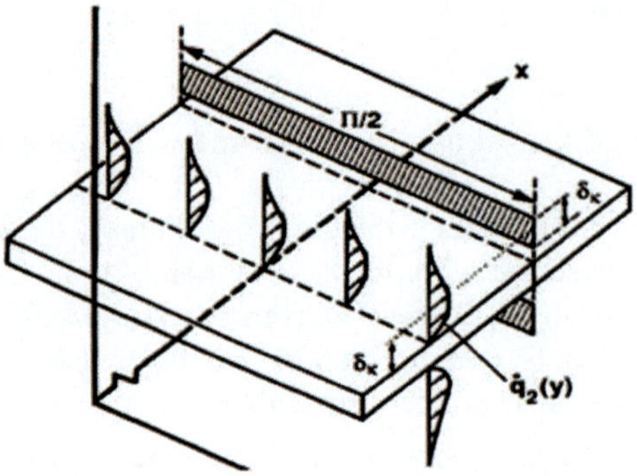

Fig. 3 Representation of the hydrodynamic heat flux density along x-direction. [6] Reproduced from

$$\dot{Q} = \Pi \int_0^\infty \dot{q}\,\mathrm{d}y \tag{13}$$

$$\dot{Q} = \frac{1}{4}\Pi\delta_k T_m \beta p_1(x) u_1(x)(\Gamma - 1)\left[\Gamma = \frac{\nabla T_m}{(\nabla T_m)_{\text{crit}}}\right] \tag{14}$$

- The heat flux depends on the area, $\Pi\delta_k$
- The product $p_1(x)u_1(x)$ implies the positioning of the stack and it vanishes if the plate is positioned at either a velocity node or a pressure node. Halfway between the nodes, this term is at its maxima.
- The temperature gradient factor is given by $(\Gamma - 1)$. At $(\nabla T_m)_{\text{crit}} = \nabla T_m$, $\Gamma - 1 = 0$, and hence heat flux is zero. For $\nabla T_m > (\nabla T_m)_{\text{crit}}$, $\Gamma - 1 > 0$ and the heat flows toward the pressure node, while for $\nabla T_m < (\nabla T_m)_{\text{crit}}$, the reverse occurs.

4 DELTAEC Simulations and Results

In the present analysis, a numerical study using DELTAEC [7] has been conducted to establish the effect of variations in the important governing parameters on the thermoacoustic effect. Design Environment for Low-Amplitude Thermoacoustic Energy Conversion (DELTAEC) is used to conduct parametric studies for obtaining the performance characteristics of thermoacoustic equipments. To model the pressure amplitude, DELTAEC utilizes the Helmholtz differential equation of second order. Two first-order differential equations for $p_1(x)$ and the complex volume flow rate amplitude $u_1(x)$ are coupled to set up the equation [6]. Solution of these equations is obtained by integrating them for each segment. The boundary conditions comprise of continuity of pressures, volume flow rates, and other variables at the junctions between different segments. The solution of the energy-flow equation along with the

Fig. 4 Configuration in DELTAEC simulation

acoustic solutions of pressures and volume flow rates in the stacks gives the mean-temperature profile. The resonator length of the experimental model is 300 mm and the stack spacing is set as 0.5 mm, with plate thickness as 1 mm. The tube diameter is 66 mm. For the study of the experimental model, the DELTAEC setup is shown in (Fig. 4) in which and no heat exchangers have been used in the simulations. The stack material is chosen as Mylar, which is a plastic sheet, closely replicating the photographic film used in the experiment. The stack type used in the simulation is STKSLAB, which comprises of parallel plates. The plate spacing is initially set as $4\delta_k$, and the other parameters have been varied. To achieve resonance condition, in DELTAEC the end condition is set as HARDENED with target impedance values set to infinity so that velocity becomes zero at this boundary. The duct is considered to be made of ideal solid so that there is negligible heat loss to the surroundings. Combined parametric variations have been done and a comparative study is done to find the operating conditions.

4.1 *Stack Spacing*

Perpendicular to the direction of gas motion, there are two important parameters which govern the temperature distribution. The thermal penetration depth, $\delta_k = \sqrt{2K/\rho\omega c_p}$ where K = thermal conductivity, ρ = density of medium, ω = frequency in rad/s, c_p = Specific heat capacity at constant pressure. The viscous penetration depth is given as, $\delta_v = \sqrt{2\mu/\rho\omega}$, where μ is the dynamic viscosity. These characteristic lengths give the extent to which heat and momentum can diffuse laterally during a time interval. The order of this time scale is of the oscillation time period. The effects of thermal conduction and viscosity due to the solid plates, on the gas are negligible beyond these lengths. If the plate spacing is much greater than the thermal penetration depth, the oscillating gas parcel would experience no temperature effect of the stack plate. This would result in no temperature difference and hence, there will be no heat transfer at the ends of the stack. If the plates are too closed to each other, the viscous effects would be high and consequently dissipate the kinetic energy of the packets damping their oscillations. Hence, there is an optimum spacing between the stack plates for which the highest temperature difference is obtained. This optimum gap is around $4\delta_k$. This variation is highlighted in (Fig. 5a). There is an optimum stack length for which the temperature difference reaches a peak, given

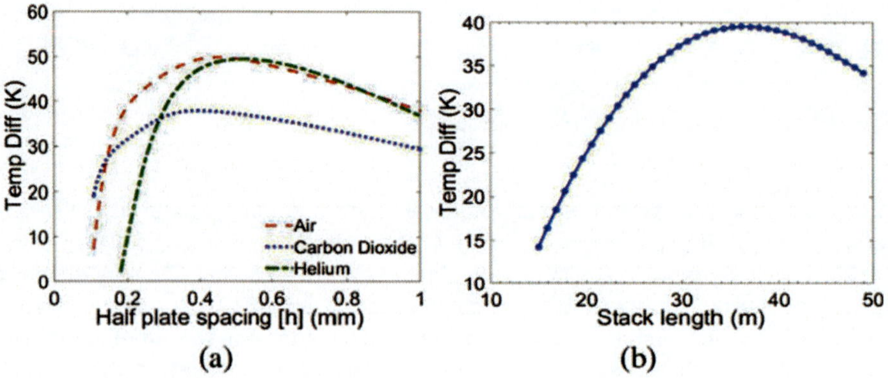

Fig. 5 **a** Variation in ΔT with stack spacing, **b** variation in ΔT with stack length (for air)

other conditions remain fixed. In (Fig. 5b), the stack is positioned at 155 mm from the speaker end, and the optimum stack length is around 50 mm.

The ends of the stack must be located such that there is highest pressure oscillation across the stack length; hence for a fixed position of the stack, there exists one length across which this is achieved.

4.2 *Frequency*

For a given length of the resonator column, gas and environmental conditions the resonant frequencies are constant. When resonance condition is achieved there is maximum energy influx since the oscillations are constructive in nature. Consequently, maximum temperature difference is achieved when this condition is achieved in the tube (Fig. 6a). Also for this particular frequency, there is maximum mass flow

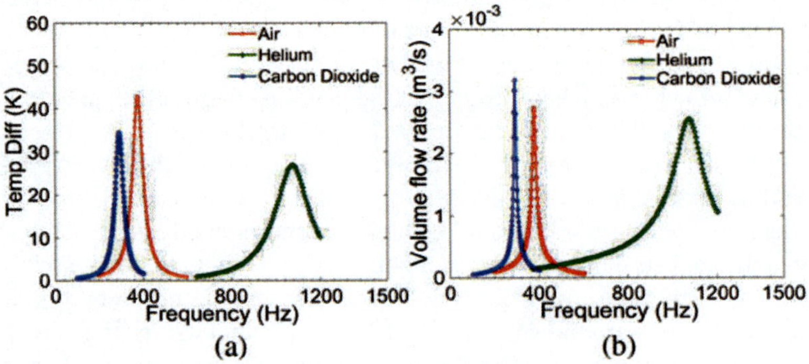

Fig. 6 **a** Variation in ΔT with frequency, **b** variation in flow rate with frequency

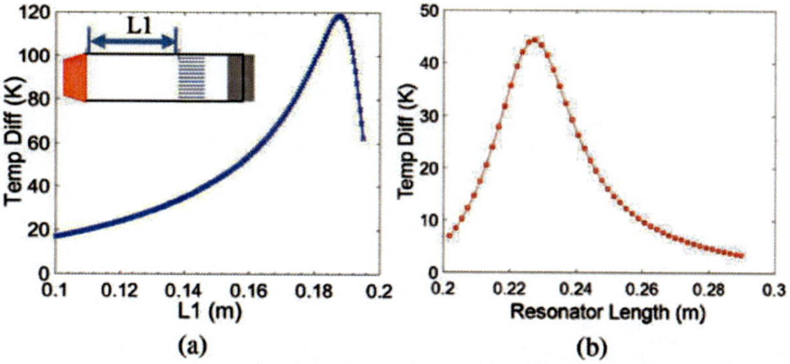

Fig. 7 **a** Variation in ΔT with stack position for air, **b** variation in ΔT with resonator length

rate since the oscillations in pressure and velocity are in phase (Fig. 6b). For achieving maximum performance for a given set of conditions, the speaker must sound waves at a particular frequency. The operating frequency for different mediums depends on the speed of sound in that medium. For Helium, the frequency is around 1200 Hz, and for CO_2, it's around 300 Hz. Generally, higher frequency sounds produce a more painful sensation; hence using CO_2 is a better option. There exists an optimum stack position (Fig. 7a) for which the temperature difference is the highest, since the gas parcel undergoes the maximum compression and expansion within this length.

According to the simulation results, the maximum temperature difference occurs when the stack is placed 180 mm away from the speaker end. To achieve resonance at this frequency the resonance column must be adjusted accordingly. For our setup of a tube having length of 300 mm, the resonant frequency of air is around 380 Hz (1st harmonic), for ambient temperatures of 300 K and pressure of 1 bar. The variation of resonator length with temperature difference for a frequency of 340 Hz is plotted in (Fig. 7b), highlighting the need of resonance to achieve the best performance.

4.3 *Pressure Amplitude*

The pressure amplitude governs the extent to which a gas parcel undergoes compression and expansion and the associated temperature fluctuations. Increasing the pressure amplitude produces a greater temperature difference across the stack since the extent over which the gas undergoes the cyclic compressions is more, but this increase is limited due to time taken for heat to diffuse in the fluid (Table 1). The length through which heat diffuses in a fluid in a given time ($1/\omega$) is governed by the thermal penetration depth. For a fixed diffusion length, higher δ_k implies faster diffusion rates. The time available for heat diffusion is limited and the gas parcel stays at either end of the stack for a time ($1/\omega$). For the same stack spacing, a higher δ_k implies higher diffusion rate. The amount of heat energy that can be transmitted from the gas to the

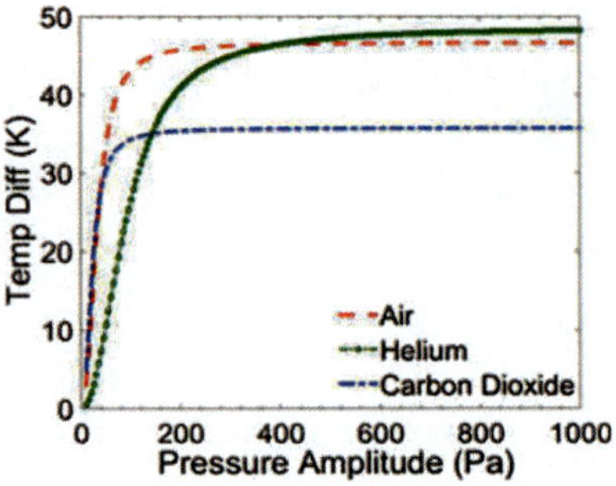

Fig. 8 Variation in ΔT with pressure amplitude for different gases

stack and vice-versa is directly proportional to δ_k. Hence, the presence of a higher temperature difference between the stack and gas parcel (due to increased pressure amplitude) does not imply higher temperature difference across the stack (Fig. 8).

The rate of increase in the temperature difference curve of helium is less as compared to the other two gases since it has a much higher specific heat capacity. The plots for the different gases for increasing pressure amplitude highlight this property. $T = 300$ K are different for each gas. Calculating δ_k for the different mediums, thermal penetration depth of helium is the highest and for CO_2 is the lowest hence the increment in the curve for CO_2 ceases at lower pressure amplitude as compared to helium.

4.4 *Mass Flow Rate and Tube Size*

The oscillating mass flow, generated by the vibration of the diaphragm of the speaker, determines the thermal capacity of the system. The pressure oscillation directly depends on the displacement of the membrane and hence the volume flow amplitude (for a fixed density) increases with increase in this pressure amplitude, linearly in this case. The temperature difference produced across the stack depends on the pressure oscillation introduced in the tube. Initially, with an increase in the amplitude of pressure, the temperature difference rises but it ultimately attains a constant value, as explained in the previous subsection. For a given pressure amplitude, the volume flow rate for the wider tube, whose area is almost 10 times of the other one, is much higher than the narrower one (Fig. 9a). The temperature difference increases with increment in pressure amplitude (implying increasing in volume flow rate), since there is an increased temperature oscillation. But beyond a particular flow rate, this temperature difference is constant. This phenomenon is explained in Section *C*. The increase in the tube diameter has minimal effect on the maximum temperature difference achieved since with increase in the cross-sectional size, a bigger stack is required and the total area to be cooled is consequently higher (Fig. 9b).

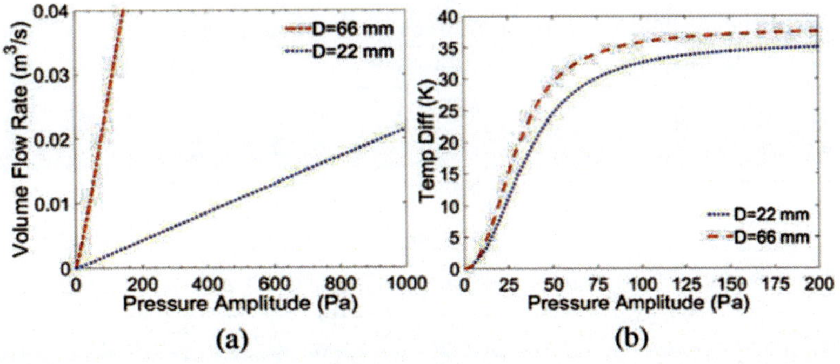

Fig. 9 **a** Flow rate versus pressure amplitude (air), **b** ΔT versus pressure amplitude for different C.S. Area

The tube diameter is an important criterion in the selection of the speaker, since the flow rate depends on the cross-sectional area. The acoustic power in DELTAEC is calculated as:

$$\dot{E} = \frac{1}{2}\mathrm{Re}\left[p_1\widetilde{U_1}\right],\quad U_1 = \text{volume flow rate } (\mathrm{m}^3/\mathrm{s})$$

$$\mathrm{d}B = 10\log(P/P_0),\ P_0 = 10^{-12}\,\mathrm{W}$$

The acoustic power required in the tube to develop the temperature difference is obtained by multiplying the area of the tube with the intensity. Considering air as the refrigerating medium, the plots have been made for two sizes of the resonator, with the diameters selected such that one of the areas is approximately 10 times that of the other.

The acoustic power necessary at the inlet of the tube for the smaller one is about 1/5th of the power required for the larger tube (Fig. 10a). The net power generated in the speaker is dissipated as electrical heating and losses from the tube, and only a

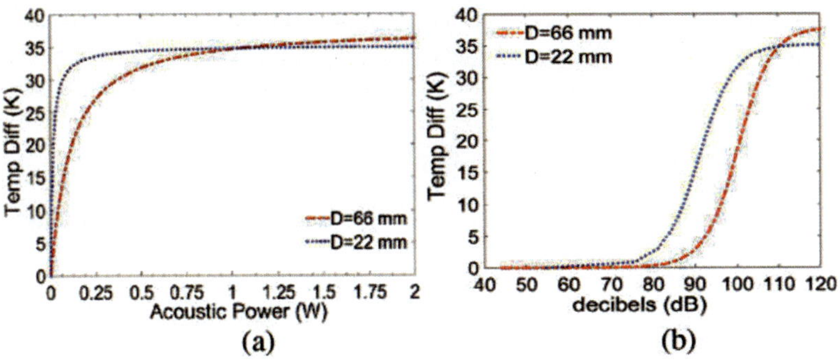

Fig. 10 **a** ΔT versus acoustic power (air), **b** ΔT versus sound intensity (in dB)

Table 1 Thermal penetration depths and diffusion rates for different gases

Medium	Resonant frequency (Hz)	c_p (J/kg − K)	δ_k (mm)	Diffusion rate (λ) (mm/s)
Air	380	1004.7	0.137	52.06
Helium	1092	5193.4	0.237	258.8
CO_2	294	860.4	0.109	33.52

small component of the gross power generated is available in developing the refrigerating effect. This is one of the reasons the COP of thermoacoustic systems is low. Improvements can be made in the resonator column like better casing and sealing of the openings. The sound generated is of the order 100–110 decibels which is similar to that of a rock concert (Fig. 10b).

5 Conclusion

This parametric study gives an insight into the parametric dependences of the thermoacoustic effect for the experimental model. The stack spacing is varied from 1–6 δ_k, and it is around 3–4δ_k, for which the maximum temperature is achieved. There is an optimum stack length which depends on the stack positioning. The operating frequency required for a particular medium is crucial in achieving performance. Using air is the simplest since there is no problem of leakage but operating in a lower frequency range is more desirable. From the study, CO_2 is one such gas in which the column has a lower resonant frequency. The length of the resonator and operating frequency are interdependent and proper selection of each is required. The pressure amplitude is directly related to the power rating of the speaker. For different mediums, the pressure amplitude for which the maximum temperature difference is achieved is unique and depends on a combination of the properties of the respective gases. The effect is constant beyond a pressure amplitude; hence an optimum speaker size is required. Varying the cross-sectional area of the resonator tube has a negligible effect on the temperature difference since the area to cool is more. An important result is the required acoustic power for different tube diameters. Acoustic power of 0.20 W is needed for the tube with a diameter of 22 mm and 1 W for a tube with thrice the diameter. Hence using a tube with a smaller diameter can significantly reduce the power consumption, but the area being cooled is also reduced.

References

1. Tijani MEH, Zeegers JCH, de-Waele ATAM (2002) The optimal stack spacing for thermoacoustic refrigeration. J Acoust Soc Am 112:128
2. Kim YT, Kim MG (2000) Optimum positions of a stack in a thermoacoustic heat pump. J Kor Phys Soc 36:279–286

3. Setiawan I, Setio-Utomo AB, Nohtomi M, Katsuta M, (2012) Experimental study on the influence of the porosity of parallel plate stack on the temperature decrease of a thermoacoustic refrigerator. In: Proceedings of 3rd international conferences on physics
4. Wantha C, Assawamartbunlue K (2011) The impacts of the resonance tube on performance of a thermoacoustic stack. Front Heat Mass Trans 2:1–8
5. Russell A, Weibulla P (2002) Tabletop thermoacoustic refrigerator for demonstrations. Daniel Science and Mathematics Department, Kettering University, Flint, Michigan 48504
6. Swift GW (1988) Thermoacoustic engines. J Acoust Soc Am 84:1146–1180
7. Ward B, Clark J, Swift G (2017) Design environment for low-amplitude thermoacoustic energy conversion. DELTAEC Version 6.4b2.7 Users Guide. Los Alamos Laboratory
8. Rott N (1975) Thermally driven acoustic oscillations, part III: Second-order heat flux. Z Angew Math Phys 26(43):49

Shifting the Focus from Macro- to Micro-waste to Energy (WTE) Plants as a Solution to the Solid Waste Management

Tenzing Dorjee Pradhan, B. B. Pradhan, and A. P. Tiwary

1 Introduction

The world-ecology is a fragile one which is an understatement. With the ever-increasing urbanization comes the problem of waste disposal. The easiest and the most common method of waste disposal is the landfill method. However, it requires a large amount of space, and also it pollutes the environment. Even if we somehow create ways and means to prevent the pollution caused due to landfills, there is the problem of space which is required. WTE (Waste to Energy) is an alternative that is being explored and successfully implemented in some developed countries and is being explored by every developing nation including India. Saini et al. [1] estimated the potential of power generation in India as 2539 MW with the total waste generated at 127 k Tons/day. Dabe et al. [2] projected that with the current trends in population growth and urbanization, by the year 2041 the MSW generation in India could go up to 4.4 lakh tons per day. There are many WTE technologies prevalent in the world, however, all the technologies require two things, (i) an extensive study of the incoming waste, which is to be used as the raw material for power generation, and (ii) proper and effective management of waste segregation. The WTE technology is developing at a rapid pace in India especially in the metros. However, waste management is still in its infancy or in some cases even nonexistent in a large number of cities and towns resulting in heavy environmental degradation. The solution could be the development of micro-level WTE plants established by small communities and institutions rather than macro-level plants run as big industries, needing a large number of management resources. This study tries to look into the feasibility of a

T. D. Pradhan (✉) · A. P. Tiwary
Department of Mechanical Engineering, Sikkim Manipal Institute of Technology, Majitar, Sikkim, India

B. B. Pradhan
Sikkim Institute of Science and Technology, Chisopani, South Sikkim, India

S. K. Ghosh et al. (eds.), *Advances in Thermal Engineering, Manufacturing, and Production Management*, Lecture Notes in Mechanical Engineering,
https://doi.org/10.1007/978-981-16-2347-9_14

micro-level WTE plant which would reduce the amount of waste being sent to the landfills, thereby reducing the burden on the municipalities and government for waste management.

2 Literature Review

Much research has been carried out in the field of WTE technology. Some important research work in India is being presented to understand the current status of WTE technology, which highlights the potential and scope for the further development. Kumar et al. [3] stated that waste management, in India, has some major issues and depends upon improper waste infrastructure and the informal sector. The authors have stated that one of the major issues with waste management in India, is the lack of responsibility shown by the community toward waste. They have also stated that the current status of SWM (Solid Waste Management) in India is abysmally poor, as appropriate methods of waste disposal are not followed. The track record of WTE in India has also been poor. Large-scale MSW (Municipal Solid Waste) incinerators have failed to work and have landed in trouble. The first MSW incinerator plant which was built in Timarpur, New Delhi in 1987 with the capacity to process waste at 300 tons per day, had to be shut down owing to poor waste segregation due to the erratic nature of the waste composition and most importantly inappropriate selection of technology for the time and other maintenance issues [4]. The latest of the WTE plants in Okhla Ghazipur and Bawana, New Delhi has also run into serious operational problems with notice being slapped by the green tribunal and also a case being registered at the Hon'ble Supreme Court of India. It was found that they were neither producing nor using RDF (Refuse-Derived Fuel), a fuel which is produced from the industrial and commercial waste and were using the untreated MSW which has a calorific value much less than the required calorific value for self-sustaining combustion and the result was toxicity in the emissions [5].

Kalyani and Pandey [6] studied the WTE status in India and highlighted the failure of the large-scale WTE plants in India. As per the authors, the country has invested large sums of capital to develop the WTE plants, 10 aerobic composting projects in the 1970s, 1 Incineration plant in the 1980s, a large-scale bio methanation project, 2 RDF projects in 2003 but have all shut down. They have stated that the failure is not owing to the technology but due to improper financial and logistical planning.

Sebastian et al. [7] focused on a WTE technology, namely Incineration and suggested that incinerability of the MSW and the amount of waste generated are crucial for the feasibility of large-scale WTE plants. This is owing to the fact that the capital cost for a large-scale WTE plant is pretty substantial. The authors have developed an incinerability map of India based upon a tool developed during the research, known as *i*-index. The choice of the WTE technology to be used depends a lot upon the composition of the waste. The composition of the waste changes from region to region and also has some change owing to seasons. Dabe et al. [2] reported the composition of the MSW in India as organics (51%), recyclables (17.5%) and inserts (31%).

Late and Mule [8] studied the open dumpsite in the city of Chennai and observed that about 35% of the waste had the potential for composting and 30% of the waste could have been recycled. These numbers however differed between different metros. The authors reported that a large quantity of the old MSW material is combustible in nature and the open dumping of such waste materials creates heavy environmental pollution. The scenario of MSWM (Municipal Solid Waste Management) in India is extremely poor. As reported by Joshi and Ahmed [9] in their study, most of the ULBs (Urban Local Bodies) do not have any formal action plan as regards to the implementation of the MSWR (Municipal Solid Waste (Management and Handling) Rules, 2000). It is also reported that no ULBs of any city can claim for 100% waste segregation at the producer level. It was stated that, out of the total waste collected, only 12.45% waste is actually processed scientifically, and the rest is also again sent for open dumping in landfills.

3 Prevalent WTE Technologies

The following is a brief description of the prevalent WTE technologies available in the world today.

3.1 Biochemical Conversion

This process comprises two options (i) digestion (production of biogas) and (ii) fermentation (in which ethanol is produced) [7]. It leads to a lower temperature and lower reaction rates. For biochemical conversion, the waste must be of high moisture content. The process includes the aerobic conversion like composting, anaerobic decomposition/digestion (which is of usual occurrence in landfills, digesters and controlled reactors), and anaerobic fermentation converting sugars which converts sugars from hydrolyzed cellulose or hemicellulose in order to generate ethanol.

3.2 Anaerobic Digestion (AD)

It is a method to convert biogenic material, anaerobically with the use of microbes. It has to be carried out in an oxygen-free environment and results in a fuel gas, called biogas. This is one of the most prevalent technologies in rural India.

3.3 Fermentation

The fermentation technique involves an operating condition in which there is an absence of free oxygen, and results in the production of biogas, consisting mainly of gases like methane and carbon dioxide and carries impurities like siloxane, H2S, ammonia, moisture and particulate matter. However, a clean upgraded AD process produces BioS methane which can be used as a biofuel [10]. MSW is one of the most complex and variable feedstock compositions and is highly dynamic in nature. The generation of the biogas depends upon numerous factors. Even the source of its collection to segregation to seasonal variation and cultural differences has a significant effect in quality of the resultant fuel [2].

3.4 Thermo-Chemical Conversion

For the production of energy from waste, Thermo-chemical conversion is the most widely used technology in the world. It consists of a combination of chemical processes with or without the production/addition of heat.

The options which are available in the thermo-chemical conversion process are combustion, gasification, pyrolysis and liquefaction [11]. This method of conversion occurs at a higher temperature and has a faster conversion rate and thus is best suited for the lower moisture feedback. However, the various technological, environmental and economic viability of the technology must be adopted before the same is practically implemented.

3.5 Combustion/incineration

It is one the most widely used processes of converting the waste into heat and further converted into electricity or mechanical power with the help of furnaces, steam turbines, stove turbo generators, etc. The process can be used to burn any type of biomass having moisture content $< 50\%$. However, it has been investigated that for biomass having high moisture content, biological conversion is the process best suited for energy conversion [11].

3.6 Gasification

This process, theoretically, lies between the processes of combustion and pyrolysis and it involves a partial oxidation of the substrate. The oxygen is added in a controlled manner such that the fuel is not completely oxidized and complete combustion does not occur. The gasification process is an exothermic process and occurs at temperature is above 650 °C. However, some heat may be required in order for the process to initiate and also to sustain it. Syngas is the main product and contains carbon monoxide, methane and hydrogen. The syngas which is generated from the gasification process has a net calorific value (NCV) of 4–10 MJ/Nm^3 [12].

3.7 *Pyrolysis*

Pyrolysis is the process of thermal degradation of substances in the absence of oxygen and is in complete contrast to the gasification and combustion. An external heat source typically between 300 and 850 °C is required to sustain the process. Pyrolysis also produces Syngas and a solid residue known as Char, which is a combination of carbon and non-combustible material. This syngas contains gases and combustible constituents which includes carbon monoxide, hydrogen, methane and a wide range of VOCs (Volatile Organic Compounds). The syngas produced from pyrolysis has an NCV between 10 and 20 MJ/Nm^3 [13].

4 Challenges in Macro-Level WTE Plants

In order to successfully manage solid waste, the selection of appropriate treatment technologies is of utmost importance. The main criteria for the selection of proper waste treatment technologies are the quantity of waste, the characteristics of waste, the physical properties, composition of waste, availability of land, capital investment, time factor for treatment, etc. An inappropriate selection of the technology may lead to failure of the entire waste management system [14]. The biggest challenges for WTE plants are the selection of appropriate technology and the segregation of waste. The WTE technologies are based upon the type and the composition of the waste, and hence, it is imperative that the one knows the type of waste generated. The amount of waste generated and the characteristics of solid waste vary from one place to the other. A large number of factors influence the quantity and composition. Average annual income, type of population, the social behavior of the larger group, climate, industries and the waste materials market all influence the type and amount of waste [8, 15]. In a country like India, which is diverse in all aspects, the quantity and characteristics of the solid waste are an immense variable. Thereby, in terms of WTE, the policy of one shoe fits all, fails miserably. Another factor for WTE plants to be successful is the proper collection and segregation of waste [7]. The same has also been highlighted by Peter et al. [16].

The system of collection of waste in India is highly unorganized, to say the least. Most of the urban areas in the country lack proper storage at the producer level. There is also a flaw in the collection system adopted in the various cities. Neither are they properly designed nor properly located. Furthermore, there are not adequate bins distributed across the cities. This has resulted in a poor collection efficiency, which is about 70% in Indian cities and states [17–22]. The most successful WTE plants are situated in the developed countries of Europe, America and Asia. The most important feature of these success stories in the WTE plants is the developed waste management and segregation plans.

Chaliki et al. [23] presented various successful WTE plants in their extensive study. The success story suggests that for WTE plants to be successful in India, it requires an overhaul of the existing waste management system and also a massive change in the attitude of the producers of waste. For a country like India having its diversity in ecology, population, culture and climate, it is a challenging task, which at the moment seems a long way off.

5 Concept of Micro-generation of Power

Juntunen and Hyysalo [24] in their study of micro-generation refer to the generation of power which is done at the place where the energy is required to be used or at the source. Micro-generation is a small-scale generation which is done at a local level, and it consists of generation of heat or electricity or both, and it also generates very small amounts of energy as compared to the large-scale centralized plants. The authors have stated that in the case of generation of electricity, it can be for the use of the buildings or neighborhoods' occupants, and the same cannot be used to feed the national grid. Some studies define micro-generation for WTE with power output below 50 kW. [25, 26]. There are many models of establishing these micro-generation plants. It can be established with the funds received by the community from the government and can also be established by the community themselves with a fee-for-credit model.

6 Necessity of Micro-level WTE Plants

With all the research and studies mentioned above, it can be understood that managing a large-scale WTE plant in a developing and diverse country as India is a very difficult task. A large-scale WTE plant requires a large amount of capital to establish the same. Furthermore, the technology selection needs to be accurate so that the appropriate technology for the WTE plant can be used. Despite all the feasibility studies and research which goes into the establishing the plant, the operation of the plants depends upon numerous factors ranging from the efficiency of its workers to the segregation of the waste. Segregation of waste, especially in India, is not an easy job, as the society as a whole does not take responsibility for waste management. In order for the plant to operate, proper segregation of the waste is of utmost importance. If the segregation of the MSW is not done at the producer level, then it will be very difficult for it to be segregated at the plant or will add to additional capital cost. Micro-level plants on the other hand can be more manageable. When the target for waste segregation is limited to a small community like a housing complex or a small village, it invariably has its own structured society. The responsibility of waste segregation to the producers of waste can be easily passed on and understood.

According to a study based upon 20 cities across 10 states of the country, it has been assessed that smaller cities have been more successful in implementing source segregation compared to bigger cities. The people involved in handling SWM of smaller cities have also been innovative in their approach and hence have performed well in the process of source segregation. In this report, it has been observed that small cities are more efficient in waste management than the large counterparts [27].

Similarly, the success story of Vengurla in Sindhurdurg district of Maharashtra, a town with a population of 18,000 has been highlighted in another report. In the same report, it is presented that this town was able to successfully manage its SWM and transformed its waste management system from a home-to-landfill scenario, to a town generating revenue from its processed waste. The town achieved a 95% segregation rate owing to strict rules and total society participation [28].

Another report highlights the success of small towns like Karjat of Raigad district in Maharashtra, with a population of close to 30,000 and Suryapet in Telangana, with a population of 115,000 [29]. Suryapet municipality transformed the entire town to a zero -landfill town with a revenue of around Rs. 100,000/- per month from vermicompost. The waste management system at Karjat was also transformed and overhauled in lines of Vengurla town.

7 Conclusions

The research and the articles from various sources highlight the micro-management of the leaders and how they transformed the waste management scenario. These scenarios would however be very difficult to replicate in large metros owing to the large population, diverse range of activities and the area it covers. The same would be comparatively easier to replicate at a micro-level with an even smaller population and area than that of Vengurla and Suryapet and Karjat. Thus, with the limited success of large-scale WTE plants in India and other developing nations, it would be wise to also explore the possibility of a micro WTE plant to reduce the burden on the organizations for efficient disposal of MSW. It is also imperative that in order to make the micro WTE plants successful, an appropriate WTE plant can be developed with appropriate selection of WTE technology. It is, however, of utmost importance that we look upon the micro WTE plant as a way of proper disposal of waste for environmental protection and not as means of generating revenue. The revenue generated from the micro WTE plants is a means of making the process sustainable.

The authors in this study have tried to bring into picture an alternative solution to the waste management problem by shifting the focus from macro to a more manageable micro-level solution.

References

1. Saini S, Rao P, Patil Y (2012) City based analysis of MSW to energy generation in India, calculation of state-wise potential and tariff comparison with EU. Procedia Soc Behav Sci 37:407–416
2. Dabe SJ, Prasad PJ, Vaidya AN, Purohit HJ (2019) Technological pathways for bioenergy generation from municipal solid waste: renewable energy option. Environ Prog Sustain Energy 38(2):654–671
3. Kumar S, Smith SR, Fowler G, Velis C, Kumar SJ, Arya S, Rena Kumar R, Cheeseman C (2017) Challenges and opportunities associated with waste management in India. Roy Soc Open Sci 4(3):160764
4. Indo-UK Seminar Report (2015) Sustainable solid waste management for cities: opportunities in SAARC countries, 2015. http://www.neeri.res.in/short%20Report_Indo-UK%20Seminar%20(25-27th%20March%202015.pdf
5. Young M (1989) The technical writer's handbook. University Science, Mill Valley. https://www.downtoearth.org.in/blog/waste/how-delhi-s-waste-to-energy-plants-are-way-off-the-mark-63026
6. Kalyani KA, Pandey KK (2014) Waste to energy status in India: a short review. Renew Sustain Energy Rev 31:113–120
7. Sebastian RM, Kumar D, Alappat B (2020) Mapping incinerability of municipal solid waste in Indian sub-continent. Clean Technol Environ Policy 22(1):91–104
8. Late A, Mule MB (2013) Composition and characterization study of solid waste from Aurangabad city. Univ J ENviron Res Technol 3(1):55–60
9. Joshi R, Ahmed SS (2016) Status and challenges of municipal solid waste management in India: a review. Cogent Environ Sci 2(1):1139434
10. Scarlat N, Dallemand JF (2011) Recent developments of biofuels/bioenergy sustainability certification: a global overview. Energy Policy 39(3):1630–1646
11. McKendry P (2002) Energy production from biomass (part 2): conversion technologies. Biores Technol 83(1):47–54
12. Department for Environment, Food and Rural Affairs [Defra] (2007) Advanced thermal treatment of municipal solid waste. London: Defra. www.defra.gov.uk/environment/waste/wip/newtech/pdf/att.pdf
13. Schmitt E, Bura R, Gustafson R, Cooper J, Vajzovic A (2012) Converting lignocellulosic solid waste into ethanol for the State of Washington: an investigation of treatment technologies and environmental impacts. Biores Technol 104:400–409
14. Naveen BP, Malik RK, Kontoni DPN (2018) Municipal solid waste management in India—challenges and feasible solutions. In: 8th international conference "from scientific computing to computational engineering, 8th IC-SCCE, Athens, 4–7 July 2018 ©LFME
15. Yadav IC, Devi NL (2009) Studies on municipal solid waste management in Mysore city—a case study. Rep Opin 1(3):15–21
16. Peter AE, Nagendra SS, Nambi IM (2019) Environmental burden by an open dumpsite in urban India 2019. Waste Manage 85:151–163
17. Saxena S, Srivastava RK, Samaddar AB (2010) Sustainable waste management issues in India. IUP J Soil Water Sci 3(1):72–90
18. Rathi S (2006) Alternative approaches for better municipal solid waste management in Mumbai, India. J Waste Manag 26(10):1192–1200
19. Siddiqui TZ, Siddiqui FZ, Khan E (2006) Sustainable development through integrated municipal solid waste management (MSWM) approach—a case study of Aligarh District. In: Proceedings of national conference of advances in mechanical engineering, Jamia Millia Islamia, New Delhi, India, 20–21 Jan
20. Gupta S, Mohan K, Prasad R, Gupta S, Kansal A (1998) Solid waste management in India: options and opportunities. Resour Conserv Recycl 24(2):137–154
21. Maudgal S (1998) Waste management in India. J India Assoc Environ Manag 22(3):203–208

22. Khan RR (1994) Environmental management of municipal solid wastes. Indian J Environ Prot 14(1):26–30
23. Chaliki P, Psomopoulos CS, Themelis NJ (2016) WTE plants installed in European cities: a review of success stories. Manage Environ Qual Int J 27(5):606–620
24. Juntunen JK, Hyysalo S (2015) Renewable micro-generation of heat and electricity—review on common and missing socio-technical configurations. Renew Sustain Energy Rev 49:857–870
25. IEA (2004) Review of small scale waste conversion systems. International Energy Agency
26. Biffa (2002) Future perfect, an analysis of Britain's waste production and disposal account, with Implications for Industry and Government of the Next Twenty Years
27. https://www.cseindia.org/forum-of-cities-that-segregate-assessment-report-2017-18-9080
28. https://www.downtoearth.org.in/news/waste/a-waste-management-model-for-small-towns-58771. (2018)
29. https://www.financialexpress.com/opinion/on-waste-management-small-towns-drive-big-change/1499461

Design and Performance Evaluation of Box-Type Solar Cooker with Energy Storage

Abid Ali and Naiem Akhtar

1 Introduction

Amongst the renewable energy technologies, solar energy is recognized as one of the most potential choices since it is free and provides environmentally clean energy. The major energy requirements of domestic consumers include lighting, cooking, and heating. Among these, cooking accounts for substantial amount of primary energy demand. There is an increasing consideration over the renewable energy options to meet the cooking requirements in developing countries. Solar cooker is a cost-effective device for harnessing solar energy. It is environment-friendly and helps in reducing deforestation and air pollution. The box-type cooker designs have been studied and modified since the 1980s and different designs and their characteristics are investigated. The method for performance evaluation of the box-type solar cooker was first proposed by Mullick et al. [1]. The main problem associated with simple box-type solar cooking system is the impossibility of cooking food during the late hours of the day. This problem can be solved by storing solar energy during the sunshine period and utilized later. There are three methods for storing heat energy, namely; sensible, latent and thermo-chemical [2]. Sensible heat storage method is used for some of the solar energy applications. However, the main limitation of drawback of a sensible heat storage unit is the large volume requirements. The latent heat storage system is a better way of storing thermal energy. The latent heat storage materials are generally referred to as phase change materials (PCMs). Although the study of PCM was initiated by Telkes and Raymond [3] in the 1940s, but, the phase change materials received large attention only after the energy crisis of the late 1970s

A. Ali (✉) · N. Akhtar
Department of Mechanical Engineering, Z. H. College of Engineering & Technology Aligarh Muslim University, Aligarh, U.P. 202002, India

S. K. Ghosh et al. (eds.), *Advances in Thermal Engineering, Manufacturing, and Production Management*, Lecture Notes in Mechanical Engineering,
https://doi.org/10.1007/978-981-16-2347-9_15

and early 1980s. Better design of Box-type solar cooker with phase change material for storage of *t* energy will be more appropriate for cooking the food during late hours of the day.

There has been a significant attention in the development of solar cookers with phase change materials. Ramadan et al. [4] constructed a simple flat-plate box solar cooker with focusing plane mirrors and using locally available sand in Tanta University to store energy. In this design, a jacket of sand (0.5 cm thick) placed around the cooking pot which improved the cooker performance. Six hours per day (3 h/day outdoor and 3 h/day indoor) of cooking time has been reported. Overall energy conversion efficiency up to 28.4% had been obtained. Domanski et al. [5] designed a multi-step inner reflector box-type solar cooker. In this design, they replaced the absorber plate with reflective surfaces; these surfaces were tilted at different angles around the cooking vessel holder. The cooking vessel was made with two concentric cylinders of aluminium. The outer wall has a diameter of 18 cm and height of 12 cm, and the inner cylinder has a diameter of 14 cm and height of 10 cm. The tests were conducted by filling the gap between the vessels/cylinder with 1.1 kg stearic acid and 2 kg magnesium nitrate hexahydrate, respectively. They reported that charging time and storage efficiency strongly depend on solar intensity and the melting point of PCM. The overall utilization efficiency was reported about 3–4 times higher than the steam and heat-pipe solar cooker. Buddhi and Sahoo [6] used stearic acid as a storage medium for a box-type solar cooker. They filled the stearic acid below the absorber plate of the cooker. The absorber plate has 28 × 28 cm dimension at the bottom, 40 × 40 cm at the top and vertical depth in the absorber plate is 8 cm, where the cooking pot is placed for cooking. They designed this cooker for a cooking capacity of 0.75 kg. They reported that after solar radiation cut-off the temperature of PCM and absorber plate decreases at a faster rate in the first two hours and thereafter at a slower rate. The rate of heat transfer from the PCM to the load during the discharging mode of the PCM is slow, and more time is required for cooking an evening meal. Vigneswaran et al. [7], Sharma et al. [8] and Buddhi et al. [9] used simple box-type solar cookers with PCM storage unit made of two concentric cylinders and the space between concentric cylinders filled with PCM. Vigneswaran et al. [7] used 2.9 kg of oxalic acid dehydrate (melting point 101 °C), for energy storage by considering the energy required to cook 0.5 kg of rice. The overall utilization efficiency was reported to be 15.74% with a single reflector for 0.4 kg of water load and this efficiency increases to 25.47% with four reflectors for 0.8 kg of water load. Sharma et al. [8] used 2.0 kg of acetamide (melting point 82 °C) and reported that the second meal of food could be cooked if it is loaded before 15:30 h. Buddhi et al. [9] used 4 kg of acetanilide as PCM and reported that 0.5 kg food was cooked if loaded at 19:00 h. Mallikarjuna Reddy et al. [10] designed two solar cookers with circular trays, one with energy storage and one without energy storage system. The diameters of trays were 68 cm with height of 9.8 cm and the wax was filled in the gap of 1.0 cm. In energy storage system the space between the two trays filled with paraffin wax (melting temperature 55 °C) with fin arrangement to increase heat transfers from top absorber plate to bottom wax surface. They reported that cookers with energy storage are more useful than without energy storage system and has higher temperature during evening time.

Sharma et al. [11] designed and tested an evacuated tube solar collector and PCM storage unit consisting of two evacuated tube solar collectors. The PCM storage unit fabricated by means of two hollow concentric cylinders of aluminium. The inner and outer diameters are 30.4 cm and 44.1 cm, respectively, with height of 42 cm. The space between both the cylinders is filled with 45 kg Erythritol. This system was able to cook two meals in a single day, noon meal and evening meal. However, the quantity of the phase change material (Erythritol) used was huge.

From the literature, it is clear that the thermal energy stored in the box-type solar cooker with the PCM is kept either in the space between two concentric cylinders or below the absorber plate with some deepness in absorber plate for cooking pot. Also, the performance of the Box-type solar cooker increases with the use of materials having higher melting temperature (specifically above 100 °C). In the present work, the design of the existing Box-type solar cooker has been modified. The Box-type solar cooker available in the market has the opening on the top side. The glass frame at the top is opened to keep the cooking pot on the absorber plate. It causes heat loss from the top. Moreover, when the food is cooked in the pot the absorber plate temperature is above 100 °C and inside air temperature is also high which can be harmful to human skin of the operator. It may become more dangerous if some amount of water spill out on the absorber plate during cooking.

In the present design, the top glass cover frame is fixed and a window is provided on the backside of the cooker for keeping the cooking pots on the absorber plate, which prevents the energy loss from the top. Moreover, energy storage using Erythritol (melting temperature 118 °C) is provided below the absorber plate for cooking during evening time.

2 Proposed Design

The box-type solar cookers available in the market generally have 0.25 m^2 aperture area, generally designed according to the BIS STANDARD, part II of "Solar cooker-Box-type-Specification *Second Revision* of IS 13429" [12]. These cookers are used for cooking one meal during the day and don't have any energy storage material. Also, the available cookers in the market have the opening for keeping/taking out the cooking pots is at the top. It causes heat loss from the top. Moreover, when the food is cooked in the pot the absorber plate temperature is above 100 °C and inside air temperature is also high which can be harmful to human skin of the operator. It may become more dangerous if some amount of water spill out on the absorber plate during cooking.

In the present design, the top glass cover has been fixed and a window is provided at the backside of the cooker for keeping the cooking pots on the absorber plate, which prevents the energy loss from the top. Apart from this modification, the second aim is to store thermal energy by using phase change material and use this energy during the late hour of the day. The selection of the material depends on the melting temperature, the latent heat of fusion, density and other properties, for example, toxicity and cost of the PCM. For the purpose of cooking using a box-type solar cooker, the PCM should have melting point above 100 °C. The phase change material; commercial-grade Erythritol ($C_4H_{10}O_4$) which is easily available in the Indian market at low cost (Rs. 325/kg) has been selected. The feasibility of charging of PCM has been checked and the quantity of PCM required is estimated. The calculation of heat losses during the charging and discharging period has been carried out.

2.1 Description of the Proposed Box-Type Solar Cooker

The PCM-cooker is designed for 0.5 m^2 aperture area. It consists of an absorber plate and a PCM-tray, constructed of GI-sheet of 0.5 mm thickness. The absorber plate and PCM-tray have an isosceles trapezoidal base. The height of the frontside of the cooker, facing the sun, is kept low, whereas the height of the backside is kept high so that the size of the opening is sufficient for keeping/taking out the pots. The frontside and backside heights of the absorber plate are 10 cm and 22 cm, respectively, and PCM-tray has 8.5 cm height. The PCM-tray is filled with a known quantity of heat storage material, employed under the absorber plate. The sides and bottom of the trays are encased in a box made of wood. The outer dimensions (length and width) of the base of the wooden box are 81.5 cm and 80.7 cm, respectively. The heights of the frontside (facing the sun) and backside are 27 cm and 41 cm, respectively. The space provided between the PCM-tray and encasing is filled with glass wool to provide thermal insulation. A double glass cover unit of glazing having length and width of 71.5 cm and 70 cm, respectively; is fitted in the wooden box at the top. A rectangular window/opening of dimensions 26 cm and 13 cm width and height, respectively, is provided at the backside of the cooker for keeping/taking out the cooking pots. A flat reflector mirror is encased in a wooden frame which serves as a reflector. To reduce the top heat loss during off-sunshine hours, the cooker is closed with reflector mirror frame by putting it on the glass cover frame. Glass wool of 3 cm thickness is provided between the wooden cover and the reflector mirror to further reduce the top heat loss during off-sunshine hours. The cooking pots are cylindrical in shape and have flat bases with 18 cm diameter and 7 cm in height. The cooking pots are provided with tight-fitting flat covers. The absorber plate and cooking pots are painted black. A secondary reflector is also used to increase incident solar radiation on the absorber plate. The isometric view of the PCM-cooker designed is shown in Fig. 1 and the backside view is shown in Fig. 2.

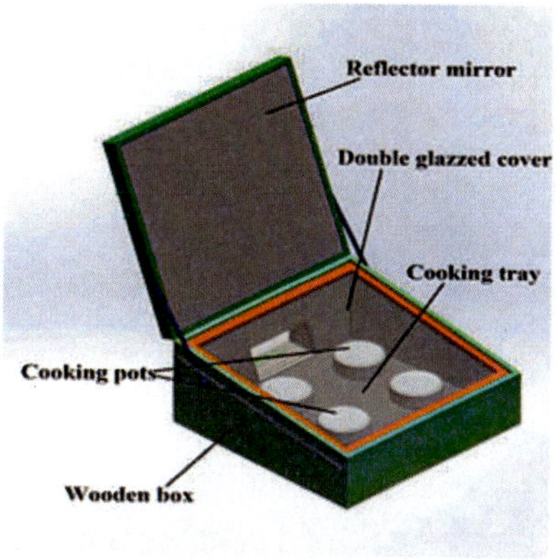

Fig. 1 Isometric view of PCM-cooker

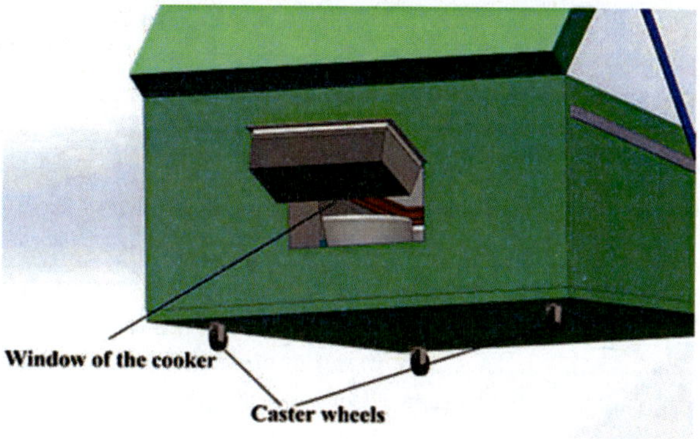

Fig. 2 Backside window view of PCM-cooker

2.2 Operation of Box-Type Solar Cooker with Phase Change Material

During the charging period, the solar radiation incident on the box-type solar cooker is transmitted through the double glass cover and reaches the absorber plate. The beam radiation reflected by the reflector mirror is also transmitted through glass cover and reaches the absorber plate. The solar radiation received on the absorber plate is converted into thermal energy and raises the temperature of the interior of the cooker. When there is no load on the absorber plate, a part of this thermal energy is utilized to increase the temperature of the interior of the cooker plus the heat supplied to the PCM and remaining energy lost to the surroundings. When there is a load on the absorber plate one part of thermal energy is also supplied to the load. As the temperature of the absorber plate increases the heat loss to the surrounding is also increased. During the late hour of the day, there is no solar radiation available and the cover of the cooker is closed (with reflector mirror). During this period only PCM is the source of thermal energy in the cooker. During off-sunshine hour, a part of the thermal energy supplied by PCM to the load, raise its temperature and a part of thermal energy is lost to the surrounding. After some time the load temperature becomes equal to the PCM temperature, the energy loss to the ambient by the cooker is the sum of the energy loss by the load and energy loss by the PCM.

2.3 Thermal Analysis of Proposed Design

The thermal performance of PCM-cooker is assessed in terms of charging-discharging time. The time required for complete melting of PCM is taken as charging time with initial temperature of PCM is equal to ambient temperature. The discharging time is defined as the time required for the solidification of PCM during off-sunshine.

A major part of the incidence solar energy is lost to the ambient specifically when the cooker interior is at high temperature. For calculation of energy loss to the ambient, the calculation of the overall heat loss coefficient is required. The overall heat loss coefficient consists of two parts; one is the top heat loss coefficient and the second side and bottom heat loss coefficient. During the charging period, there is incidence solar radiation and the cover of the cooker is open. During discharging period there is no solar radiation and the cover of the cooker is closed. So, top heat loss coefficient has two cases. In charging mode, the top heat loss is from the absorber plate to the surrounding through the glass cover. The top heat loss coefficient of the cooker is calculated by using the analytical procedure proposed by Akhtar and Mullick [13]. In discharging mode, the top heat loss coefficient is calculated by applying energy balance between the absorber plate and first glass cover, first glass cover to second glass cover, second glass cover to the wooden cover and from the wooden cover to ambient. An iterative procedure is used for solving the heat balance equations until the convergence is achieved. For fixed dimensions of the cooker and a constant velocity of wind, the top heat loss coefficient is a function of the plate and ambient temperature. The side heat loss coefficient and bottom heat loss coefficient are calculated with the help of Fourier's law of heat conduction. The overall heat loss coefficient is the sum of top heat loss coefficient and side heat loss coefficient and bottom heat loss coefficient. After calculation of the overall heat loss coefficient as a function of absorber plate temperature, the time taken to charge the PCM, duration of discharging and quantity of PCM required is calculated by applying energy balance on the absorber plate.

The thermal analysis of cooker with 6 kg of Erythritol as phase change material, considering the intensity of solar radiation as 700 W/m^2 is shown in Table 1.

The above results show that it takes approximately 7 h to charge 6 kg PCM. And it takes more than 3 h to solidify PCM along with heating the 4 kg of water up to 100 °C. Without load, complete PCM will solidify in more than 6 h.

Table 1 Result of thermal analysis of proposed design

Pre-charging time (ambient temperature to 118 °C)	1.05 h
Melting time (at 118 °C)	5.90 h
Charging time (pre-charging time + Melting time)	6.95 h
Total discharging time with load (4 kg of water)	3.23 h
Total discharging time without load	6.63 h

3 Experimental Setup and Instrumentation

The cooker was designed, fabricated and experiments were performed in the Department of Mechanical Engineering, Aligarh Muslim University Aligarh, India. The location of Aligarh city is (27.89° N, 78.08° E) and 178 m above mean sea level. Experiments were performed during the month of January to April-2019.

The experimental setup consists of two box-type solar cookers. One cooker is purchased from the market named as standard cooker, shown in Fig. 3. The second cooker is developed in the lab integrated with energy storage material named as PCM-cooker, shown in Fig. 4. Pyranometer (Kipp and Zonen CMP-11Secondary standard) is used for the measurement of solar radiation and T-type thermocouples with data-logger are used for measurement of temperatures.

Fig. 3 Standard cooker

Fig. 4 PCM-cooker

Table 2 Dimension of box-type solar cookers

		Standard cooker	PCM-cooker
Aperture area (m^2)		0.25	0.5084
Reflector mirror area (m^2)	Primary	0.348	0.5476
	Secondary	N.A	0.7972

The standard cooker has only a primary mirror reflector and PCM-cooker has a primary as well as secondary mirror reflector; both the cookers have black painted absorber plate. The dimensions of both the cookers are shown in Table 2.

3.1 Experimental Procedure

The experiments were conducted for: (i) Comparison of PCM-cooker with standard cooker and (ii) Performance evaluation of PCM-cooker.

3.1.1 Comparison of PCM-Cooker with Standard Cooker

In this test, PCM-cooker's performance is compared with a standard cooker to observed the effect of PCM used in the cooker. This test is carried out between 10:00 A.M. and 2:30 P.M. The test is carried out for no-load test and load test with two continuous loads. In standard cooker load is 1.5 L of water and the PCM-cooker load is 3 L of water in two and four pots, respectively.

3.1.2 Performance Evaluation of PCM-Cooker

The performance of PCM-cooker is evaluated with maximum load of 4 L of water. This test is conducted from 10:00 to 20:00 h. In this test two loads are placed; one in the morning which completed by the time of 13:00 h after that only charging is carried out of the PCM up to 15:40 h after 15:40 h PCM-cooker is placed inside the room and the second load is placed about 15:50 h and the test is carried out up to 20:00 h.

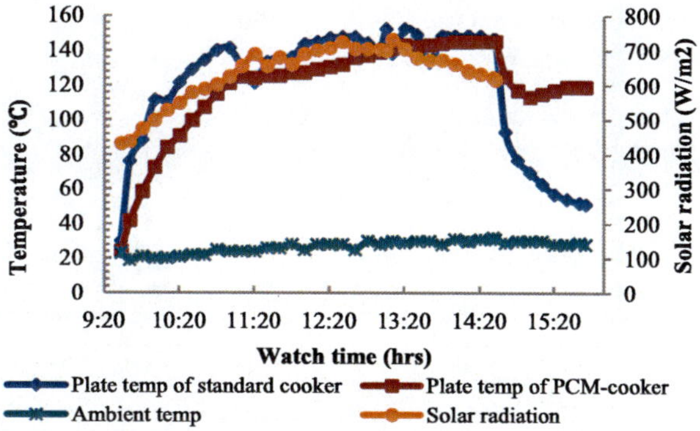

Fig. 5 Variation of plate temperature, ambient temperature and solar intensity with time of the day in no-load test on 23-February-2019

3.2 Experimental Result and Discussion

3.2.1 Comparison of Performance of PCM-Cooker with Standard Cooker in No-Load Test

Figure 5 shows the variation of plate temperature, ambient temperature and intensity of solar radiation with time of the day for standard cooker and PCM-cooker in no-load test performed on 23-February-2019. The test starts at 9:33 h, the cover of the cooker is closed at 14:32 h and cookers placed inside the room and test continue till 15:45 h.

From the experiment, it is observed that the absorber plate of standard cooker attained maximum temperature, 152.5 °C, at 13:20 h and the absorber plate of PCM-cooker attained maximum temperature, 145.1 °C, at 14:20 h.

3.2.2 Comparison of Performance of PCM-Cooker with Standard Cooker in Load Test

Figure 6 shows the variation of plate temperature, water temperature and ambient temperature with time of the day in two continuous loads test conducted on 26-March-2019. The first load placed at 10:27 h in both the cookers. The loads on both the cookers are replaced with the second load at 13:17 h after attained maximum temperature. The test continues till second load in both the cookers attained maximum temperature.

From the experiments, it is observed that in standard cooker the first load attained maximum temperature, 96 °C, at 12:59 h. Then first load is replaced with second load at 13:17 h and the second load attained maximum temperature, 94.2 °C, at 15:33 h. In PCM-cooker first load attained maximum temperature, 95.9 °C, at 13:03 h, now first load is replaced with second load at 13:17 h and second load attained maximum temperature, 98.9 °C, at 15:33 h.

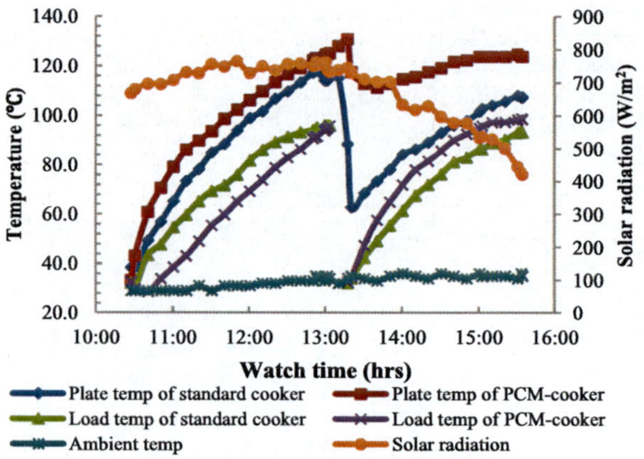

Fig. 6 Variation of plate temperature, load temperature, ambient temperature and solar intensity for two continuous load test on standard cooker and PCM-cooker on 26-March-2019

3.2.3 Performance of PCM-Cooker with Primary and Additional Reflector Without Load

Figure 7 shows the variation of plate temperature and ambient temperature with time of the day for PCM-cooker with additional reflector in no-load test performed on 01-April-2019. The test started at 10:08 h, the cover of the cooker is closed at 15:47 h and cooker placed inside the room. The observations are continued till 20:50 h.

From the experiment, it is observed that during charging mode the absorber plate attained maximum temperature, 162.1 °C, at 12:26 h. During off-sunshine mode the absorber plate temperature drop to 109.9 °C at 16:40 h, then the absorber plate temperature starts increasing and attained 118.9 °C at 17:20 h. After that, the absorber plate temperature continuously decreases and the test is concluded at 20:50 h with absorber plate temperature of 90.0 °C.

Discussion on variation of plate temperature after solar radiation cut-off

When solar radiation is cut-off and the cooker is placed inside the room the plate temperature falls below the melting temperature of PCM and after some time plate temperature starts increasing and reach at melting temperature of PCM. This

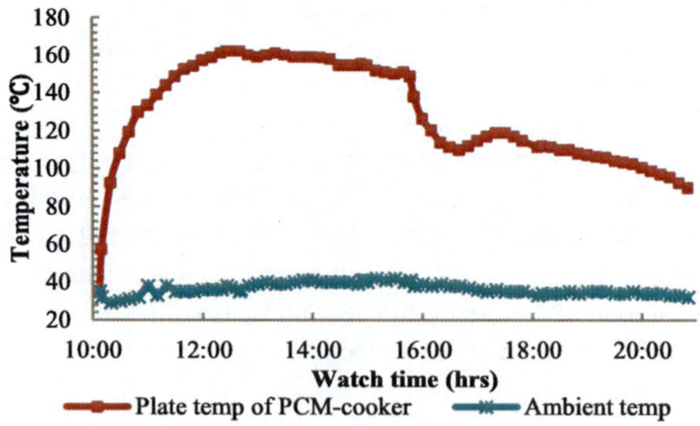

Fig. 7 Variation of plate temperature with time of the day in no-load test on PCM-cooker on 01-April-2019

happened because when solar radiation is cut-off by closing its cover the overall heat loss coefficient decreases to almost half.

During the charging mode of PCM-cooker the top heat loss is through top glass covers to the atmosphere. This heat loss depends on the temperature difference between glass temperature and ambient temperature. Due to heat balance between two surfaces the heat loss from plate to inner glass is equal to heat loss from inner glass to outer glass and this also equal to heat loss from top glass to atmosphere. After solar radiation is cut-off by the closing cover, the heat loss from cooker is from outer surface of the wooden cover to atmosphere, this heat loss depends on temperature difference between temperature of wooden cover and atmosphere. And this heat loss also depends on the temperature difference between plate and inner glass. Since heat loss becomes almost half after closing the cover, the temperature difference between plate and inner glass decreases after solar cut-off. Hence, the temperature of inner glass increases. For this, some amount of heat is required. Also, the air medium requires some amount of heat to increase its temperature. This heat is supplied by the PCM. When the PCM starts supplying the heat, the temperature of PCM and the plate temperature start decreasing. Since the heat is supplied by the upper layer of the PCM adjacent to the absorber plate the top layer of PCM will initially solidify. As the heat capacity of PCM in a solid-state is low, the plate temperature goes below the melting point of PCM. Once equilibrium between plate and inner glass is achieved at a lower temperature the layer of solid PCM below the absorber plate will start taking heat from the liquid PCM below the solid layer of PCM. When solid layer gets heat from liquid PCM, it also supplies heat to the plate and the plate temperature increases and reaches the melting point of PCM. Thus, the absorber plate temperature increases and equilibrium at a higher temperature (melting temperature of PCM) is achieved and after that plate temperature starts decreasing gradually due to the continuous heat loss.

3.2.4 Performance of PCM-Cooker with Primary and Additional Reflector with Load

Figure 8 shows the variation of plate temperature, water temperature and ambient temperature with time of the day with the use of additional reflector performed on 27-April-2019. The test is performed with two loads. The first load of 4 kg water kept at 10:10 h in PCM-cooker, which is removed from the cooker after the maximum temperature, is achieved at 12:51 h and then the cooker is left for charging of PCM. The charging is continued till 15:43 h, then the cover is closed and the cooker is placed inside the room. After that, the second load of 2 kg water is kept in the PCM-cooker and observations are continued till 20:20 h.

From the experiment, it is observed that during charging mode the first load of 4 kg water attained maximum temperature, 97.1 °C, at 12:51 h, now first load is removed from the cooker at 12:51 h. After charging of PCM, the cover of the cooker is closed and placed inside the room. The second load of 2 kg water is placed at 15:51 h in the

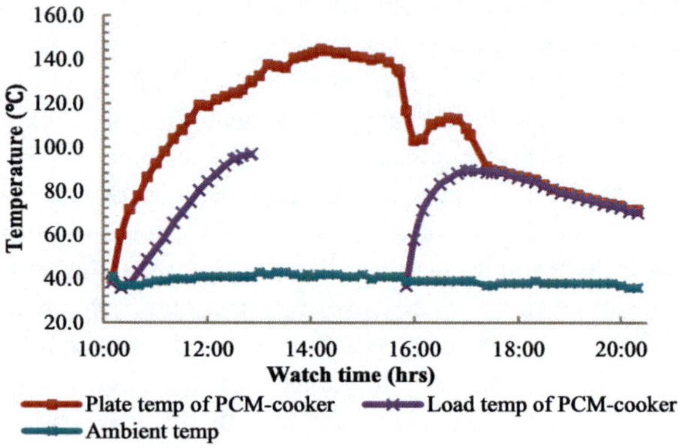

Fig. 8 Variation of plate temperature, load temperature and ambient temperature with additional reflectors with time of the day in load test on PCM-cooker on 27-April-2019

PCM-cooker. The second load attained temperature, 89.7 °C, at 17:04 h. The test is continued up to 20:20 h with load (water) temperature achieved is 70.1 °C.

Temperature of second load (during off-sunshine hours) in PCM-cooker at different load conditions

The cooker is charged during the sunshine hours and then placed in the room with the cover of the cooker in closed condition. The performance of the cooker is evaluated in terms of the rise in temperature in different load conditions. The variation of 2 and 4 kg of water on different days is shown in Fig. 9.

It can be seen from the plot that the temperatures achieved during off-sunshine hours for 2 kg and 4 kg of water are 92 °C and 85 °C, respectively.

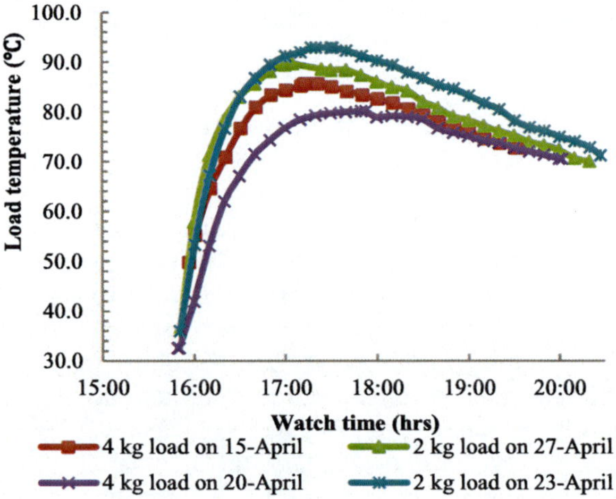

Fig. 9 Variation of load temperature during off-sunshine hours on 15-April, 20-April, 23-April, and 27-April

3.3 Cooking of Foods in Standard Cooker and PCM-Cooker

The experimental study shows that the first load placed in the PCM-cooker during sunshine hour attained the boiling point of water, but when the second load is placed in cooker it does not achieve boiling temperature of water. The maximum temperature achieved by the second load is 85.6 °C and 92.2 °C when the cooker is loaded with full load (4 kg) and half load (2 kg), respectively. These results show that the second load does not attain the boiling temperature of water, which creates doubt regarding the cooking of the meal in the late hour of the day. A matter of curiosity, to clarify this doubt the PCM-cooker is tested by cooking two meals on the same day.

On 29-April and 30-April load tests with food were performed by placing rice and pulse in both the cookers. On 29-April 550 g load (100 g rice + 200 g water in one pot and 50 g pulse + 200 g water in the second pot) on the standard cooker and 1.1 kg load (200 g rice + 400 g water in one pot and 100 g pulse + 400 g water in the second pot) on PCM-cooker are placed at 10:00 h, which are prepared/cooked by 12:00 h. After that PCM-cooker is left for charging up to 15:50 h. Then the charged cooker is placed inside the room and then second meal of 1.1 kg is placed on PCM-cooker before 16:00 h and we get well cooked and hot food around 19:30 h.

Similar test again conducted on 30-April with increased loads: 1.3 kg load (200 g rice + 600 g water in one pot and 100 g pulse + 400 g water in the second pot) on the standard cooker and 2.6 kg load (400 g rice + 1200 g water in one pot and 200 g pulse + 800 g water in the second pot) on PCM-cooker at 10:00 h, the food was cooked by 12:40 h. Then the PCM-cooker is left for charging up to 15:50 h and after charging the PCM-cooker is placed inside the room and then second meal 1.95 kg load (300 g rice + 900 g water in one pot and 150 g pulse + 600 g water in the second pot) is placed on PCM-cooker before 16:00 h and at 19:30 h well cooked and hot food was ready.

4 Conclusions

On clear days with high intensity of incoming solar radiation, the absorber plate temperature of the standard cooker is higher than the temperature of PCM-cooker. This is due to the fact that when the absorber plate temperature of the PCM-cooker reaches higher than the melting point of the phase change material, it transfers the heat to the PCM and the rise in temperature is less. Although the overall efficiency of both the cookers is approximately same, the standard cooker is unable to cook the food in the late hours of the day. The cooker with phase change material can be used for cooking the food during the daytime as well as during the late hours of the day. The use of an additional (secondary) reflector during the charging time is beneficial for cooking the food when the solar radiation is moderate. During off-sunshine hours, the second load (second meal of the day) attains a temperature around 90 °C in the PCM-cooker. As the cover of the cooker is closed during the off-sunshine hours, the heat loss decreases and the temperature drop is slow and the temperature of the load is approximately 70 °C at 20:00 h. The cooking of rice and pulse is possible during the off-sunshine hours of the day using the PCM-cooker.

References

1. Mullick SC, Kandpal TC, Saxena AK (1987) Thermal test procedure for box-type solar cookers. Sol Energy 39(4):353–360
2. Sharma A, Chen CR, Murty VVS, Shukla A (2009) Solar cooker with latent heat storage systems: a review. Renew Sustain Energy Rev 13(6–7):1599–1605
3. Telkes MRE (1949) Storing solar heat in chemicals -a report on the Dover house. Heat Vent 46(11):80–86
4. Ramadan MRI, Aboul-Enein S, El-Sebaii AA (1988) A model of an improved low cost-indoor-solar-cooker in Tanta. Sol Wind Technol 5(4):387–393
5. Domanski R, El-Sebaii AA, Jaworski M (1995) Cooking during off-sunshine hours using PCMs as storage media. Energy 20(7):607–616
6. Buddhi D, Sahoo LK (1997) Solar cooker with latent heat storage: design and experimental testing. Energy Convers Mgmt 38(5):493–498
7. Vigneswaran VS, Kumaresan G, Sudhakar P, Santosh R (2017) Performance evaluation of solar box cooker assisted with latent heat energy storage system for cooking application. IOP Conf Ser Earth Environ Sci 67(1)
8. Sharma SD, Buddhi D, Sawhney RL, Sharma A (2000) Design, development and performance evaluation of a latent heat storage unit for evening cooking in a solar cooker. Energy Convers Manage 41(14):1497–1508
9. Buddhi D, Sharma SD, Sharma A (2003) Thermal performance evaluation of a latent heat storage unit for late evening cooking in a solar cooker having three reflectors. Energy Convers Manage 44(6):809–817
10. Reddy SM, Sandeep V, Sreekanth M, Daniel J (2017) Development and testing of a solar cooker with thermal energy storage system. Int Energy J 17:185–192
11. Sharma SD, Iwata T, Kitano H, Sagara K (2005) Thermal performance of a solar cooker based on an evacuated tube solar collector with a PCM storage unit. Sol Energy 78(3):416–426
12. BIS Standards on solar cookers, IS13429 part II components, Bureau of Indian Standards, Manak Bhavan, New Delhi-110002
13. Akhtar N, Mullick SC (2007) Computation of glass-cover temperatures and top heat loss coefficient of flat-plate solar collectors with double glazing. Energy 32(7):1067–1074

A Proposal for Harnessing an Alternative Source of Renewable Energy

Nabanita Paul

1 Introduction

Renewable energy, as the name suggests, is obtained from special types of sources which have high contributing nature and recovered into original state within a time span comparable to human time scale. In the twenty-first century, civilization is heading for a techno-savvy world, and to meet the burgeoning demands of the global village, it requires an almost constant supply of huge amount of energy. The conventional source of energy is primarily fossil fuel, which falls under non-renewable category as its developing time is much beyond human life time. Nowadays, technology is built up to access energy from several available renewable energy resources. Solar energy being a perennial source is considered as the best alternative source of energy. Next comes the case of wind energy. From the dawn of civilization, these energy sources made the planet worth for living. Energy being constant from the beginning of creation, solar and wind energies were continuously flowing, but in order to harness these energies, suitable engines are required. Using solar panel or turbine, people brought those green energies at their control and applied ever-flowing treasure according to their need. Nowadays, rooftop solar panels are connected to electric grid and made our homes self-sufficient [1]. The grid-connected system allows storage of energies for future use either on a daily basis or on interseasonal duration. From the dawn of creation, wind is blowing along with its adhered energy but intervention of turbine made people able to use this clean energy according to their own needs. Mechanical energy associated with natural flow of water is accessed as electric energy by installing suitable engines. These pollution-free green energies are available until the sun shines, the wind blows, and the water flows. Thermal energy residing inside earth's interior as geothermal energy though less common is also harnessed to avail electric energy. Temperature difference between the earth's

N. Paul (✉)
Department of Chemistry, A.P.C. Roy Government College, Siliguri, India

S. K. Ghosh et al. (eds.), *Advances in Thermal Engineering, Manufacturing, and Production Management*, Lecture Notes in Mechanical Engineering,
https://doi.org/10.1007/978-981-16-2347-9_16

crust and interior allows the passage of thermal energy towards the earth's surface. Radioactive disintegration in the interior part of the earth liberates energy mostly in the form of heat. This, when localized over a zone, is able to melt rocks nearby; during volcanic eruption, molten mass comes out in the form of lava. At the present age, drilling is done into volcano to access thermal energy of earth's interior. Steams coming from underground hot water resources are utilized for producing electricity in geothermal power plant. In ocean thermal energy conversion (OTEC), the temperature difference between surface of the ocean and deep inside the ocean is utilized in a power plant for transforming thermal energy as electric energy. Almost three-fourth of world is covered with ocean, so harnessing this energy has the potential to provide maximum benefit. Landfill gas mainly contains methane, and this gas is obtained by microbial decomposition of organic wastages [2]. In a landfill, once methane gas is generated, it will be available for a prolonged period. Moving turbine within the flow of this gas, electric generators are constructed. This saves the present world from two dimensions. Along with mitigating demands of energy, it reduces green-house gases from our environment. Burgeoning population dumped more garbage in landfill and induction of suitable engine assures steady supply of methane, which is a main component of natural gas. In municipal solid-waste management, municipal solid-waste management generators are introduced to open a new sector for developing a source of renewable energy and curbing potential environmental danger. In short, basic target is spread to avail energy from all possible sources. Standing at this crucial juncture, this article is tossing a plan which is supposed to offer little relief to people in mitigating impending energy crisis.

2 Underlying Basis of the Proposal

Energy of *strained molecule is a probable source of alternative energy.* Reversible photo-induced cyclization–decyclization reaction forms the underlying basis of the current proposal [3]. Photochemical cyclization of two ethylene molecules by UV-B radiation (310–320 nm) produces cyclobutane molecule. Cyclic dimer on irradiation with 270–280 nm photon breaks up to reproduce ethylene [4] (Scheme 1).

Despite the bonds between all carbons of cyclobutane in saturated state, it is an extremely strained molecule. During cyclization, orbitals have to undergo a wide deformation from original orientation. In addition, overlap of orbitals takes place in a very poor manner. All these together are responsible for developing an inherent

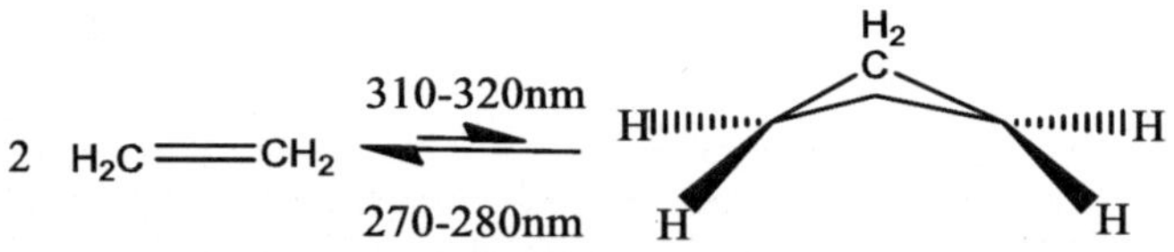

Scheme 1 Photo-induced cyclization-decyclization cycle

strain in the molecule [5]. So cyclobutane has a natural tendency to breach connective linkage between adjacent carbons in order to reproduce initial ethylene molecule. This decyclization is also a photochemically feasible exothermic transformation [6]. Thermal energy liberated from strained molecule during the backward reaction is likely to be utilized in storing for its convenient use. It is this reversibility of the cyclization–decyclization process that appears as the essence of current proposal and frames the basis of the renewable source of energy [7].

2.1 Materials and Arrangements

Material required to materialize the scheme is bioethylene which is likely to be generated by catalytic dehydration of bioethanol. Bioethylene obtained from catalytic dehydration of bioethanol when exposed to 310–320 nm radiation undergoes efficient cyclic dimerization. Cyclobutane on exposure to photochemical excitation at 270–280 nm promotes decyclization-producing precursor ethylene and releases considerable amount of energy in the form of heat. Technology should be devised to continue the cycle of ethylene consumption and reproduction. Once a suitable technology is introduced, it would possibly turn as an efficient machine. Bioethylene required to materialize the scheme is planned to produce from biomass of ripened banana skin, sugarcane, etc., and the device needed to set up to avail energy is costly. However, one time installation would suffice the requirement.

2.2 Biofuel: an Alternative of Petrochemical Industry

With urban lifestyle, production of biofuel from biomass opens a new avenue for production of energy. Recently in Punjab, biomass made from stubble of both paddy and wheat crops shows a new dimension in developing new source of renewable energy [8]. Bioproduction of bioethanol from sugarcane, corn and cellulose-based materials (lignocelluloses) is a challenging task for modern scientists to keep the transportation system running. Yeast-catalysed fermentation process of sugar obtained from sugarcane is materialized widely in Brazil for production of bioethanol. Leaves and wastages left after sugar extraction (bagasse) are used for heating and other necessary actions. India being a high sugarcane-producing country has a possibility of developing bioethanol production plant. At present, bioethylene plant in Brazil and India produces ~ 0.3% of global ethylene capacity. However, United States utilizes corn for producing bioethanol, ~ 63% of the global production. The production cost of sugarcane-bioethylene in India is not very high—~ USD 1200/t in comparison with global average of petrochemical ethylene which is ~USD 1,100/t. India Glycols Limited, established in India in 1989, produces major bioethylene from molasses. Bioethylene is considered as a green fuel since it reduces production of 40% greenhouse gases (GHG).

2.3 *Theoretical Estimation of Energy*

Ethanol-to-ethylene (ETE) technology reports that one tonne bioethylene requires 1.74 tonnes of bioethanol and these conversion demands 1.6 GJ per tonne of bioethylene [9].

2.4 *Banana Peel and Bioethylene*

Yellow skin of banana is enriched with ethylene. From the thrown banana peel, one can easily extract bioethylene. People should be made aware of the potential of banana skin, and policies should be introduced for proper disposal of garbage [10].

3 Results and Discussion

A scheme is planned to harness renewable energy, in which an arrangement is made with two chambers made up of aluminium sheet. Side walls of the first tank are set with many UV-B lamps, and these lamps are connected with solar panels. Two types of UV lamps are set. One type would emit radiation near 310–320 nm range, while another group of lamps would emit 270–280 nm radiation. Inner walls of top and bottom layers are coated with black or navy blue colour and highly shielded. Through the inlet of the first tank, bioethylene is fed. This arrangement is connected with a thermo-electric generator (TEG) which converts thermal energy into electric energy. Batteries are charged with this generated electricity and stored for future use.

3.1 *Working Principle*

Bioethylene is first exposed to irradiation with 310 nm photons coming from UV-B lamps. These lamps are lighted using solar energy. Photochemically excited ethylene initiates cyclic dimerization and produces cyclobutane [11]. A sensor is kept inside the arrangement which allows 300 nm lamps glowing until ethylene concentration is above a critical level. When the concentration falls below critical level, 300 nm lamps get disconnected, but 275 nm lamps start glowing. This radiation would initiate breakage of cyclobutane [12]. Cyclobutane to ethylene conversion is exothermic. Liberated heat is allowed to enter Seebeck generator which transforms thermal energy into electric energy, and it is utilized to charge different types of batteries for future use. This procedure is run with either a critical concentration of ethylene or more than this. Decyclization of cyclobutane though regenerate ethylene, yet with prolonged use when concentration of ethylene falls below critical concentration, the machine would send a message to user either by blinking indicator light or by producing typical sound. Ethylene tank is refilled with required amount of starting materials for the next spell of reaction. However, this plan would appear as an excellent technology if UV-B radiations coming towards the earth's surface be can be trapped for triggering photo-induced reversible conversion [13].

3.2 Possible Application Sites

As a start-up phase, this technology is planned to employ for domestic purposes—for running home appliances. Second phase can be applied on street light and small sectors. Other than transportation, small private cars can use it for lightening and other purposes. In the next phase, it is planned to charge inverters; the energy released is stored inside battery for future use.

4 Conclusion

People engaged themselves in harnessing energy from all possible sources to continue the journey started from the dawn of civilization. Following same target, the current proposal is awaiting the suitable technology to intervene so that it can touch the taste of reality. The proposal discussed in the article talks about a scheme which is likely to offer energy for a prolonged period once the arrangement is materialized. Ethylene is the main feedstock of the machine; if this raw material is targeted to acquire from solid waste management sector, then the venture would provide dual impact on society. Release from strained state is a premier target of any system. In ethylene–cyclobutane conversion, as the latter breakdown its structure, releases energy which must be stored and utilized on requirements like running small home appliances, lightening small sectors, using water geyser and charging cell phone batteries. The most advantageous fact associated with the current proposal is recycling of ethylene during transformation. This specific feature would provide a special character to the technology in harnessing renewable energy. Price required to implement the current proposal would be expected to be high. But with initial investment on necessary arrangement, it would run for a considerable length of time since the basis of the technology is connected with a reversible reaction initiated by same agitating agency. Introduction of this technology would curtail our dependency on fossil fuel. At this moment, it is a sheer proposal but anticipated energy that it is likely to give in the future is expected to be high. The article discussed a proposal which is supposed to be a potent alternative energy resource, and hence, the current proposal offers a new dimension to mitigate the cutting-edge problem of energy.

Acknowledgements The author is indebted to Dr. T K Dasgupta for his classroom lectures which help in building the article.

References

1. Nwaigwe KN, Mutabilwa P, Dintwa E (2019) An overview of solar power (PV systems) integration into electricity grids. Mater Sci Energy Technol 2(3):629–633
2. Themelis NJ (2007) Methane generation in landfills. Renew Energy 2(7):1243–1257
3. Woodward RB, Hoffmann R (1971) The conservation of orbital symmetry. Verlag Chemie (Academic Press), Germany, pp 113–70
4. Normann ROC, Coxon JM (2012) Principles of organic synthesis. CRC Press, India, pp 280–283
5. Claiden J, Greeves N, Warren S, Wothers P (2001) Organic chemistry. Oxford University Press, New York, pp 905–909
6. Carey FA, Sundberg RJ (2012) Advanced organic chemistry. Springer, India, pp 888–892
7. Fukui K (1997) Frontier orbitals and reaction path. In: Fukui K, Fujimoto H (ed) World's scientific series in 20th century chemistry, vol 7. World scientific, Singapore, pp 203–207
8. Kumar P, Kumar S, Joshi L (2015) Alternative use of crop stubble. In: Socioeconomic and environmental implications of agricultural residue burning, Springer Briefs in environmental science. Springer, New Delhi, pp 69–89
9. Haro P, Ollero P, Trippe P (2013) Technoeconomic assessment of potential processes for bioethylene production. Fuel Process Technol 114:35–48
10. Chillet M, Bellaire LL, Hubert O, Mbeguie DM (2008) Measurement of ethylene production during banana ripening. Fruits 63:253–254
11. Coffey P, Jug K (1974) Semiempirical MO calculations on symmetry governed reactions. Theor Chim Acta 34(3):213–224
12. Jug K, Kruger HW (1979) A comparison between ethylene dimerization and Diels-Alder reaction. Theor Chim Acta 52:19–26
13. Diara C, Vernieri P, Gorini S, Lercari HP (2007) Damaging and non-damaging effects of UV-B in South American bean genotypes under controlled environment condition. Adv Hort Sci 21(1):28–34

Energy-Efficient Cooling Scheme of Power Transformer: An Innovative Approach Using Solar and Waste Heat Energy Technology

Tapash Kr. Das, Ayan Banik, Surajit Chattopadhyay, and Arabinda Das

1 Introduction

A transformer can be defined as the most integral and influential device in applied power engineering. Power system optimization and reliability can be achieved by maintaining the transformer in its normal working. Transformers are highly robust and designed accordingly to work continuously under safe parameters. Transformers serve as a tie-line between GENCO, TRANSCO, and DISCOMS. The transformer is employed in different geographic locations and is exposed to vast temperature variation. As a result, the maximum hotspot winding temperature of the transformer reflects its standard aging limit. Continuous high temperature inside the transformer may result in quick aging and often require replacement, which is uneconomical and problematic.

2 Literature Survey

M. Sorgic et al. (2010) have studied and presented a case study on oil-directed cooling of power transformers versus oil forced [1]. An overview of water-based lithium bromide (LiBr) absorption cooling mechanism [2] has been introduced in (2002) by

T. Kr. Das · S. Chattopadhyay
Department of Electrical Engineering, GKCIET (Under MHRD, Government of India), Malda, India

A. Banik (✉)
Department of Electrical Engineering, National Institute of Technical Teachers' Training & Research, Kolkata, India

A. Das
Department of Electrical Engineering, Jadavpur University, Kolkata, India

S. K. Ghosh et al. (eds.), *Advances in Thermal Engineering, Manufacturing, and Production Management*, Lecture Notes in Mechanical Engineering,
https://doi.org/10.1007/978-981-16-2347-9_17

Castro et al. In (1995), Aphornratana et al. have executed a detail thermodynamic-based evaluation associated with cycle absorption refrigeration system (VARCs) [3]. Daikin Industries, Ltd. [4] in (2002) has successfully awarded by the Patent Authority of United States to develop specially design refrigerating equipment using air-coolant absorption technique, which proves a milestone for industrial application of VAS. A double-effect air conditioner heater (DEACH) [5] has been developed by De Vuono et al. A unique, comprehensive review work has been prepared by D.H. Shroffet et al. (1985) on paper aging in power transformers [6]. Later, a detailed report was presented on power transformer aging and life extension by Muhammad Arshad et al. [7]. Lin Chaohui et al. (2017) have introduced a new method for aging calculation and prediction of life span for oil-type power transformers [8]. Activation energy tool has been invented to estimate lifetime for large rating transformer [9] by P. V. Notingher et al. in (2011). M. K. C. Martin et al. have proposed all-new micro-technology-based chemical heat pumps. Author in their previous work has studied and analyzed different nature of PV microgrid under various physical conditions and faults using a complex mathematical tool which includes FFT and DFT. Recently, authors have been working on waste heat recovery technology and the application of thin-film solar PV for a better tomorrow and to promote green energy [10–13].

3 Necessity of Cooling for Transformer

The method for cooling of transformers means keeping the generated temperature within a safe limit. During different power flow stages, losses appear as heat and dissipated. The losses in the form of heat energy increase proportionate with transformer loading. The factors that accelerate heat generation in the transformer are core and load losses. The winding loss plays the most significant role in unwanted heat generation as compared to other losses. Transformer operations lead to several faults, i.e., internal and external faults. An internal fault is considered to be more severe and complicated than any other fault. The basis of internal faults in a transformer is due to overheating, which gives rise to acute faults, i.e., *insulation breakdown, turn to turn fault, phase-to-ground, and lastly earth fault*.

3.1 *Mathematical Form to Determine Transformer Temperature Level*

$$\Delta t = (\Sigma PL/As)0.833 \tag{1}$$

As stated in above equation,

Temperature rise in degree Celsius (°C) = $\mathbf{\Delta t}$.

Total losses in transformer/equivalent heat loss in mW = $\mathbf{\Sigma PL}$.

Effective surface area of main tank in cm^2 = **As**.

Table 1 Conventional transformer cooling techniques

1. Air cooling (dry-type trans.)	1. Oil cooling (oil-immersed trans.)
• *Air natural (AN)* • *Air blast (AB)*	• *Oil natural air natural* • *Oil natural air forced* • *Oil forced air natural* • *Oil forced air forced*
2. Oil and water cooling • *Oil natural water forced (ONWF)* • *Oil forced water forced (OFWF)*	**2. Special cooling** • *Vapor compression system (VCS)* • *Ammonia-based refrigeration systems*

4 Evolution in Transformer Cooling Techniques

In 1886, William Stanley built the first reliable commercial transformer prototype, which had further been developed in the early 90s, and since then, several upgradations have been made till now [14]. The unwanted heat generated in the transformer due to losses in core and windings inside the tank may result in the downfall of efficiency. To minimize these losses, different cooling arrangements have been considered. Popular methods of cooling for utility transformers have been categorized in accordance with coolant types:

- *Air natural and forced cooling (ANFC).*
- *Air and oil cooling (AOC).*
- *Oil and water cooling (OWC).*
- *Special cooling (SC)* (Table 1).

5 Design and Fabrication of the Proposed System

The very first concept of refrigeration had been started with the introduction of the vapor ammonia system (VAS). With the upgradation of time, technological advancement has developed many new refrigerating technologies, which further fails to achieve sustainable goals. A modernized VAS system may prove beneficial in wide aspects during this crisis hour, including domestic and large industrial applications. The VAS stands superior over any other existing refrigeration system as thermal power is being utilized, rather than conventional power sources. The compressor's role has totally been upgraded by a unit consisting of four major components: generator, ammonia-absorber, servo pump, and pressure release valve. The waste heat of transformers has been employed for the heating up strong compound of aqua NH3 within the specially featured generator. Throughout heat transferring action, NH3 coolant has been forced out under heavy pressure and transformed as a weak compound, i.e., aqua-NH_3, into the generator. The analyzer has been built as an integral part of the generator and functioning to remove unwanted water particles, which have been further revised by a rectifier (dehydrator). On the contrary, a compound of weak NH_3 has allowed for returning into absorber under light pressure, after passing through the pressure release valve. Then, under very high pressure, the generator's NH_3 fumes have converted into droplets inside the condenser under excessive pressure liquid NH_3. The above solution needs to be passed through the expansion valve to the evaporator for ample cooling. The chilled coolant has been circulated by the servo pump all across the tank of the transformer through surrounded copper capillaries and completed cycle finally. Thermodynamic properties in each phase of the cycle have varied on corresponding refrigerant flow rates and largely depend on the input parameters. This cycle operates smoothly until smooth heat exchanging is ensured. The above-discussed parameters are listed as follows:

T_a: Absorber temperature in Celsius (°C)
T_g: Generator temperature in Celsius (°C)
T_c: temp. within condenser in Celsius (°C)
T_e: temp. inside the evaporator in Celsius (°C)
Q_e: Refrigeration load in kilo-watt (kW)

In the above set, pressure drop in individual components is neglected to avoid mathematical complexity. The theoretical coefficient of performance of vapor absorption system (VAR) is (Fig. 1):

$$\textbf{Theoretical COP} = \frac{Tg - Tc}{Tg} \times \frac{Te}{Tc - Te} \tag{2}$$

An all-new transformer cooling system is introduced and developed with proper implementation. Special emphasis has been given to design, and a complete outline of the system is presented in a simplified manner. A thin-film PV solar module has been used for secondary power generation, which helps to charge lithium-ion battery,

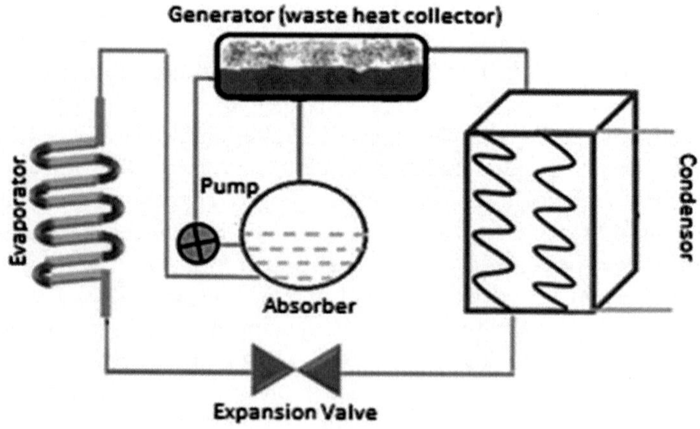

Fig. 1 Simplified layout of vapor absorption cycle (VAS)

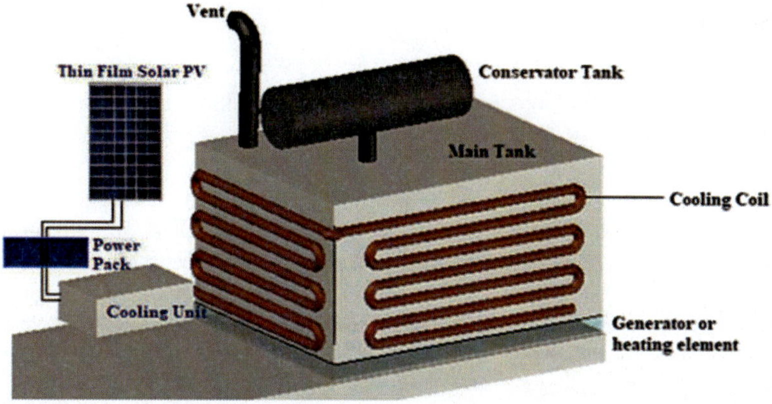

Fig. 2 Design and fabrication of proposed model

placed in the power pack with an additional specialized charge controller. Pumps are employed for uniform circulation of the coolant through a dedicated copper duct distributed over the power transformer's main tank. Precision temperature monitoring and protective devices have been incorporated within the power pack. A special design valve has been used to regulate the coolant's discharge rate based on the temperature variation. Moreover, computer process control has been adopted by using Arduino-based integrated thermal sensors for real-time data transfer to the remote control room to achieve better monitoring. If something unusual happens, it can be tripping the whole cooling system without disturbing the transformer's normal operation. It also helps synchronize bilateral operation between thin-film PV, charging, and discharging of the lithium-ion battery as shown in Fig. 2.

6 Algorithm

- ***Generator:****The liquefied ammonia solution presenting the generator is being heated by the external source of heat, i.e., transformer waste heat.*
- ***Analyzer:****It is mounted over the generator to ensure proper healing to liquid ammonia.*
- ***Condenser:****Refrigerant, i.e., liquefied ammonia solution, enters the condenser at high pressure and temperature and gets condensed.*
- ***Expansion Valve:*** *It helps in the reduction of high temperature and pressure.*
- ***Evaporator:****Final cooling is achieved here; using a copper duct, the cooling is further distributed over the transformer's surface.*
- ***Absorber:****It consists of a weak refrigerant solution for the absorption of ammonia; water has been used as an absorbent.*

7 Future Scope

Along with several global agencies, the United Nations have set their new vision or objective to transform the world energy requirement in strategic partnership with the government of various nations. The above-discussed technology has enormous potential for future implementation and toward the way of promoting sustainable development and waste to energy conversion. It has been appraised that energy requirement statistics indicating a peak in the graph and annually have been projected about 778 η-J and expected to be reached in the year 2035. These energy deficiencies can only be fulfilled by emphasizing waste to energy and other green technologies. Waste heat utilization using vapor absorption system (VAS) can be modified and implemented in various sectors all over the globe [14].

- Power generation or cooling in automobiles using exhaust gas.
- Industrial waste heat utilization for achieving cooling in cold storage.
- Waste heat recovery of thermal power plant flue gas to achieve cooling in the control room and cooling tower.
- Ice plants may utilize the waste heat from nearby industries in order to attain very low cryogenic temperature.

7.1 Salient Features

- Moving part absent, i.e., no wear and tear with simple and compact construction.
- Silent operation and does not involve high initial cost and maintenance.
- Reduced vaporization pressure and temperature, i.e., low risk of an accident.
- Energy-efficient system thus load variation does not affect performance.
- Self-regulated system and Arduino kit helps in real-time monitoring of temperature
- As it does not require electric power, thus, power failure does not affect its operation.

7.2 Future Challenges

- The system operator must have expertise and skilled in VAS.
- Large floor space area is required, and system cooling can be achieved slowly.
- The refrigerant used is lightly toxic in nature so adequate precaution needs to be adopted.

8 Conclusion

An exhaustive case study of the literature on vapor absorption systems (VAS) has been carried out over a wide range of energy retreat mechanisms. With regard to vapor ammonia cycles (VACs), it has been found that very few studies have often been carried out on large capacity systems from design approaches and applications. In order to improve the average life of transformers, it becomes necessary to ensure an effective cooling system, which would be beneficial to maintain internal thermal stability. Natural and forced cooling has insufficient capacity to deal with adverse atmospheric conditions for higher rating transformers. In the above study, an attempt has been made to introduce and demonstrate an all-new transformer cooling technique using vapor absorption system (VAS), powered by solar and waste heat. The proposed effective idea of cooling has entire control over the transformer winding temperature within the safe limit so that it might not affection oil-paper insulation. The life span of a large power transformer with normal operation consistency can be extended with these optimized cooling arrangements. This unique system has enormous potential which can be explored and utilized to promote system efficiency, including several other dimensions like understanding ever-increasing power demand and environmental restrictions. An Arduino-operated real-time data tracker and thermal sensors have been incorporated to monitor transformer thermic levels. Researchers are continuing to bring the vapor absorption system in a portable hybrid form, with ultra-modern energy utilization features over any other existing refrigeration and air conditioning system.

Glossary

PV: Photovoltaics.

CPC: Computer process control.

VAS: Vapor ammonia system.

VCS: Vapor compression system.

ACR: Automatic compression system.

AIR: Advance industrial refrigeration.

VARC: Vapor absorption refrigeration cycles.

Acknowledgements Authors sincerely acknowledge the support received from Ghani Khan Chowdhury Institute of Engineering and Technology (A CFTI under MHRD, Govt. of India) West Bengal and Industrial Training Institute (ITI) in this regard.

References

1. Sorgic M, Radakovic Z (2010) Oil forced versus oil directed cooling of power transformers. IEEE Trans Power Del 25(4):2590–2598
2. Castro J, Leal L, Pozo P, Oliet PSC (2002) Development and performance of an air-cooled water-LiBr absorption cooling machine. In: Universität Politecnica de Catalunya; International Forum on Renewable Energies

3. Aphornratana S, Eames IW (1995) Thermodynamic analysis of absorption refrigeration cycles using the second law of thermodynamics method. Int J Refrigeration
4. Daikin Industries, Ltd., U.S. Patent 6109060 (2000) AirCooled Absorption Type Refrigerating Apparatus
5. De Vuono AC, Hanna WT, Osborne RL, Ball DA (1991) Development of a Double-effect Air-Conditioner Heater (DEACH)
6. Shroff DH et al (1985) A review of paper ageing in power transformers. In: IEEE proceedings, vol 132, Pt. C, no 6
7. Arshad M, Islam SM, Khalique A (2004) Power transformer aging and life extension. In: 8th international conference on probabilistic methods applied to power systems, Iowa University, Ames, Iowa, pp 498–501
8. Lin C, Zhang B, Yuan Y (2010) The aging diagnosis of solid insulation for oil-immersed power transformers and its remaining life prediction. In: Asia-Pacific Power and Energy Engineering Conference (APPEEC), vol 3, issue 2, pp 1–3, 2017 IJARIIE-ISSN(O)-2395-43963741
9. Notingher PV, Badicu LV, Dumitran LM, Setnescu R, Setnescu T (2011) Transformer board lifetime estimation using activation energy. In: XV International Symposium on Electromagnetic Fields in Mechatronics, Electrical and Electronic Engineering
10. Das TK, Banik A, Chattopadhyay S, Das A (2019) Sub-harmonics based string fault assessment in solar PV arrays. In: Chattopadhyay S, Roy T, Sengupta S, Berger-Vachon C (eds) Modelling and simulation in science, technology and engineering mathematics, MS-17 2017, Advances in Intelligent Systems and Computing, vol 749. Springer, Cham. https://doi.org/10.1007/978-3-319-74808-5_25
11. Banik A, Sengupta A (2021) Scope, challenges, opportunities and future goal assessment of floating solar park. In: 2021 innovations in energy management and renewable resources (52042), Kolkata, India, 2021, pp 1–5. https://doi.org/10.1109/IEMRE52042.2021.9386735
12. Das TK, Chattopadhyay S, Das A (2019) Remote solar panel temperature monitoring in microgrids. Int J Eng Appl Manage Sci Paradigms (IJEAM) 54(3): 289–296 (Thailand, ISSN 2320-6608)
13. Das TK, Banik A, Chattopadhyay S, Das A (2019) FFT based classification of solar photo voltaic microgrid system. In: ICACCP-2019 International conference on advanced computational paradigms, Sikkim, India. https://doi.org/10.1109/ICACCP.2019.8882995. Publisher: IEEE, IEEE Digital Xplore. Gangtok, India Paper ID-1570504042
14. Stanley Jr W (November 28, 1858–May 14, 1916) Patents https://en.wikipedia.org/wiki/William_Stanley_Jr

Investigation of a Solar Concentrator for Water Distillation

Sunita Mahavar, Ankit Goyal, and Boris V. Balakin

Nomenclatures

I_s solar insolation	W/m^2
T_w water temperature	°C
T_a ambient temperature	°C
T_s receiver surface temperature	°C

Abbrevations

AC	Activated Charcoal
CLC	Cylindrical Light Copper
CPA	Cylindrical Pressurised Aluminium
SBG	Spherical Borosilicate Glass
TDS	Total Dissolved Solids

S. Mahavar (✉) · A. Goyal
Department of Physics, University of Rajasthan, Jaipur, India

B. V. Balakin
Western Norway University of Applied Sciences, Bergen, Norway

NRNU Moscow Engineering Physics Institute, Moscow, Russia

S. K. Ghosh et al. (eds.), *Advances in Thermal Engineering, Manufacturing, and Production Management*, Lecture Notes in Mechanical Engineering,
https://doi.org/10.1007/978-981-16-2347-9_18

1 Introduction

According to the UN World Water Development Report, 3.7 billion people are currently affected by drinking water scarcity. In 2050, this number could increase up to 5.7 billion. For the people living in remote arid areas in countries like India, no water purification devices are available at an affordable cost to supply potable water [1]. It is estimated that up to 575,000 Indian villages alone face the problem of brackish and contaminated water [2, 3]. Although the availability of freshwater is much greater in Norway, the Norwegian industry consumes large volumes of freshwater: in 2017, Statoil used 14.8 Mt freshwater; Norsk Hydro together with daughter companies' report used 57.3 Mt even after 27% recycling [4, 5]. Major developments in water purification technologies are needed in the current scenario at an affordable price. The abundant solar radiation, the clean character of solar energy, the high cost of fossil fuels and negative emission consequences of fossil fuel consumption, along with large requirements for water desalination, are the key drivers of a strong focus on the development of solar desalinator via this collaborative work. Among various types of solar desalinators, the concentrating solar collectors are of most interest as these can produce useful heat in medium- to high-temperature levels. There are several studies been done in solar distillation using concentrating technology. Muraleedharan et al. [6] have designed and constructed a modified active solar distillation system (MSDS) and have compared its performance parameters with a conventional solar still (CSS). The MSDS consists of a Fresnel lens concentrator with an evacuated receiver tube and serpentine loop-type heat exchanger [6]. A solar dish concentrator has been built with a simplified distillation system whose yield per square metre provided sufficient drinking water to meet the daily needs of at least two adults by Prado et al. [7]. A novel light concentration and direct heating solar distillation device embedded underground are proposed by Zhu et al. [8]. They have made a theoretical model of heat and mass transfer by which estimated theoretical efficiency is up to 57%. An experimental device is also developed and tested by Zhu et al. [8]. The process of photothermal evaporation in nanofluids is a promising application in solar energetics. Ulset et al. [9] have established the process in nanofluids with less expensive carbon black (CB) and iron oxide (IO) nanoparticles. They have also developed an empirical model of photothermal steam generation. A solar dish system was used for brackish water desalination under Iraqi condition by Chaichan et al. [10]. In the storage tank paraffin wax was used as phase change materials, that increased the system efficiency about 3 times than the without PCM storage. A theoretical and experimental study of a parabolic concentrator is done by El-Kassaby [11] for sea water distillation. The effect of water flow rates on performance of apparatus is investigated. A water desalination study was also performed by Gowtham et al. [12] on a parabolic trough concentrator. The paraffin wax as latent heat storage material was used and tested as energy storage material for better performance. Various solar concentrator parameters are studied and effect of those are analysed by Rafeeu and Kadira [13]. The effect of lauric acid as phase change material (PCM) on a solar still was tested by Al-Hamadani and Shukla [14]. A considerable improvement in water yield is found in the study. A solar concentrator working with water as working fluid produces steam. That steam can be primarily used for electricity generation or steam cooking and after that can be condensed to get distil water. Hence, distilled water can be treated as secondary output of a solar concentrator system. Considering this, in the present work water distillation through a solar concentrator developed at laboratory is tested. The effect of different receivers on the water yield is tested. So far, the effect of different concentrations of activated charcoal on distilled water output has not been studied, which is targeted in the present work. For this, an indoor set-up is designed and experiments are conducted. The results of indoor testing may be further confirmed using solar concentrator as extension of the present work.

2 Dish Solar Concentrator and Receivers

A prototype solar concentrator is developed in the University of Rajasthan, Jaipur, India. The diameter of the dish is 121 cm, and depth is 20 cm. The paraboloid solar dish is made up by joining thin GI sheets of triangular geometry to produce the desired curvature. The dish is mounted via jack system on an iron shaft, resting and turning on a bearing. The bearing is fitted into the quadruple iron stand which has wheels for locomotion. The movable shaft is equipped with a disc-type gear which receives motions from the stepper motor. The system is shown in Fig. 1. The polymeric acrylic sheets are used as promising reflector material due to their lightweight, flexibility, non-fragility and moderate-temperature performance in the severe weather conditions. For water distillation, different types of receivers are used: cylindrical light copper (CLC), cylindrical pressurized aluminium (CPA) and spherical borosilicate glass (SBG); the dimensions are mentioned in Table 1.

3 Experimental Study

3.1 Set-up

On a number of days, the experimental studies have been conducted at the Solar Energy Research Laboratory (SERL), University of Rajasthan, Jaipur (26.92°N, 75.87°E), on different types of receivers. During the experiments, the water temperature (T_w) and surface temperature (T_s) are recorded by 85XX + Masibas data logger (0.1 °C least count) in every 2-min interval (sensor type K) and weather parameters (solar insolation (Is), ambient temperature (T_a), wind speed, etc.) are measured by weather station (Virtual Pvt. Ltd) installed at SERL.

The pH and total dissolved solids (TDS) of collected water samples are measured by EcoTestr instruments. A glass condenser of length 30 cm and diameter 14 cm is used for vapour/steam condensation. For steam flow, silicon pipe of (inner) diameter

Fig. 1 Dish solar concentrator and experimental set-up

Table 1 Different types of receivers and their dimensions

Receiver type	Height (m)	Diameter (m)	Base area (m^2)	Weight (kg)	Capacity (L)	Concentration ratio
CLC	0.14	0.11	0.010	0.20	1.3	114
CPA	0.11	0.13	0.013	0.66	1.5	86
SBG	-	0.13	0.013	0.19	1.0	86

1 cm is used. The cold water in condenser is circulated through an electric pump of capacity 40 W.

The experimental arrangements are shown in Fig. 1 for CPA receiver. The system is manually tracked as per the requirement during the experiment hours. In order to test transparent receiver as distiller, some laboratory experiments are also conducted with a conventional electric heater of capacity 1 kWh. The effect of ratio of activated charcoal in water is tested.

3.2 Thermal Profiles

In this paper, representative temperature profiles of CPA and CLC receivers are shown in Figs. 2 and 3, respectively. Experiment with both the receivers (CPA and CLC) is carried out for 500 ml water distillation. First set of CPA experiment is started at

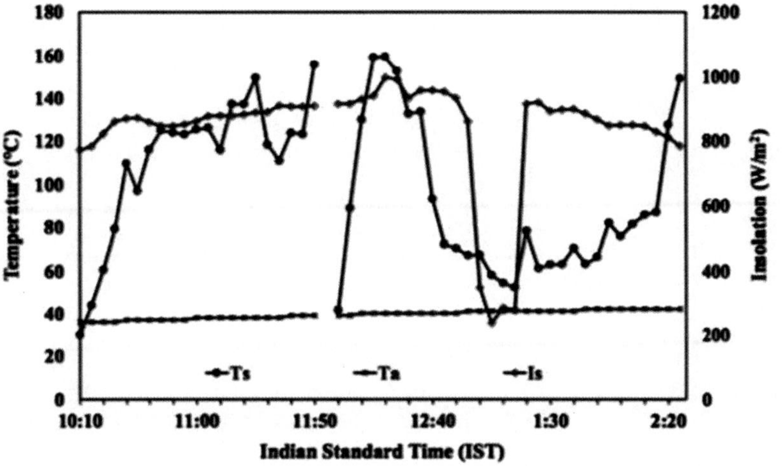

Fig. 2 Thermal profile of cylindrical pressurized aluminium (CPA) receiver on 27 May 2019 with Indian Standard Time (IST) (T_a and T_s are ambient and surface temperatures, respectively, and I_s is solar insolation W/m^2)

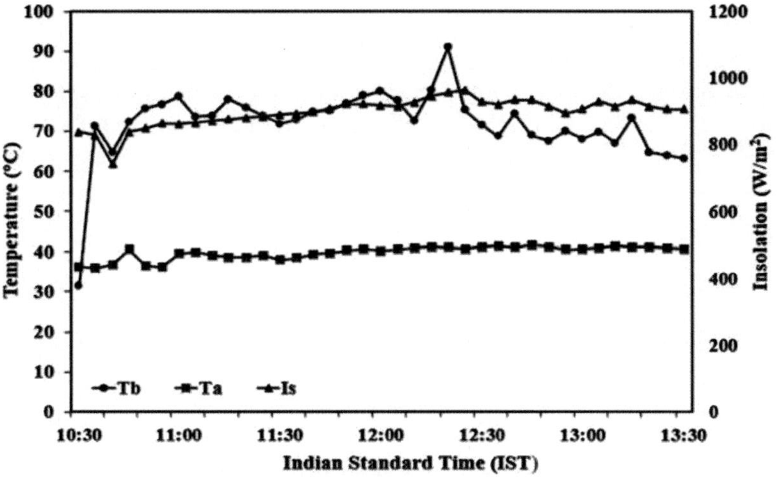

Fig. 3 Thermal profile of cylindrical light copper (CLC) receiver on 25 May 2019 with Indian Standard Time (IST) (T_a and T_b are ambient and water temperatures, respectively, and I_s is solar insolation W/m^2)

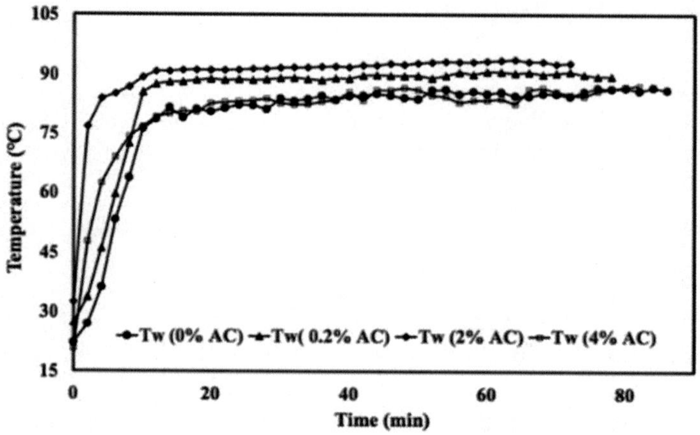

Fig. 4 Water temperature (T_W) with time in spherical borosilicate glass (SBG) receiver with AC concentration 0%, 0.2%, 2% and 4%

10:10 IST, and second set is started at 12:00 IST. To increase distilled water yield, an indoor study is also carried out with different ratios (0%, 0.2%, 2% and 4% of activated charcoal (AC) with 500 ml water in SBG). This profile is shown in Fig. 4 (Table 2).

4 Results and Discussion

The surface temperature profile shown in Fig. 2 reveals that in the CPA receiver the surface temperature remained high enough, and the maximum temperature is about 155 °C in first set. In second set, surface temperature reached around 158 °C in 15 min but due to insolation fall could not sustain further. In first set of experiment, 500 ml water distillation occurred in 1 h 40 min, while in second set this time is 2 h 25 min due to sudden clouds and insolation (Is) falls between 12:05 and 01:20 IST (i.e. 25 min). The average insolation is about 865 W/m^2, and collected yield is 400 ml for first set; in second set, these values are 817 W/m^2 and 430 ml, respectively. The average collection of distilled water remained 275 ml/h. The TDS and pH of collected sample are found to be 10 ppm and 6.5, respectively, while the initial values were 300 ppm and 9.1, respectively. Figure 3 shows the temperature profile of CLC receiver. The maximum temperature of water is about 90.8 °C. In this experiment, 500 ml water distillation occurred in 3 h. The average insolation is about 897.42 W/m^2, and collected yield is 350 ml. The TDS and pH of collected sample are found to be 10 ppm and 6.5, respectively, while the initial values were 400 ppm and 8.4, respectively. Figure 4 depicts that ratio of activated charcoal in water plays a vital role in water distillation. For concentration ratio 0, 0.2% and 2%, the inside temperature increases significantly, and time of distillation reduces. But,

Table 2 Performance comparison of CPA and CLC receivers

Receiver	Amount of water taken (ml)	Water condensed (ml)	Time taken (min)	Avg. insolation (W/m^2)	Max temp. (°C)	pH		TDS	
						Initial	Final	Initial	Final
CPA	500	400	100	865	155	9.1	6.5	300	10
CLC	500	350	180	897	90.8	8.4	6.5	400	10

Table 3 Test results with variable concentration of activated charcoal (AC) in 500 ml water

AC %	Water yield (ml)	Power consumption (kW)	pH	TDS
0	330	1.43	7.4	10
0.2	225	1.30	6.5	10
2.0	330	1.20	4.9	20
4.0	360	1.37	4.0	70

for high concentration this increase in temperature is not observed in the initial testing as infers from Fig. 4. Moreover, for highest concentration time consumption is also remained higher than 2% concentration. These findings are also summarized in Table 3. This initial study with activated charcoal (AC) indicates that an optimized ratio will reduce distillation time with an appropriate water yield. The outdoor experimental study infers that under similar insolation conditions the performance of CPA receiver is found better than CLC receiver. The thickness of CPA receiver is much than CLC receiver, yet the rate of steam generation remained high in CPA due to pressurized outlet point. Moreover, copper is a superior heat conductor than aluminium, yet in the present study, the aluminium receiver showed better performance over the copper one. It indicates that designing of receiver is utmost important including other material aspects in solar concentrator. The indoor testing confirms that activated charcoal increases the rate of steam generation and hence reduces time of water distillation. This study also reveals that there exists an optimum concentration of AC for distillation.

5 Conclusion

The abundant solar radiation, the clean character of solar energy, the high cost of fossil fuels and negative emission consequences of fossil fuel consumption, along with large requirements for water desalination, are the key drivers of a strong focus on the development of solar desalinator. In the present study, a dish solar concentrator developed at laboratory is tested for water distillation. For this purpose, two receivers are tested. The distilled water yield in less time period is found higher for the cylindrical pressurized aluminium (CPA) receiver, while the weight of this receiver is greater than the other cylindrical light copper (CLC) receiver. It may conclude that over material and weight of a receiver designing is more crucial as per the application of a solar concentrator. Some indoor tests on conventional electric heater to observe the effect of activated charcoal on water distillation are also performed. These tests confirm that with the less expensive material the distilled water yield rate can be increased and an optimum value of concentration also exists to reduce distillation time period.

Acknowledgements Authors, Sunita Mahavar and Ankit Goyal, are thankful to DST (New Delhi) and SERB (New Delhi) for project funds under those this experimental study is conducted.

References

1. WWAP (United Nations World Water Assessment Programme)/UN-Water (2018) The United Nations world water development report 2018: nature-based solutions for water. UNESCO, Paris
2. Murugavel K, Srithar K (2011) Performance study on basin type sloe solar still with different wick material and minimum mass of water. Renew Energy 36:612–620
3. Tabrizi FF, Dashtban M, Moghaddam H, Razzaghi K (2010) Effect of water flow rate on internal heat and mass transfer and daily productivity of a weir type cascade solar still. Desalination 260:239–247
4. Sustainability report, Statoil, 2017
5. Norsk Hydro, Annual Report, 2017
6. Muraleedharan M, Singh H, Udayakumar M, Suresh S (2019) Modified active solar distillation system employing directly absorbing Therminol 55–Al_2O_3 nano heat transfer fluid and Fresnel lens concentrator. Desalination 457:32–38
7. Prado G, Vieira LGM, Damasceno JJR (2016) Solar dish concentrator for desalting water. Sol Energy 136:659–667
8. Zhu Z, Zheng H, Wang Q, Chen M, Li Z, Zhang B (2018) The study of a novel light concentration and direct heating solar distillation device embedded underground. Desalination 447:102–119
9. Ulset ET, Kosinski P, Balakin BV (2018) Solar steam in an aqueous carbon black nanofluid. Appl Therm Eng 137:62–65
10. Chaichan M, Abass K, Kazem H (2015) Design and assessment of solar concentrator distillating system using Phase Change Materials (Pcm) suitable for desertic weathers. Desalination Water Treatment 57
11. El-Kassaby MM (2011) Parabolic type solar still. Renewable Energy 1:449–454
12. Gowtham M, Chander MS, Mallikarujanan KVS, Karthikeyan N (2011) Concentrated parabolic solar distiller with latent heat storage capacity. Int J Chem Eng Appl 2(3):185–188
13. Rafeeu Y, Ab Kadira MZA (2012) Thermal performance of parabolic concentrators under Malaysian environment: a case study. Renew Sustain Energy Rev 16(6):3826–3835
14. Al-Hamadani AAF, Shukla SK (2011) Water distillation using solar energy system with lauric acid as storage medium. Int J Energy Eng (IJEE) 1(1):1–8

Floating Photovoltaic Plant in India: Current Status and Future Prospect

Debajit Misra

1 Introduction

Rising power demand, rapid reduction of fossil fuels, increase of greenhouse gas emission, etc. are the major concern in the development of renewable energy technology. At present times, concerned countries all around the world are focusing on renewable energy technology for sustainable power generation, whereas the prime attention is given on solar energy, particularly in solar PV technology. Solar PV technology is being implemented in large scale compared to other renewable energy technology, and this technology is now growing rapidly. However, installation of solar PV requires enough land, which is, nowadays, very limited due to high population growth. Solar PV plants usually require four to five times more land than traditional power plants. In view of this, water surfaces are being utilized for power generation. Water reservoir, canal, lake, etc. are being used where floated structures are being placed on water surface to mount solar panels for generating electricity.

In India, FPV technology is a new and emerging concept. India is a place of huge population. Land in India is scarce, and it has multiple uses like land which is used for agriculture, for grazing, for industries and certain settlements. Land accusation is big problem here. Further, for installation of land-based PV plant, places should be plain and shadow free. Thus, in most of the cases, waste infertile lands are being used but again to convert these wastelands into power plant, lots of effort are required to convert these lands into making plain. Floating power plant on the other hand does away with these difficulties. Floating PV technology could be a very promising option as it requires water body, which is flat and even.

In regular (rooftop and land based) solar PV plant, one of the biggest adversary is the dust accumulation of the solar panel. In case of FPV plant, dust accumulation does not occur because it is placed on water body. Again, in India, most of the large-scale

D. Misra (✉)
Department of Mechanical Engineering, Techno India Group, Kolkata, West Bengal, India

S. K. Ghosh et al. (eds.), *Advances in Thermal Engineering, Manufacturing, and Production Management*, Lecture Notes in Mechanical Engineering,
https://doi.org/10.1007/978-981-16-2347-9_19

solar power plants exist in southern and western regions which receive high solar radiation. Solar PV provides optimum power at a temperature of 25 °C, and there is a constant relation between temperature and electrical efficiency. After 25 °C, in every 1 °C rise of temperature, the efficiency of a solar panel decreases by almost 0.5% [1]. Thus, solar PV at high ambient temperature means that its efficiency has been reduced by some extent. At present, we can get solar module in market with maximum 20% efficiency. Solar system in water body has natural evaporative cooling as water evaporates and cools the solar modules and keeps the temperature lower than that of the surroundings. This can increase the efficiency of a floating solar plant compared to the regular solar plant. Recently, India has decided to include solar and wind power in hydropower projects to create combined source of renewable energy. India has many lakes, large water reservoirs and hydropower plants, and thus, floating PV power generation can be an attractive alternative of renewable power generation by utilizing the surface of water bodies.

2 FPV Technology

2.1 Major Components

Typically, floating PV system is almost same as regular PV system, except that the PV panels and often inverters are fitted on a floating structure. The DC current generated by PV modules is collected by combiner boxes and converted to AC current by inverters. Generally, in small plants which are very close to shore, inverters can be placed on land. Some extra arrangements are very much necessary for floating PV system compared to regular PV system. The primary components of a floating PV system are the following: (Fig. 1).

(i) Pontoon: For installation of floating PV plant, firstly, pontoon (floating structure) is constructed which easily and safely holds solar panels. It is a sturdy structure which is placed on water surface carrying panels and other electrical equipments.
It also provides safety from wind velocity to a certain extent. Numerous hollow lightweight plastic floats made of fibre-reinforced plastic (FRP), high-density plastic (HDPE) or medium density plastic (MDPE) which are placed over the floating structures to form a pontoon. PV panels are fitted on these pontoons maintaining suitable inclination to receive highest possible solar radiation.

(ii) Mooring and anchoring system: The pontoon or floating structure is held securely with mooring and anchoring system. With the help of this system, floating structure can freely move or stay in a suitable position.

(iii) Waterproof cable and connector: All cables remain on water, and the cabling is designed to be shock and waterproof.

(iv) PV system components: It includes PV panels, string inverter and other electrical equipments which are essential for transmitting electricity.

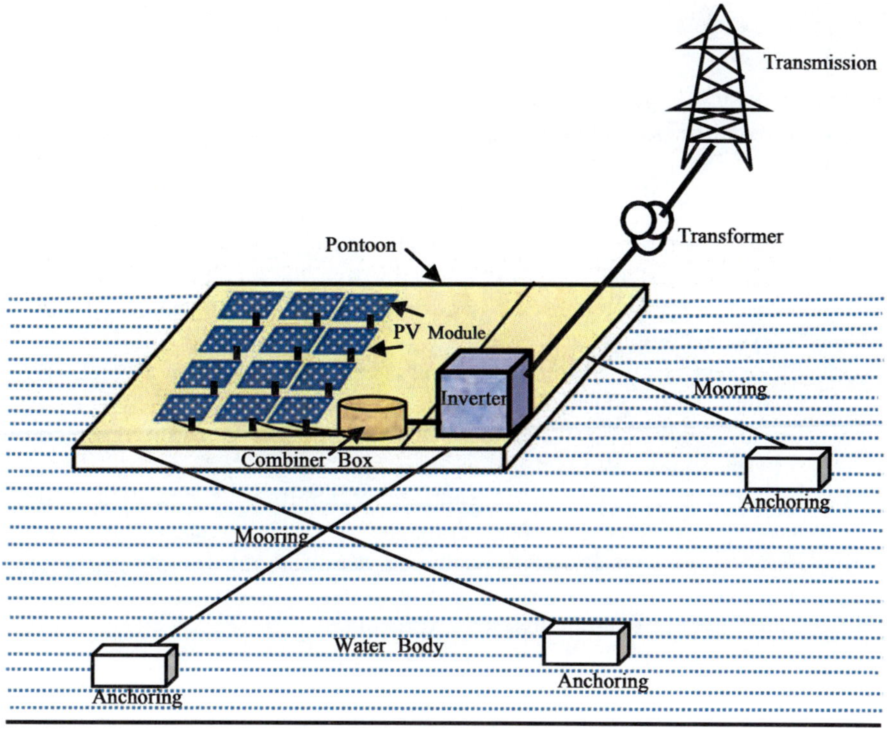

Fig. 1 Simplified schematic diagram of an FPV system

2.2 *Benefits of FPV Plant*

(i) Arable land can be conserved by using water body.
(ii) Unused water body can be converted into useful profitable renewable energy generation area.
(iii) There is no land acquisition issue.
(iv) More power can be generated by increasing the efficiency of PV panel.
(v) Floating PV panels require low maintenance cost.
(vi) Cleaning process of PV panels is very easy.
(vii) Water is conserved by reducing evaporation.
(viii) Reduced water evaporation decreases the possibility of drought in the near area.
(ix) Water clarity can be improved by reducing algae formation due to the constant shade on water surface.

2.3 Challenges

FPV technology is in growing stage, and lots of researches and studies are required to analyse its prospects and challenges ahead. Though it offers lots of advantages, there may be following challenges which should be considered with great attention before its implementation.

(i) As the FPV system design is complicated, its installation requires much more investment cost and long payback period than the regular PV system.
(ii) As entire FPV system remains on water surface, moisture content can affect on system components.
(iii) FPV plant is always prone to natural calamities like heavy wind, storm, cyclone, flood and tsunami.
(iv) Securing profitability is a challenging task as per unit generation cost is more than the regular PV plant.
(v) FPV system component's life span and durability can be affected by stress and vibration caused by water along with other environmental conditions.
(vi) Selected site's water surface should be clean and dirt free else it can affect the system.
(vii) The system installation in a sea is difficult.
(viii) Aquatic system's biodiversity may be affected.

3 Global Scenarios

FPV technology was started in Europe, Japan, South Korea and the USA for research purpose within the time periods from 2008 to 2014. The prime focus was to acquire knowledge that how this technology could utilize for large-scale power generation. Deployment of the global floating solar is growing very rapidly post 2014. According to the report by World Bank, this growth is almost exponential. The reason behind this growth is attributed to its vast expansion in South Asia, Europe and the USA. The total capacity of installed FPV globally was 1097 MW by end of 2018 [2].

At present, Asia possess leading position in terms of FPV installation. The implementation of FPV plant in Japan is the highest (130.59 MW) in the world owing to its dense population and low land availability. China, South Korea, Taiwan and India are the others major countries in Asia. Recently, Singapore, Malaysia, Indonesia, Vietnam and Bangladesh prepare plan to develop FPV plant. In Europe UK, France, Italy, the Netherlands, Belgium and Portugal are the major countries who are implementing FPV plants. Presently, world's largest FPV plant is situated in China on 86 hectare water body, and its capacity is 40 MW [3]. This plant was commissioned in the year 2017. In October 2019, France opened 17 MW FPV plant on 17 hectare water reservoir for electrifying 4733 households [4]. It is Europe's largest floating solar plant. In the same time, USA opened 4.4. MW FPV plant in Sayreville, New Jersey, the largest in North America [5]. In South America, Brazil, Chile and Mexico are investing on FPV plant. Brazil had just 305 KW FPV plant on Goias agricultural firm [2]. Presently, one 1.01 MW plant has been completed on the São Francisco river [6]. South Africa built continents first FPV plant in 2019, which was 60 KW in capacity and located at Marlenique fruit farm in Western Cape [7]. Ivory Coast also sets target for implementation of FPV plant. Australia soon launches their first floating plant in Jamestown, and it serves power for wastewater treatment facility.

4 FPV Plant in India

4.1 Present Status

In India, the concept of floating PV system was first initiated by Tata Power in 2011 with a small pilot project, and then, in 2012, a second pilot project was developed on the banks of the Sabarmati River in the state of Gujarat.

In 2014, India's first 10 KW floating PV plant was installed by Vikram-solar at Rajarhat in Kolkata [8]. This project was developed under supervision of Arka Renewable Energy College in Kolkata and New Town Kolkata Development Authority. The PV system comprises of 40 polycrystalline modules with 250 Wp each. The plant can produce more than 14 MWh power in a year (Fig. 2).

India's second floating PV plant was installed at Banasura sagar reservoir in January 2016 at Wayanad, Kerela. It spreaded on 111.5 m^2 water surface. The plant was built on hollow concrete platform which could carry 4500 tonnes load. In late 2017, it became the India's first large floating PV plant bearing 500 kWp capacity covering 1.25 acres of water surface. The PV plant comprises of 1938 solar modules which have been installed on 18 hollow ferro-cement floaters. The plant can generate 7.5 lakh units of power annually (Fig. 3).

Fig. 2 10 KW FPV plant at Rajarhat in Kolkata [8]

Fig. 3 500 KW FPV plant at Wayanad in Kerala [9]

In 2017, NTPC installed 100 kWp floating PV plant at Kayamkulam in Kerala, on the adjacent lake of NTPC's Rajiv Gandhi Combined Cycle Power Plant (RGCCPP) [10]. The system was installed at 0.32 acre water surface by Swelect Energy Systems Ltd., Chennai with support from NETRA and NTPC within a very short time of 22 days (Fig. 4).

In July 2016, Chandigarh Renewal Energy and Science and Technology Promotion Society (CREST) installed a 10 KW floating PV plant at Dhanas lake. Yellow 2 Gen Power, a Gurgaon-based company has completed this project aiming at solar power generation cum lake water aeration through fountains to improve the oxygen levels in the lake. The floating plant has been set up with 34 modules each of 300 W with dual axes tracking technology to get maximum solar insolation. The plant has been built on a platform that consists a large outer ring that floats on water (Fig. 5).

The Bhurbandha village, which is located at the Morigaon district Assam, has not been connected to the grid up to 2016. In 2017, 10.5 KW floating PV plant was built

Fig. 4 100 KW FPV plant at Kayamkulam, Kerala [10]

Fig. 5 10 KW FPV plant in Chandigarh [11]

to supply electricity to the villagers. This project was implemented by Quant Solar technology in association with Assam Energy Development Agency (AEDA). It is the first floating PV plant in North-Eastern region of India. In this FPV plant, the PV panels were fitted over bamboo platform with the help of metallic rods and floating surfaces (Fig. 6).

Currently, Greater Visakhapatnam Smart City Corporation Limited (GVSCCL) has initiated country's largest floating PV plant of 2 MW capacity on the Mudasarlova reservoir in Visakhapatnam. The PV plant has been built over 20 acres area (Fig. 7).

Thermal power plant management authority Vidarbha Industries Private Limited (VIPL) at Butibori, Nagpur set up 4.8 KW floating solar PV plant which is an initiative of carbon-free green energy. The plant has 16 solar modules of 300 Wp each (Fig. 8).

Fig. 6 10.5 KW FPV plant in Assam [12]

Fig. 7 2 MW FPV plant in Visakhapatnam [13]

Fig. 8 4.8 KW FPV plant in Nagpur [14]

Indian Oil Corporation Ltd. (IOCL) developed grid-connected 100 kWp fixed tilt FPV Plant at Panipat in Haryana [11]. The power plant was a floating structure that supports multiple arrays of solar PV modules. The plant is housed in a raw water reservoir (RWH), located in the Naphtha Cracker unit of IOCL (Fig. 9).

4.2 *Projects Under Development*

In 2018, India has earned the fifth position in the world in the capacity of installation of solar power [15]. Solar Energy Corporation India Limited (SECI) reported that India set a big target in the development of floating solar plant in the states like Andhra Pradesh, Uttar Pradesh, Gujarat, Kerala and Rajasthan and is moving rapidly

Fig. 9 100 KW FPV plant at Panipat in Haryana [11]

to achieve this target. Those targets have become operational or are going through various developmental stages. Following are the projects in the country which are still under construction.

(i) NHPC is working to develop a 50 MW floating solar plant using a 350 acre water body at West Kallada in Kerala.
(ii) The Shapoorji Pallonji Group in collaboration with the Solar Energy Corporation of India (SECI) is initiating to develop a 50 MW plant which is a part of the proposed 150 MW floating solar power plant planned at Rihand Dam, Sonbhadra district, Uttar Pradesh.
(iii) National Thermal Power Corporation (NTPC) has taken initiative for the development of 25 MW floating solar PV project at NTPC Simhadri in Andhra Pradesh.
(iv) NTPC has also commissioned for development of 1 MW floating solar power plant on Kawas Reservoir in Aditya Nagar, Surat in Gujarat.
(v) Maharashtra State Electricity Distribution Company Ltd. (MSEDCL) has just started for the development of 1 GW of floating solar plant in the Ujjani Dam in Solapur district of Maharashtra.
(vi) In Gujarat, the government-owned Sardar Sarovar Narmada Nigam (SSNNL) has started to build a 100 MW canal-top solar power project atop the branch canals of river Narmada. It would cover nearly 40-km canal.
(vii) Hindustan Zinc, a group firm of the Vedanta Group, has almost completed their 1 MW floating solar power plant at Ghosunda Dam, near Chittorgarh, Rajasthan. The solar plant could be able to supply annually 1993 MWh of energy.
(viii) NHDC, a joint venture between NHPC and Madhya Pradesh State Government, has started to set up a 25 MW of floating solar plant on the Omkareshwar reservoir in Khandwa district of Madhya Pradesh.

(ix) NHPC has been working to develop 40 MW of floating capacity at the Chiplima hydropower facility in Odisha.

4.3 Recent Projects Proposal

Recently, ten floating solar projects proposal has been launched with capacity ranging from 100 KW to 1 GW. World Bank, through its Clean Technology Fund, has financed up to 10 MW for the development of some of these projects. The German Development Bank (KfW) is funding for the development of two projects of total capacity of 40 MW. The project proposals are listed below:

(i) In 2018, NTPC planned to expand their power generation capacity by grid-connected floating solar at its Rajiv Gandhi Combined Cycle Power Plant (RGCCPP) Kayamkulam in the southern state of Kerala. A tender was released of 70 MW capacity.

(ii) BHEL is constructing a 25 MW floating solar power plant at NTPC Simhadri Super Thermal Power Station in Deepanjalinagar, 40 km from Visakhapatnam. Once completed, this floating solar power plant would be the largest in Andhra Pradesh.

(iii) NHPC plans to set up a 600 MW floating solar cum hydroelectric project in the Satara district of Maharashtra.

(iv) In September 2019, the Bihar Renewable Energy Development Agency (BREDA) released tender inviting bidders to develop two floating solar PV power projects. One 2 MW project will be developed on private water bodies and another 2 MW to be developed on a government-owned water body.

(v) Solar Energy Corporation of India Ltd. (SECI) invites online bids for 100 MW floating solar power plant at the reservoir of Getalsud Dam, Ranchi, Jharkhand and 50 MW floating solar power plant at the reservoir of Dhurwa Dam, Ranchi, Jharkhand.

(vi) In 2019, government released a project proposal to develop 1GW floating solar plant on India's largest reservoir, the Indira Sagar Dam in Madhya Pradesh.

(vii) In January 2020, SECI requests for selection (RfS) for setting up a 4 MW grid-connected floating solar PV power project at Kalpong dam, Diglipur, North Andaman.

(viii) A tender has been accorded for 100 MW floating solar PV project at Ramagundam in Telangana.

(ix) In 2020, SECI has sent a proposal to State Government of Odisha for developing 500 MW FPV plant at the Hirakud reservoir.

(x) 100KW floating PV plant is still under process at Loktak Lake in Manipur, which is the largest fresh water lake in North-East India. The project for installation has been commissioned by Manipur Renewable Energy Development Agency (MANIREDA).

5 Future Prospect

India is in the third position in the World after China and the USA in terms of number of Dams. National Register of Large Dam (NRLD) reported that India has 5264 dams with an area of approximately 1.4 million hectares and 437 dams which are under construction. Again, according to Reservoir Fisheries of India, it has 56 large reservoirs covering 1,140,268 hectares, 180 medium reservoirs covering 527,541 hectares and 19,134 small reservoirs covering 1,485,557 hectares areas. Besides these, India has countless water bodies. Thus, India could utilize few portions of those water sources in the development of the huge hubs for FPV projects.

Currently, floating solar plants in the world acquire within 10–25% area of the water body. Thus, there is no major impact on aquatic habitats. The location of the plant should be selected according to the depth of the water body so that it keeps floating when water level falls during summer. To evaluate power generation capacity of 35 large water reservoirs in India, 10% area of water body has been considered. Again, it has been considered that 1 KW floating plant requires 40 m^2 area of water surface [13]. Indian large water reservoirs, lakes and reservoirs adjacent to dams are considered here for evaluation [16–18]. In this evaluation, only large water body, whose area is equal to 100 km^2 or above, has been taken under consideration. It is also considered that 1750 L/m^2/year of water could be saved due to reduction in evaporation, and the yearly saving of water has been calculated accordingly [19] (Table 1).

It is found that by utilizing FPV plants on 35 large water reservoirs, India potentially can generate 38.88 GW power. Besides this, it has the ability to save 2730 billion litres of water in every year. Normally, for coal-based thermal power plant, if G10 category coal is used having calorific value of 2615 kcal/KWh, 4528 tonnes coal would require generating 1 MW power per annum [20]. Thus, for the present case to generate 38.88 GW, 176.04 megatonnes coal could be saved yearly. Also, its equivalent greenhouse gas emission could be reduced in every year.

6 Conclusion and Remarks

It may be concluded that FPV system is a very effective renewable power generation system in the current scenario when fossil fuel is constantly reducing. This system, unlike the regular PV systems, is an eco-friendly power generation system for its efficiency in reducing water evaporation rate and thus helping in water conservation. It also can solve the persistent problem of land availability for installation of solar PV. Though its initial cost is higher than the regular PV system, a floating PV system generates almost 25% more power compared to the regular PV system. Its application can also be possible in aquaculture farms and effluent treatment centres, especially in remote areas which are not connected to the power grid. With proper government support and public awareness, the FPV technology could make a revolutionary

Table 1 Energy potential and water saving using FPV plant of major lakes, reservoirs and dams in India

Name	Reservoir area (km^2)	Location	Potential (GW)	Water saving (mL/year)
Vembanad lake	2033	Kerala	5.08	355,775
Chilika lake	1165	Odisha	2.91	203,875
Indira Sagar	913.4	Madhya Pradesh	2.28	159,845
Shivsagar lake	891.78	Maharastra	2.22	156,060
Hirakud dam	743	Odisha	1.85	130,025
Gandhi Sagar dam	723	Madhya Pradesh	1.8	126,525
Getalsud dam	717	Jharkhand	1.79	125,475
Idukki dam	649.3	Kerala	1.62	113,626
Srisailam dam	616.4	Andhra Pradesh	1.54	107,870
Polavaram dam	600	Andhra Pradesh	1.5	105,000
Rihand dam	468	Uttar Pradesh	1.17	81,900
Sriram Sagar	450.82	Telengana	1.12	78,892
Pulicat lake	450	Andhra Pradesh	1.12	78,750
Tungabhadra dam	378	Karnataka	0.94	66,150
Sardar Sarovar	375.3	Gujarat	0.93	65,676
Jayakwadi dam	350	Maharastra	0.87	61,250
Ujjani dam	337	Maharastra	0.84	58,975
Linganamakki dam	316.65	Karnataka	0.79	55,413
Loktak lake	287	Manipur	0.71	50,225
Nagarjun Sagar	284.9	Andhra Pradesh	0.71	49,856
Bargi dam	267.97	Madhya Pradesh	0.66	46,894
Pong dam	260	Himachal Pradesh	0.65	45,500
Almatti dam	242.3	Karnataka	0.6	42,401
Maharana Pratap Sagar	240	Himachal Pradesh	0.6	42,000
Tawa reservoir	225	Madhya Pradesh	0.56	39,375
Somasila	221.28	Andhra Pradesh	0.55	38,724
Bisalpur dam	218.36	Rajasthan	0.54	38,213
Somasila dam	212.28	Andhra Pradesh	0.53	37,149
Ranapratap Sagar	198.3	Rajasthan	0.49	34,701
Bhakra dam	168.35	Himachal Pradesh	0.42	29,460
Mettur dam	153.46	Tamil Nadu	0.38	26,854
Kangsabati	124.32	West Bengal	0.31	21,756
Bhadra dam	112.5	Karnataka	0.28	19,686
Indravati	110	Odisha	0.27	19,250

(continued)

Table 1 (continued)

Name	Reservoir area (km^2)	Location	Potential (GW)	Water saving (mL/year)
Kolab dam	100	Odisha	0.25	17,500

change in renewable power generation in India. Recently, Indian cabinet addressed a long pending concern, The Dam Safety Bill, 2019, which could ensure safety to the currently existing dams and could pave the way for many more FPV projects in India. SECI, a government undertaking organization, took the responsibility for implementing the FPV projects in India. The organization has developed schemes 'Expression of Interest' (EoI) all over the country. The main feature of EoI is to understand the viability of developing large-scale FPV plant and to encourage the developers to develop FPV plants. The organization has addressed the government's intentions and objectives in front of the developers.

Owing to the large investment of the FPV projects, financial support is needed from government or private sectors. In India, as the concept of FPV technology is new, finding skilled workforce could be a problem. So, proper training programmes are necessary in the utilization of human resource to drive the project efficiently. Since it is a new technology, there is no historical evidence to acquire knowledge about the impact of water on PV system for long-term application. For large-scale installation, building public awareness is needed through seminar and workshop. In India, lots of research and development are required, and better financial infrastructure and manufacturing facilities are needed to develop large-scale projects. Again, in remote places, adequate transmission network is required. India is capable to manage above-mentioned challenges to make a huge FPV power hub in coming future.

References

1. Moharram KA, Abd-Elhady MS, Kandil HA, El-Sherif H (2013) Enhancing the performance of photovoltaic panels by water cooling. Ain Shams Eng J 4(4):869–877
2. World Bank Group, ESMAP, and SERIS (2018) Where sun meets water: floating solar market report—executive summary. Washington, DC: World Bank. Available at: http://documents.worldbank.org/curated/en/579941540407455831/pdf/Floating-Solar-Market-Report-Executive-Summary.pdf
3. PV Magazine (2017) 40 MW floating PV plant in China connected with sungrow's inverters (Online). Available at: https://www.pv-magazine.com/2017/05/19/floating-pv-plant-in-china-connected-with-sungrows-inverters/
4. PV-Tech (2018) Akuo energy starts construction on 17 MW floating solar plant in France [Online]. Available at: https://www.pv-tech.org/news/akuo-energy-starts-construction-on-17mw-floating-solar-plant-in-france
5. PV Magazine (2019) America's largest floating solar project completed (Online). Available at: https://pv-magazine-usa.com/2019/10/23/americas-largest-floating-solar-project-completed/
6. PV-Tech (2019) Ciel & Terre's 1.01MW floating plant completed in Bahia, Brazil (Online). Available at: https://www.pv-tech.org/news/ciel-terres-1.1mw-floating-plant-completed-in-bahia-brazil

7. PV-Tech (2019) New Southern Energy completes first floating PV project in South Africa (Online). Available at: https://www.pv-tech.org/news/new-southern-energy-completes-first-floating-pv-project-in-south-africa
8. Floating solar power plant in West Bengal, India. Available at: https://www.vikramsolar.com/case-studies/floating-solar-power-plant-in-west-bengal-india/
9. Solarplaza, (2019) Top 100 floating solar projects. Available at: https://www.solarplaza.com/channels/markets/11968/top-100-floating-solar-projects/#accesstop100
10. NTPC installs India's first largest solar PV plant in Kerala. Available at: https://economictimes.indiatimes.com/industry/energy/power/ntpc-installs-indias-largest-floating-solar-pv-plant-in-kerala/articleshow/57577004.cms
11. A bellwether for floating solar industry in India (Online). Available at: http://www.yellow.org.in/project.html
12. Clean Future (2017) Assam village uses floating solar platform for electric generation (Online). Available at: http://www.cleanfuture.co.in/2017/08/02/assam-village-uses-solar-for-electric-generation/
13. PV-Tech (2018) GVSCCL inaugurates 2 MW floating solar plant in Andhra Pradesh (Online). Available at: https://www.pv-tech.org/news/gvsccl-inaugurates-2mw-floating-solar-plant-in-andhra-pradesh
14. Energetica (Power generation magazine, India) Nagpur's first floating solar PV plant installed using delta inverter. Available at: http://www.energetica-india.net/news/nagpurs-first-floating-solar-pv-plant-installed-using-delta-inverters
15. Ministry of new and renewable energy. Available at: https://mnre.gov.in/
16. National register of large dams. Available at: http://www.indiaenvironmentportal.org.in/files/file/NRLD%202018.pdf
17. Reservoir fisheries of India. Available at: http://www.fao.org/3/V5930E/V5930E01.htm#ch1
18. Dam in India (Online). Available at: https://www.mapsofindia.com/maps/dams/tungabhadra-dam.html
19. CWC (2011) Evaporation control in reservoirs. Basin Planning and Management Organisation, Central Water Commission, New Delhi
20. Norms for coal consumption. Available at: http://www.cea.nic.in/reports/others/thermal/tppd/coal_cons_norms.pdf
21. Solar energy corporation India Limited (SECI). https://seci.co.in/tenders.php
22. National Thermal Power Corporation (NTPC) Floating solar. https://www.ntpc.co.in/en/search/node/floating%20language%3Aen
23. BHEL to commence work on floating solar plant at NTPC. Available at: https://www.thehindu.com/news/cities/Visakhapatnam/bhel-to-commence-work-on-floating-solar-plant-at-ntpc/article29858452.ece
24. India planning 1 GW floating solar project in Madhya Pradesh. Available at: https://ieefa.org/india-planning-1gw-floating-solar-project-in-madhya-pradesh/
25. Central Water Commission. Available at: www.cwc.gov.org
26. Hartzell T (2016) Evaluating potential for floating solar installations on Arizona water management infrastructure. University of Arizona, Tucson (Arizona)
27. PV-Tech (2018) NTPC tenders 70MW of floating solar at combined cycle power plant in Kerala. Available at: https://www.pv-tech.org/news/ntpc-tenders-70mw-of-floating-solar-at-combined-cycle-power-plant-in-kerala

Friction Coefficient Analysis of Nano-crystalline TiO_2-Added Alumina Ceramics

Partha Haldar, Tapas Kumar Bhattacharya, and Nipu Modak

1 Introduction

Economic manufacturing of excellent quality product is one of the main targets of research in material science. Generally, some new materials are fabricated in the laboratory; then, the characterization and further testing are done. So as a result, huge experimentation is required before successfully commercialization of a product, and the process is associated with time and cost. Now, if a predictive algorithm is designed which is trained on the basis of some experimental data and can predict the other required parameters, then it will surely cause reduction of time and cost of experimentation.

Al_2O_3-based ceramics are becoming popular as mechanical moving contact materials which motivate intensive research in understanding of their wear behavior [1]. They are used as parts of rockets, jet engines, gas turbines, heat shields for space vehicles, fusion reactors, heat treatment furnaces [2, 3], cutting tools [4], etc. As newer advanced ceramics are coming in the market rapidly, so evaluation of their coefficient of friction (COF) with respect to other counter body is one of the prime concerns for practicing engineers before they are tested in a tribometer. As a result, researchers are trying to develop theoretical models to predict the COF of newer materials. Variation in wear behavior is reported by researchers by addition of various secondary phases in alumina like zirconia [5], SiC [6], CuO [7–12] and TiO_2 [13, 14]. To understand

P. Haldar (✉)
Department of Mechanical Engineering, Government College of Engineering and Ceramic Technology, Kolkata, India

T. K. Bhattacharya
Department of Ceramic Technology, Government College of Engineering and Ceramic Technology Kolkata, Kolkata, India

N. Modak
Department of Mechanical Engineering, Jadavpur University, Kolkata, India

S. K. Ghosh et al. (eds.), *Advances in Thermal Engineering, Manufacturing, and Production Management*, Lecture Notes in Mechanical Engineering,
https://doi.org/10.1007/978-981-16-2347-9_20

the wear behavior and tribological application of a material, an important factor is the COF. Recently, we have studied [15] the reciprocating frictional behavior of nano-crystalline TiO_2 addition in alumina ceramics and found that with increase in TiO_2 weight percentage in alumina (up to 4 wt.%), the COF reduces for any constant sliding frequency and normal load. In this work, we added 0, 0.5, 1, 2 and 4 wt.% of titania in to alumina and fabricated the briquettes for testing under a reciprocating tribotester. It was pointed out that with increase in sliding frequency keeping a constant normal load, the COF increases. Similarly, with increase in normal load, keeping a constant reciprocating frequency, the COF increases. The whole tribotest was done under certain level of experimentation like, when the normal load was varied as 300, 500, 700 and 1000 gmf, then the reciprocating frequency was kept constant at 30 Hz; whereas when the sliding frequency was varied as 15, 30, 45 and 60 Hz, then the normal load was kept constant at 500 gmf. These experimental COF values are taken to make the curve fitting, and the resulting curve tries to predict the weight percentage of TiO_2 required in pure alumina system to obtain a specified COF under some known sliding frequency and normal load. To validate the results in this work, we have fabricated another two types of batches of sintered sample with 1.5 and 3 wt.% of TiO_2 added alumina ceramics. Prediction of COF of any ceramic material by some computational work is still a gray area of research.

Therefore, the main objective of this work is to: Determine the weight percentage of TiO_2 required in pure alumina system to obtain a specified COF under some known sliding frequency and normal load.

2 Materials and Methodology

2.1 *Materials*

Reactive alumina (Al_2O_3) is from M/S Alcoa; chemical grade titanium (IV) isopropoxide ($Ti[OCH(CH_3)_2]_4$), citric acid ($C_6H_8O_7$) and ethylene glycol ($C_2H_6O_2$) (all of these precursors are 99.9% pure) from M/S Merck are purchased and used without further purification.

2.2 *Synthesis of Nano-crystalline TiO_2Powder*

Citric acid and titanium (IV) isopropoxide are mixed first in stoichiometric ratio, and an aqueous solution is made to synthesis TiO_2 powder. The solution is then heated at about 60 °C until complete dissolution takes place. Ethylene glycol is then added under constant stirring with the same volume proportion and heated at about 80°C until gelation occurs. After that, the gel is dried at 120 °C for 24 h in a drier. Finally, the dried material is ground in an agate mortar and heat treated at 800 °C for 3 h in a muffle furnace for proper crystallization.

2.3 *Densification*

1.5 and 3 wt.% of nano-crystalline TiO_2 powder are mixed in alumina, and a slurry is made, and also the batches are marked as T1.5 and T3, respectively. Next, the slurry is dried in a drier at 110 °C for 24 h. The densification study is conducted on 13 mm diameter and 3.8 mm high briquette which are formed in a uniaxial hydraulic press under pressure of 100 MPa with 2 min dwell time. 6 wt.% polyvinyl alcohol solution is used as binder. Briquette after drying at 110 °C for 24 h is subjected to sinter at 1600 °C for 2 h soaking at the peak temperature followed by normal furnace cooling.

2.4 *Tribological Testing*

The dry friction test on the T1.5 and T3 samples is conducted in a reciprocating friction tester (TR-282, DUCUM, India). The counter surface used is a spherical Si_3N_4 ball. The samples and the counter surface are cleaned properly before starting the test. The tribotester records the real-time COF value generated due to the motion in the interface in each second. The test duration for each case is 10 min. Two set of experimental scheme is adopted. In the first set, the normal load of 0.5 kgf and sliding time of 10 min are kept constant, and the sliding frequency is varied as 15, 30, 45 and 60 Hz. In the second set, the sliding frequency of 30 Hz and sliding time of 10 min are kept constant, and the normal load is varied as 0.3, 0.5, 0.7 and 1 kgf.

3 Results and Discussion

TiO_2 addition in alumina ceramics reduces the COF, and it is strong function of sliding frequency and normal load. As stated in the introduction section that in our previous work [15], we have recorded the COF values of T0, T0.5, T1, T2 and T4 materials under two set of experimental scheme as stated in the tribological testing section. Figure 1 indicates that the effect of sliding frequency on COF is more than the normal load.

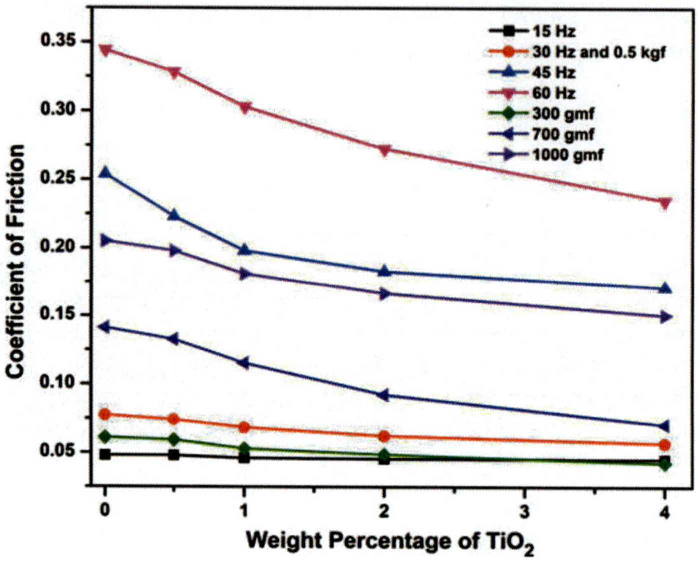

Fig. 1 Variation of coefficient of friction with respect to weight percentage of TiO_2

3.1 Computation of Frictional Coefficient in Relation with Sliding Frequency and Normal Load

The discrete stable COF values for sintered samples are used to form smooth curves in MATLAB following the best fit topology. Minimum root mean square error (RMSE) has been adopted as the governing criteria for the selection of curve among the several variants available. The COF values have been tested with different curve fitting methods, e.g., polynomial fit, used with second order, exponential fit with first and second order, smoothing spline, Weibull fit and cubic spline fit methods, out of which cubic spline fit produced the minimum RMSE and hence selected for the proposed work. The detailed flowchart of algorithm for this computational work is shown in Fig. 2. The experimental COF data are plotted in MATLAB using the best fit curve techniques in two separate models in which sliding frequency and the normal load are given independently as the abscissa and COF values are kept as the ordinate in both the cases. This process is carried out for each of the sintered samples varying in addition level of TiO_2, hence forming five sets of independent curves in each of the models. In each case, the four experimental data points corresponding to one particular sample have been joined using the predicted COF values obtained from the cubic spline analysis for the intermediate abscissa values, thus simulating the predicted COF values over the entire experimental span of either sliding frequency or the normal load to form a smooth curve. The smooth plot using the cubic spline fit method with COF and sliding frequency is shown in Fig. 3, and the similar plot between COF and normal load is shown in Fig. 4.

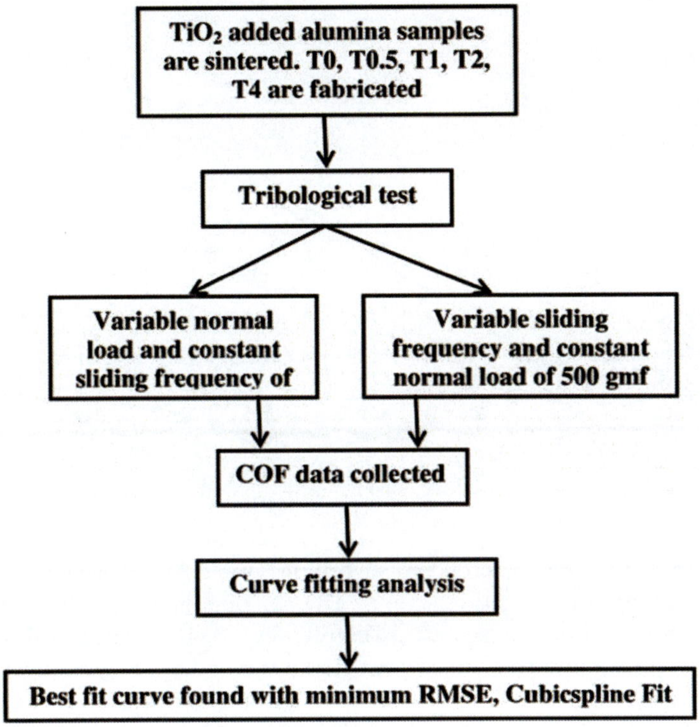

Fig. 2 Flowchart of the predictive algorithm

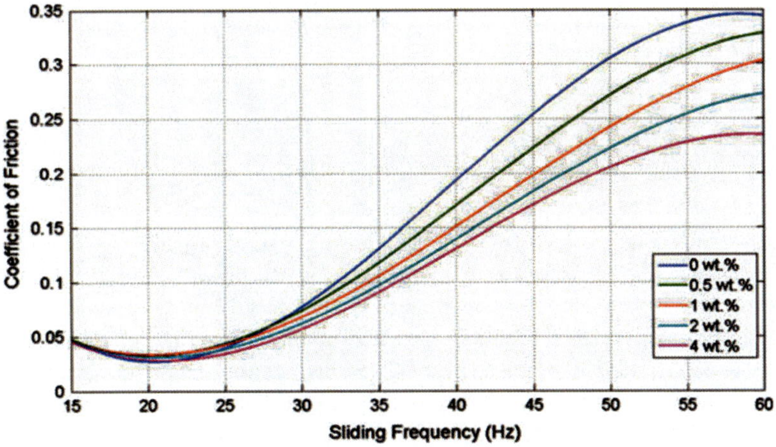

Fig. 3 MATLAB generated cubic spline fitted plot between COF and sliding frequency

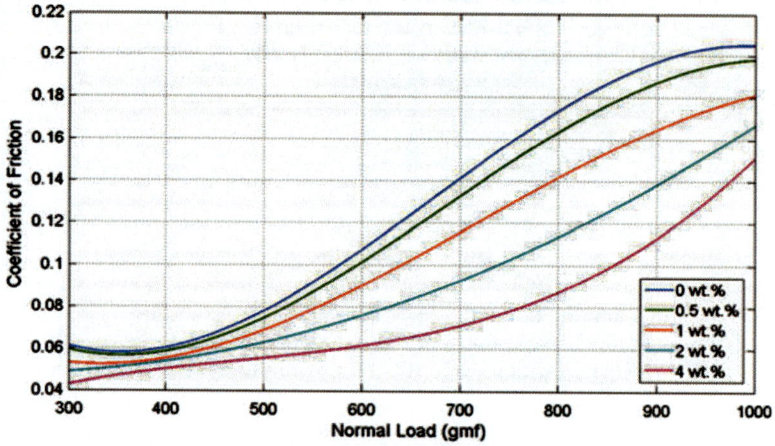

Fig. 4 MATLAB generated cubic spline fitted plot between COF and normal load

In order to validate the proposed algorithm, two different samples have been prepared with different addition level of TiO_2 in alumina other than that used for designing the curve fitting analysis. The prepared samples are tested with same sliding frequency and normal load which were used for designing the algorithm, and the COF values are recorded. The experimentally obtained data are separated to form two sets; the first set comprising of sliding frequency and corresponding COF and the second one with normal load and the COF, to feed into the two different curve fitting models, respectively, to compute the predicted weight percentage independently. These values are verified with the original addition level of the samples to carry out validation of the proposed scheme.

3.2 Validation of the Proposed Algorithm

The flowchart of algorithm for the validation is shown in Fig. 5. The proposed algorithm is validated using experimental results as shown in Tables 1 and 2. For validation, two separate batches are prepared with 1.5 and 3 wt.% of titania in alumina following the steps as discussed in the densification section, and the samples are labeled as T1.5 and T3, respectively. Table 1 shows the results obtained from the experimental setup where T1.5 and T3 are tested under variable frequency conditions of 15, 30, 45 and 60 Hz, and the COF values are recorded. The same frequencies are tested using the proposed algorithm which gave the weight percentage of addition level of TiO_2 in alumina very close to 1.5 and 3, respectively, which are included in Table 1. It is also noteworthy to mention that a maximum error of +4.43% is observed for cubic spline fit. Similar experiment with the same pair of samples has been carried out with variable normal loading of 0.3, 0.5, 0.7 and 1 kgf which gave a

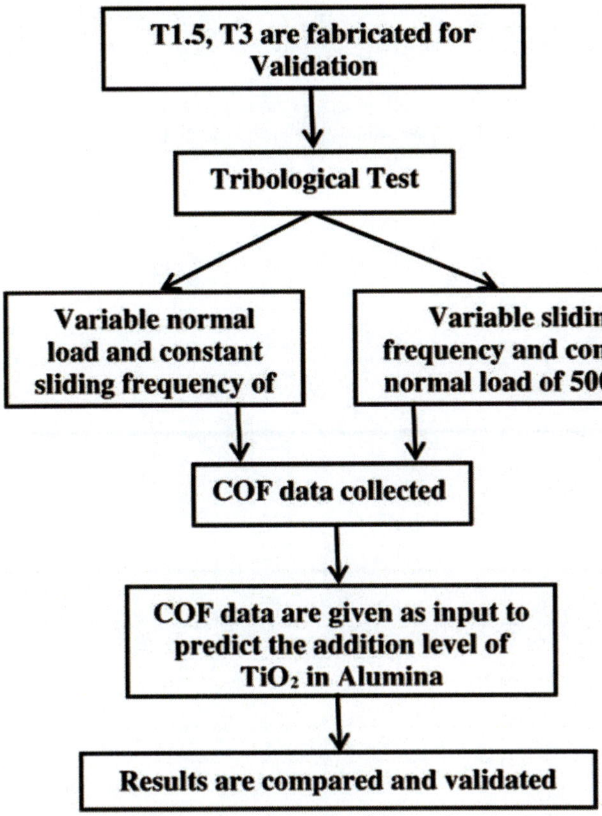

Fig. 5 Algorithm for validation of the proposed work

set of COF values (see Table 2). While testing with the proposed algorithm, weight percentage very close to 1.5 and 3 is obtained with a maximum error of −2.56% in case of cubic spline fit, thus validating the algorithm both with respect to variable frequency as well as variable normal loading conditions. The errors generated due to use of other curve fitting algorithm like polynomial of order two, smoothing spline, exponential of order one and two, Weibull fit are also shown in Tables 1 and 2. The error values lying with in ±5.00% are shown in bold format, and it can be observed that no other curve fitting techniques fit well for this data set except the cubic spline.

Table 1 COF of T1.5 and T3 at different sliding frequency and the errors generated by different curve fitting algorithms

Sample ID	SF (Hz)	Experimental COF value	Additive % using cubic spline fit	Error % due to cubic spline fit	Error % due to poly2	Error % due to exp1	Error % due to exp2	Error % due to smoothing spline fit	Error % due to Weibull fit
T 1.5	15	0.04581	1.5614	**+4.09**	-2.98×10^4	$+1.38 \times 10^3$	-3.40×10^5	-1.22×10^7	-8.93×10^7
	30	0.06483	1.4896	**−0.69**	$+4.58 \times 10^3$	$+2.34 \times 10^3$	+554.73	$+1.11 \times 10^3$	$+1.14 \times 10^9$
	45	0.18857	1.5034	**+ 0.23**	−53.98	−70.82	−10.47	−32.28	$+1.86 \times 10^9$
	60	0.285182	1.5155	**−1.04**	+ 17.51	+20.53	**+1.04**	**+2.33**	-1.98×10^{10}
T3	15	0.04555	3.1328	**+ 4.43**	-1.33×10^4	+ 817.55	-1.60×10^5	-5.52×10^6	-7.85×10^9
	30	0.059489	2.9771	**−0.76**	$+4.20 \times 10^3$	$+1.89 \times 10^3$	+554.09	+976.12	$+8.88 \times 10^9$
	45	0.176345	2.9699	**−1.00**	−63.66	−75.33	−15.30	−42.94	$+3.03 \times 10^{10}$
	60	0.251211	3.0497	**+1.66**	+15.59	+17.92	**+1.66**	**+4.81**	-5.94×10^{11}

Table 2 COF of T1.5 and T3 at different normal loads and the errors generated by different curve fitting algorithms

Sample ID	NL (gmf)	Experimental COF value	Additive % using cubic spline fit	Error % due to cubic spline fit	Error % due to poly2	Error % due to exp1	Error % due to exp2	Error % due to smoothing spline fit	Error % due to Weibull fit
T 1.5	300	0.050628	1.4616	**−2.56**	−31.16	−20.20	+450.08	−39.41	$+3.85 \times 10^{22}$
	500	0.065276	1.4738	**−1.74**	+14.87	+30.17	+8.75	+24.95	$+1.40 \times 10^{12}$
	700	0.101852	1.4869	**−0.87**	−8.99	−10.96	**−2.72**	**+0.52**	$+1.69 \times 10^{20}$
	1000	0.172436	1.5108	**+0.72**	**+2.87**	**+1.13**	**+0.71**	**−4.59**	$+1.56 \times 10^{62}$
T3	300	0.045824	2.9811	**−0.62**	+22.55	−39.89	-1.47×10^{3}	−22.69	$+4.74 \times 10^{16}$
	500	0.058391	3.0717	**+ 2.39**	−14.81	**+0.75**	**−3.74**	**−0.18**	$+2.67 \times 10^{11}$
	700	0.079892	3.0109	**+ 0.36**	+11.21	+36.04	**+1.29**	+17.65	-2.74×10^{60}
	1000	0.15806	2.9905	**−0.32**	**−2.67**	−8.13	**-0.34**	−9.43	-6.57×10^{13}

4 Conclusions

The following inferences are drawn from the above study.

- COF of TiO_2 system is directly proportional to sliding frequency, normal load and inversely proportional to weight percent of titania addition.
- For the discrete stable COF values obtained from experimentation, minimum RMSE governing criteria refers cubic spline fit method.
- The prediction the cubic spline fit behaves well for this system and has a maximum error of +4.43%.

If such kind of prediction is used for similar kind of work, then it will result in huge saving in terms of raw material, research time, and in turn, the cost of the product to the customers will reduce.

References

1. Parchovianský M, Balko J, Švančárek P et al (2017) Mechanical properties and sliding wear behaviour of Al_2O_3-SiC nanocomposites with 3–20 vol% SiC. J Eur Ceram Soc 37(14):4297–4306
2. Medvedovski E (2006) Alumina–mullite ceramics for structural applications. Ceram Int 32(4):369–375
3. Puchy V, Hvizdos P, Dusza J et al (2013) Wear resistance of Al_2O_3–CNT ceramic nanocomposites at room and high temperatures. Ceram Int 39(5):5821–5826
4. Kumar AS, Durai AR, Sornakumar T (2006) Wear behaviour of alumina based ceramic cutting tools on machining steels. Tribol Int 39(3):191–197
5. Elsen SR, Ramesh T (2016) Analysis and optimization of dry sliding wear characteristics of zirconia reinforced alumina composites formed by conventional sintering using response surface method. Int J Refract Met Hard Mater 58:92–103
6. Rodríguez J, Martín A, Pastor JY et al (1999) Sliding wear of alumina/silicon carbide nanocomposites. J Am Ceram Soc 82(8):2252–2254
7. Pasaribu HR, Sloetjes JW, Schipper DJ (2003) Friction reduction by adding copper oxide into alumina and zirconia ceramics. Wear 255(1–6):699–707
8. Song J, Valefi M, De RM et al (2012) The effect of an alumina counterface on friction reduction of CuO/3Y-TZP composite at room temperature. Wear 274:75–83
9. Ran S, Winnubst L, Blank DH et al (2007) Effect of Microstructure on the Tribological and Mechanical Properties of CuO-Doped 3Y-TZP Ceramics. J Am Ceram Soc 90(9):2747–2752
10. Pasaribu HR, Reuver KM, Schipper DJ et al (2005) Environmental effects on friction and wear of dry sliding zirconia and alumina ceramics doped with copper oxide. Int J Refract Met Hard Mater 23(4–6):386–390
11. Haldar P, Bhattacharya TK, Modak N (2019) The effect of normal load and sliding frequency on the reciprocating friction behavior of Nanocrystalline CuO-based alumina ceramics. In: Advances in micro and nano manufacturing and surface engineering. Springer, Singapore, pp 673–681
12. Haldar P, Bhattacharya TK, Modak N (2021) Effect of nano CuO addition on the tribo-mechanical behavior of alumina ceramics in non-conformal contact. Int J Appl Ceram Tec 18(1):110–118
13. Lee SW, Morillo C, Lira-Olivares J et al (2003) Tribological and microstructural analysis of Al_2O_3/TiO_2 nanocomposites to use in the femoral head of hip replacement. Wear 255(7–12):1040–1044

14. Bagde P, Sapate SG, Khatirkar RK et al (2018) Friction and abrasive wear behaviour of Al_2O_3-$13TiO_2$ and Al_2O_3-$13TiO_2$+ Ni Graphite coatings. Tribol Int 121:353–372
15. Haldar P, Bhattacharya TK, Modak N (2019) Effect of nano-crystalline TiO_2 addition on reciprocating frictional behaviour of alumina ceramics. In: IOP conference series: materials science and engineering, vol 653(1), pp 012010

Thermal and Rheological Behaviour of a Wax Based Binder and an Alumina Feedstock for Micro-PIM Application

S. K. Tanbir Islam, Sudip K. Samanta, Aditya K. Lohar, Santanu Das, and Asish Bandyopadhyay

1 Introduction

Powder injection moulding (PIM) draws significant attention in the field of near-net-shape manufacturing processes form the early 90s of the twentieth century. Micro-PIM (μ-PIM) is the miniaturised form of conventional PIM having four similar steps namely mixing, injection, debinding and sintering [1, 2]. It becomes cost-effective when mass production is the prime concern for complex micro-components [3–5]. At first, the metal or ceramic powder is mixed with thermoplastic polymers known as a binder. Often mixing is done with the help of a sigma blade mixture or twin-screw extruder. Effects of different mixing processes and homogeneity of feedstock (a mixture of binder and powder) were studied by several researchers [6–8]. Injection stage that provides the shape to the feedstock is one of the prime research areas in the powder injection moulding process. In this stage, the feedstock is heated to a certain temperature to melt the binder that offers enough fluidity to the feedstock. Then the melt is injected into the mould cavity by applying pressure [9]. Liu et al. [6] gave their attention to develop a 316L feedstock for the micro-PIM application. Powder size, particle size distribution and surface roughness were found to have affected the final product. Sometimes, the critical aspect ratio of the micro-mould was found to affect complete filling of the mould cavity during an injection [10]. However, the low viscosity of the binder system is desirable at a moulding temperature along with good green strength after moulding. On the other hand, the binder should be

S. K. T. Islam (✉) · A. Bandyopadhyay
Department of Mechanical Engineering, Jadavpur University, Kolkata 700032, India

S. K. Samanta · A. K. Lohar
CSIR-Central Mechanical Engineering Research Institute, Durgapur 713209, India

S. Das
Department of Mechanical Engineering, Kalyani Government Engineering College, Kolkata 741235, India

S. K. Ghosh et al. (eds.), *Advances in Thermal Engineering, Manufacturing, and Production Management*, Lecture Notes in Mechanical Engineering,
https://doi.org/10.1007/978-981-16-2347-9_21

removed completely without deforming the product during debinding stage. Effects of different binder compositions on mould feeling and final product were also studied [11]. To get quality, micro-component with good surface finish, characterizing the powder and the binder become unavoidable part of the PIM process [12].

Fine powder particles are required to achieve a detailed structure. Effects of powder particle size in micro-injection moulding were studied by some of the researchers [13, 14] in terms of mouldable viscosity. Their findings revealed that fine particles did have high surface contact to the binder and at the same time, it increased the chance of particle to particle contact. As a result, feedstock having fine particles was observed to show higher viscosity rather than feedstock having a coarse one.

Besides viscosity, thermal characteristic is equally important for a successful injection moulding. Thermal characterization provides the melting and degradation point of the feedstock and as well as thermal conductivity. Based on this thermal character, the injection temperature, the debinding and sintering cycle was set up [15]. The thermal debinding process was reported to affect the surface roughness of the PIM product [16, 17]. Sintering temperature, time, heating rate and sometimes environment were reported to affect product strength and surface finish to some extent [18–20]. Nowadays, flow simulation of the injection stage has been largely popularised to predict the process parameters that help to avoid defects like a short shot, wrinkle, powder-binder separation, etc. [21]. It is very important to provide thermal and rheological properties to get a realistic output from the flow simulation.

In the present investigation, a new binder combining paraffin wax (PW) and low-density polyethylene (LDPE) is designed and it is mixed with alumina powder to prepare a feedstock for micro-PIM. The alumina powder is characterized by particle shape, size and distribution. The thermal and rheological character of the binder and the prepared feedstock are studied independently. Some basic process parameters of the μ-powder injection moulding are predicted from the characterized data, particularly for this powder binder combination.

2 Materials Used

In the present work, thermally reactive alumina (Al_2O_3) powder (Make: Almatis Alumina Priv. Ltd, Kolkata) is chosen for its low sintering temperature and fine particle size. The chemical composition (as per the manufacturer's datasheet) of the alumina powder is shown in Table 1. The alumina powder is characterized by particle shape and size distribution. A two-dimensional image of the alumina powder is taken by FESEM (Make: ZEISS, Model: Sigma HD) to measure the shape factor (SF), as shown in Fig. 1.

The ImageJ software is used to measure the area and perimeters of the alumina particles and shape factor is calculated using a relation as shown in Eq. 1.

$$\mathrm{SF} = \left(4\pi A / P^2\right) \tag{1}$$

Table 1 Chemical composition of the alumina powder

Compound	Weight %
Na_2O	0.08
Fe_2O_3	0.02
SiO_2	0.03
CaO	0.02
MgO	0.07
Al_2O_3	Balance

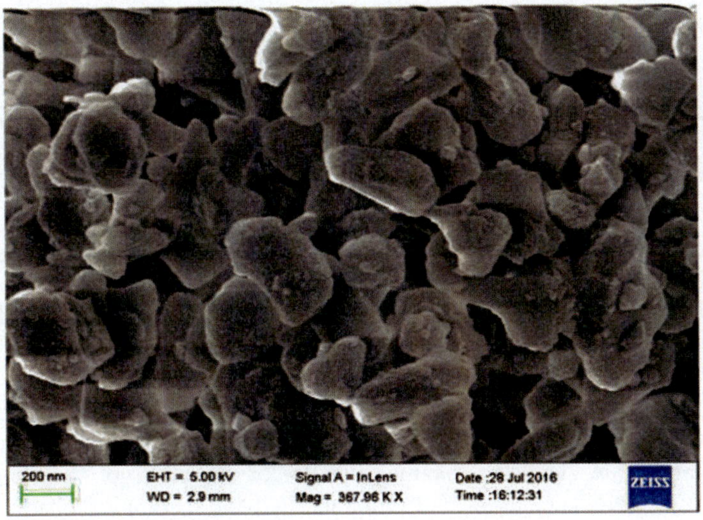

Fig. 1 FESEM image of the alumina powder

Here, *A* and *P* represent area and perimeter of the alumina particles, respectively. The average calculated SF of the said alumina powder is 0.828.

On the other hand, the particle size distribution of the said powder is measured using laser diffraction spectrometry (LDS) method and the results obtained are shown in Table 2.

A binder is designed combining paraffin wax (Make: Acros Organics, Germany) and low-density polyethylene (Make: Goonvean Fibres, UK) in a volume ratio of 65:35, respectively. From the manufacturer's datasheet, it is found that the PW and LDPE have the same density of 0.91 g/cm^3 at room temperature and their corresponding melting points are 60 °C and 90 °C, respectively.

Table 2 The tabular form of particle size distributions of the alumina powder

Percentile	Size (μm)
10.0	0.391
20.0	0.431
30.0	0.470
40.0	0.503
50.0	0.534
60.0	0.557
70.0	0.606
80.0	0.656
90.0	0.758
95.0	0.994

3 Experimental Procedure

It is important to find out the critical powder loading of a feedstock to justify the suitable powder binder ratio to be chosen. The critical powder loading is measured by comparing the theoretical and measured density of the feedstock. Here, the powder and binder are mixed using a sigma blade mixer at a temperature of 393 K (120 °C). The powder loading is increased step by step and corresponding densities are measured. Comparing the theoretical and experimental density, it is found that critical powder loading for the present powder binder combination is 49 vol. %, as shown in Fig. 2. So, a feedstock having a powder loading of 45 vol. % (<49 vol. %) is prepared and abbreviated as A-45. Further thermal and rheological characterization of the binder and the alumina feedstock (A-45) are carried out individually.

First, the point of thermal degradation of the feedstock is figured out using a Thermogravimetry analyser (Make: Perkin Elmer, USA) at a heating rate of 10 °C/min. The weight of the sample is monitored continuously during the experiment to find the weight loss with the increasing temperature.

On the other hand, a Differential Scanning Calorimetry (Make: Perkin Elmer, USA) is used to find the melting and solidification temperature of the binder as well as of the feedstock. In this case, heat flow is measured with temperature variation while heating and cooling rate of 10 °C/min is maintained precisely. In the present work, thermal diffusivity is measured using a thermal analyser (Make: NETZSCH, Model: LFA 467) and thermal conductivity is calculated from the diffusivity data. The sample is placed inside a temperature-controlled chamber on an aluminium crucible having a diameter of 12.6 mm. The chamber temperature is raised to the desired test temperature and thermal diffusivity is measured using a flashlight source method.

Another vital part of the binder and feedstock characterization is a rheological study. Here, a rotational rheometer (Make: Anton Paar, Model: MCR501) having a parallel plate (diameter 15 mm) arrangement is used to carry out the rheological experiment. The gap between the two plates is maintained at 0.1 mm and temperature

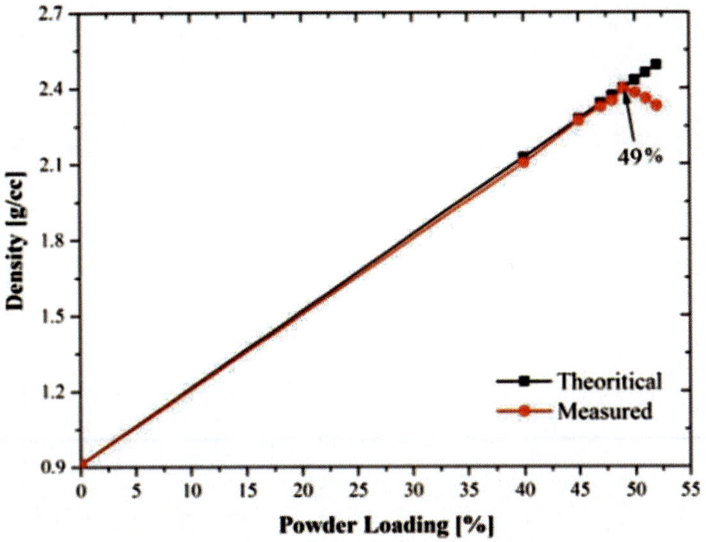

Fig. 2 Critical solid loading of the said powder binder combination

is controlled precisely by a furnace chamber during the rheological tests. The gap between the plates is intentionally reduced to increase the achievable shear rate considering a thumb rule. That thumb rule says that the gap should be at least ten times the largest particle size of the powder of a particle suspended fluid [22].

4 Results and Discussion

Analysis of the test results is essential to understand the material behaviour in various real-life situations. In this section, the thermal and rheology data are analysed to explain the physics behind the material behaviour. Some predictions are made to set up different process parameters (i.e. injection pressure and temperature, debinding and sintering cycle) for a successful powder injection moulding.

4.1 Thermogravimetry Analysis (TGA)

The sample weight with respect to temperature is plotted from the TG analysis of the feedstock (A-45), as shown in Fig. 3. It is found that the feedstock starts losing weight just above 466 K (193 °C) due to the degradation of the PW. Then the rate of degradation increases and becomes maximum near 575 K (302 °C). The assumption can be made that the PW and LDPE both degrade rapidly at this temperature. Above

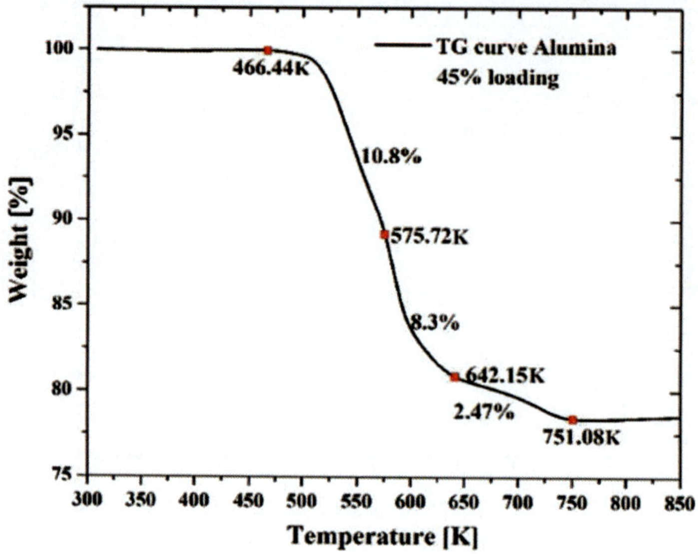

Fig. 3 Thermogravimetry Analysis (TGA) curve of feedstock A-45

642 K (369 °C), weight loss rate slows down while it becomes horizontal above 751 K (478 °C). It signifies that there are no binder ingredients left to degrade above 751 K. The weight percentage of the binder in the said feedstock A-45 is around 22.12% of the feedstock weight. The zone wise weight percentage loss of the feedstock is shown in Fig. 3 and commutatively 97.5% of the binder is removed above 478 °C. From this analysis, it can be concluded that the thermal debinding should be done above 751 K to remove the binder completely. On the other side, the injection should be done at a considerably lower temperature than the degradation point of the PW that is 466 K.

4.2 *Differential Scanning Calorimetry (DSC)*

A Differential Scanning Calorimetry (DSC) analysis of the binder and feedstock A-45 is carried out individually. The heat flow curve with respect to temperature of the binder is shown in Fig. 4a. Two endothermic peaks are found at 336.21 K and 365.52 K during heating of the binder due to melting of PW and LDPE, respectively. In the case of powder injection moulding, the cooling curve is more important as the solidification of the binder occurs after mould filling. Two exothermic peaks are found on the cooling curve of the binder at 357.22 and 334.31 K due to the solidification of the binder ingredients. It is observed that melting and solidification points of LDPE and PW are not the same. Solidification points are shifted to a lower temperature than the melting points.

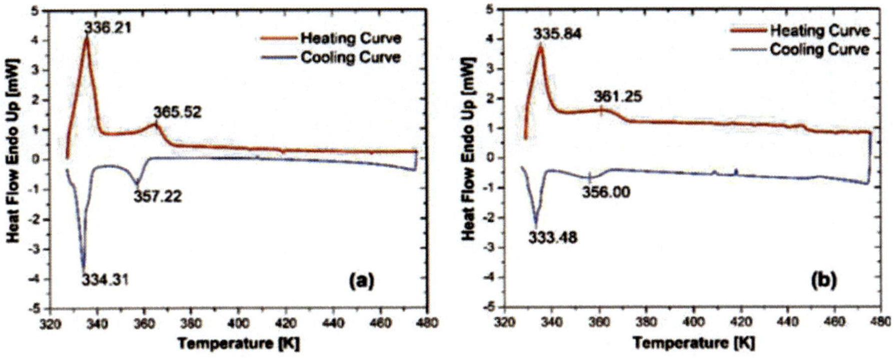

Fig. 4 DSC curve of **(a)** Binder and **(b)** Feedstock A-45

On the other side, a similar curve is plotted from the DSC results of the feedstock, as shown in Fig. 4b. In this case, the binder ingredients get melted and solidified and the powder remains as it is. That is why the A-45 feedstock shows two endothermic and two exothermic peaks just like the binder. Presence of alumina powder shifted the melting and solidification temperature of the binder ingredients to a lower temperature than the pure binder. Melting and solidification temperature of the PW in A-45 feedstock are 335.84 K and 333.48 K while it is 361.25 K and 356 K for LDPE, respectively.

4.3 *Specific Heat (C_p) and Latent Heat of Solidification*

Specific heat and latent heat sometimes play a vital role in solidification of an injected component. Here, the specific heat of solidification is evaluated from the DSC analysis and plotted against temperature. Specific heat curves for solidification of the binder and A-45 feedstock are shown in Fig. 5. In the present work, the area of the peaks on the specific heat curve is measured which gives the latent heat of solidification of the respective materials. Then the measured area is divided by the temperature range of solidification to get the average C_p of the corresponding materials. The calculated latent heat of solidification of the binder is 91,659.91 J/kg, whereas it is only 16,990.36 J/kg for the feedstock. The calculated average specific heat of the binder system and feedstock A-45 are 91,659.91/(364.3–327.05) = 2460.669 J/(K-kg) and 16,990.36/(364.8–327.45) = 454.896 J/(K-kg), respectively.

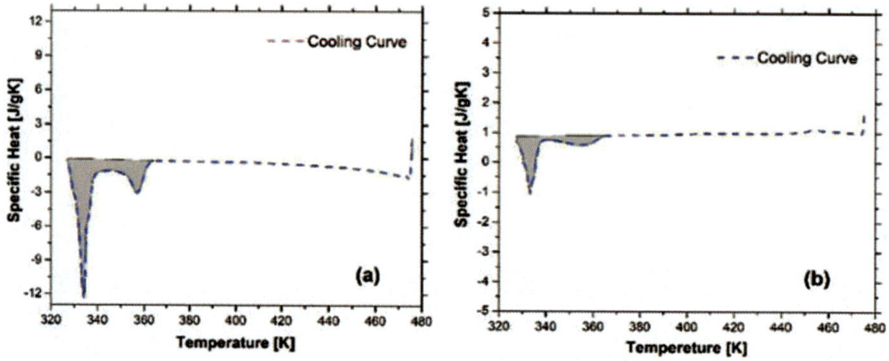

Fig. 5 Specific heat curve during solidification of the **(a)** Binder and **(b)** Feedstock A-45

4.4 *Thermal Diffusivity and Conductivity*

Thermal diffusivity is measured at a wide range of temperatures (150–45 °C) to cover the transmission of binder from the liquid state to solid-state, as shown in Fig. 6. First, the temperature is raised to 150 °C to start measuring thermal diffusivity. Thermal conductivity is calculated using a simple relation as shown in Eq. 2.

$$\lambda(T) = \alpha(T).\rho(T).C_p(T) \tag{2}$$

where, T, λ, α, ρ and C_p denote temperature, thermal conductivity, thermal diffusivity, density, and specific heat respectively. Here, density and specific heat are taken

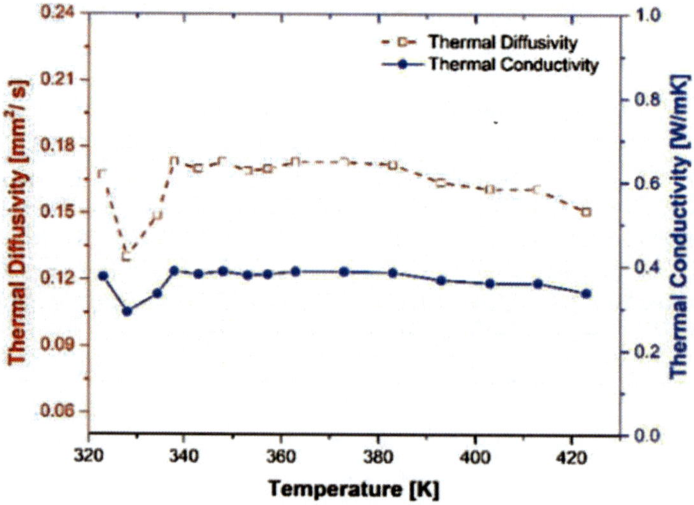

Fig. 6 The plot of the binder's thermal diffusivity and conductivity with respect to temperature

as constant to make the calculation of thermal conductivity simple. Results show that the thermal diffusivity of the binder gradually increases with a temperature drop and it suddenly drops at 328 K. On the other hand, it is found that the thermal conductivity does not vary significantly up to 338 K and it follows the trend of thermal diffusivity below 338 K.

4.5 Rheology of Binder

The degradation point of the binder is already known from the TG analysis as discussed above. Here, all the rheology tests are carried out below the degradation point of the binder. Isothermal tests are carried out at different temperatures to incorporate the effect of temperature on flow behaviour.

The controlled shear rate method is adopted to increase the shear rate from 10 to 10,000/s logarithmically (6 points/decade) and the corresponding viscosity of the binder is measured. The Fig. 7a shows the plot of the binder viscosity with respect to the shear rate at three different temperatures of 373, 383 and 393 K.

Experimental curves are well fitted with a modified Cross model and temperature dependency is incorporated into the expression [23]. A generalised expression for the modified Cross model is shown in Eq. 3 and curve fitting parameters are given in Table 3.

$$\eta(\dot{\gamma}, T) = \frac{\eta_0}{1 + C(\eta_0\dot{\gamma})^{1-n}} \tag{3}$$

where,

$$\eta_0 = B.Exp\left(T_b/T\right)$$

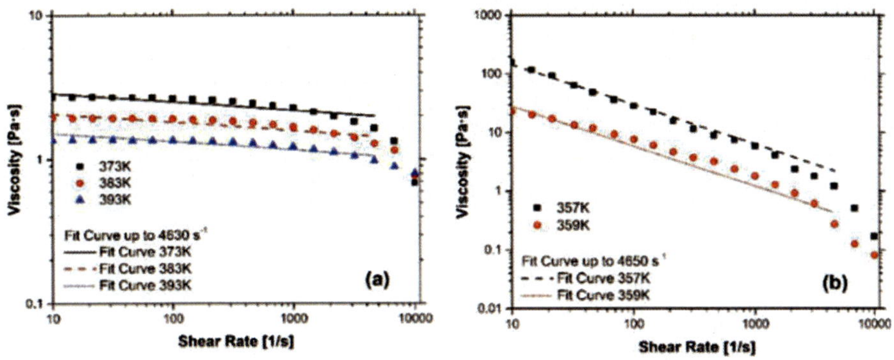

Fig. 7 Flow behaviour of the binder at **(a)** Liquid condition and **(b)** Semi-solid condition

Table 3 Curve fitting parameters of the liquid binder (at 373, 383 and 393 K)

	Shear rate (s^{-1})						
	100	1000	2150	3170	4630	6790	10,000
B	3.83E −05	2.15E−05	5.17E−06	6.05E−06	5.18E−06	1.87E−07	1.27E−06
T_b	5077.831	5081.504	5290.669	5295.124	5334.049	6268.908	5744.492
p	11.00199	5.235006	2.643666	3.228898	3.150257	1.149856	3.517997
q	−0.00246	−0.00193	−0.00373	−0.00452	−0.00487	−0.00278	−0.00787
n	0.9931	0.9658	0.9349	0.9293	0.9161	0.7631	0.8434
R^2	0.9995	0.9921	0.9799	0.9685	0.9543	0.9561	0.8176

$$C = p + qT$$

Here, feedstock viscosity, zero-shear-rate viscosity, shear rate and power-law index are denoted asη,η_0, $\dot{\gamma}$ and n respectively. B, C and T_b are curve fitting constants and the temperature in absolute scale is represented by. The temperature change incorporates some changes in viscosity as observed from Fig. 7a. In the present case, the fitting constant C is replaced by a linear temperature-dependent equation to fit the experimental data simultaneously. The intercept and coefficient of the linear equation are p and$q$, respectively. It is found that shear-thinning behaviour of the binder is not significant in this temperature range. The power-law index at the low shear rate (100/s) is 0.9931 whereas it is 0.9116 at 4630/s shear rate. However, a sharp drop of the power-law index (from 0.9116 to 0.7631) is observed at 6790/s may be due to migration of fluid from the test region. So, the curve fitting parameter up to 4630/s can be considered for mould filling simulation in future work.

Similar kind of rheological experiment is carried out at two different temperatures in the binder's solidification (mushy zone) range and the result is shown in Fig. 7b. Experimental results are well fitted with the same modified Cross model as mentioned above in Eq. 3 and the curve fitting parameters are shown in Table 4. However, a close look at Table 4 reveals that there is a comparatively larger change in powder law index (from 0.3202 to 0.3160) that is occurred above 4650/s shear rate. It signifies

Table 4 Curve fitting parameters of the semi-solid binder (at 357 and 359 K)

	Shear rate (s^{-1})						
	100	1000	2160	3170	4650	6810	10,000
B	1.28E −06	1.50E −06	1.55E −06	1.58E −06	1.62E −06	1.78E −06	2.17E −06
T_b	10,480.82	10,507.89	10,510.99	10,513.77	10,521.18	10,527.75	10,536.75
p	−178.053	−178.028	−178.026	−178.026	−178.026	−178.026	−178.026
q	0.499399	0.499395	0.499394	0.499392	0.499391	0.499389	0.499380
n	0.3235	0.3231	0.3221	0.3214	0.3202	0.3160	0.3089
R^2	0.9886	0.9841	0.9839	0.9836	0.9832	0.9800	0.9725

that the particle might migrate towards edges of the parallel plate above 4650/s shear rate. This particle migration due to centrifugal force on the suspended particle can be explained by slip phenomenon [22]. The migration of the suspended fluid can be overcome to some extent using a serrated/ rough plate instead of a smooth parallel plate.

In the present study, rheology data up to 4650/s shear rate is accepted to model the flow behaviour of the binder. It is found that there is a significant difference in power-law index between the liquid zone (about 0.9116) and mushy or semi-solid zone (about 0.3202) of the said binder. This difference signifies that the binder viscosity is highly dependent on the shear rate at the mushy zone relative to the fully liquid condition.

4.6 Rheology of Feedstock

The rheological characterization of the said feedstock A-45 is done in a similar manner as described above at two different temperatures (145 and 155 °C) and the results are plotted in Fig. 8. In this case, the same modified Cross model fits the flow curves with a coefficient of determination above 0.98 up to 3170/s shear rate. The curve fitted results up to 3170/s shear rate are shown in Table 5. The power law index (0.436505) indicates a strong shear thinning nature of the feedstock.

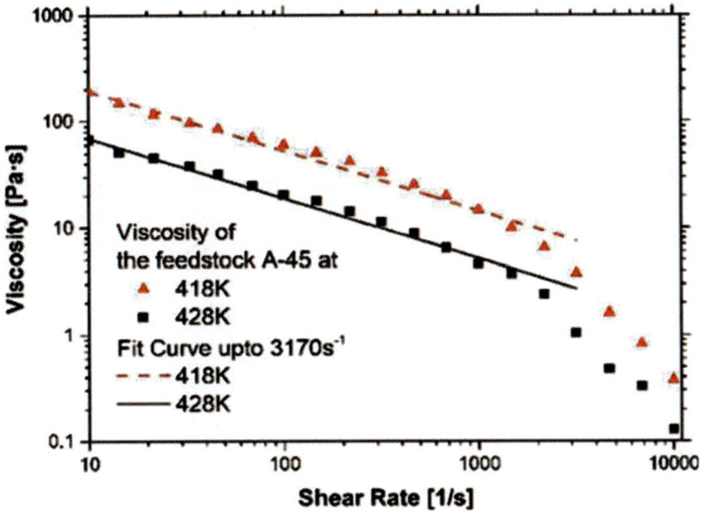

Fig. 8 Flow behaviour of the feedstock A-45 at two different temperatures

Table 5 Curve fitting parameters of the feedstock A-45 at 3170/s shear rate

Curve fitting parameters					
B	T_b	p	q	n	R^2
0.9886	0.9841	0.9839	0.9836	0.9832	0.9800

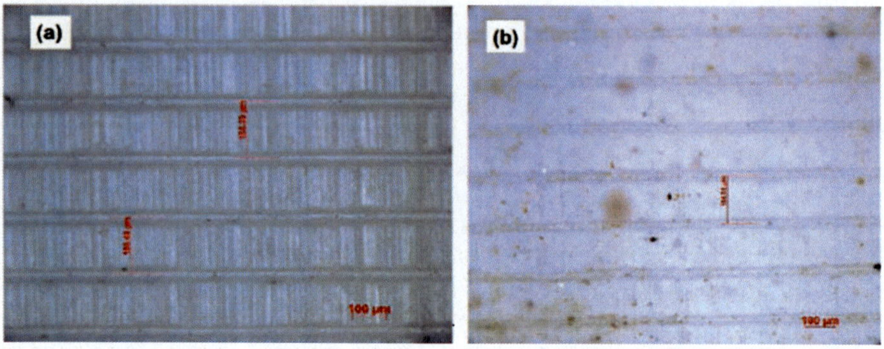

Fig. 9 Optical microscopic image of micro-channels out of alumina feedstock **(a)** Green and **(b)** Sintered

4.7 Component Development

The prepared feedstock A-45 is injected into a micro channelled mould using an injection moulding machine (Make: BabyPlast, Model: 10/6p). The injection pressure and temperature are set to 70 bar and 140 °C respectively. After that, the micro-channel specimens are thermally debound at 500 °C for 3 h and sintered at 1540 °C for 30 min. Optical images of the green and sintered micro-channels are shown in Fig. 9. An Optical microscope is used to capture those images at 10 × magnification and the distance between the peaks is measured in it. The distance between two peaks in the green part is around 186 µm (avg.) and it shrinks to 150 µm (avg.) after sintering. After sintering, the sample density of 3.813 g/cm^3 is achieved which is 97.40% of the actual powder density. A satisfactory level of shape retention is observed in the sintered component.

5 Conclusion

Each step of the micro powder injection moulding process is quite challenging to develop a defect-free product. In the present work, a binder system is designed to prepare an alumina feedstock and they are characterized by means of thermal and rheology analysis. DSC analysis shows that the binder will be completely melted down above 92 °C while TG analysis shows the degradation point (193 °C) of the feedstock. So, it can be predicted that the injection temperature will be set in between 92 to 193 °C. TG analysis shows complete removal of the binder is possible by thermal debinding above 478 °C. Thermal conductivity of the binder remains constant above 90 °C and it shows peaks on solidification temperature of PW and LDPE. On the other hand, the binder and feedstock A-45 show a temperature-dependent shear thinning behaviour in rheology analysis. A modified-Cross model fits the rheology data of the binder as well as the feedstock with a satisfactory level of coefficient of determination. Latent heat and specific heat of the binder and the feedstock are calculated considering that the characterized data can be used for simulation in the future work. The micro-channel sample is injected, debound and sintered successfully with the prepared feedstock.

Acknowledgements Authors would like to acknowledge the support from the individual member of AdMaC, CSIR-CMERI. Authors are thankful to Council of Scientific and Industrial Research (CSIR) for financial support to carry out the project.

References

1. German R, Bose A (1997) Injection molding of metals and ceramics. Metal Powder Industries Federation, Princeton, NJ
2. Piotter V, Honza E, Mueller T (2012) Recent developments in metal and ceramic micro injection moulding. Eur. Congress
3. Petzoldt F (2008) Micro powder injection moulding-challenges and opportunities. Powder Injection Moulding International 2(1):37–42
4. Ni XL, Yin HQ, Liu L, Yi SJ, Qu XH (2013) Injection molding and debinding of micro gears fabricated by micro powder injection molding. Int J Miner Metall Mater 20(1):82–87
5. Oh JW, Lee WS, Park SJ (2018) Investigation and modeling of binder removal process in nano/micro bimodal powder injection molding. Int J Adv Manuf Technol 97(9–12):4115–4126
6. Liu L, Loh NH, Tay BY, Tor SB, Murakoshi Y, Maeda R (2005) Mixing and characterisation of 316L stainless steel feedstock for micro powder injection molding. Mater Charact 54(3):230–238
7. Nurun Nahar ABM, Hossain A, Choudhury IA (2014) Quality feedstock preparation for metal injection molding using Taguchi design of experiment. In: Proceeding of the Second International Conference on Advances in Mechanical, Aeronautical and Production Techniques, pp 55–58
8. Sommer F, Walcher H, Kern F, Maetzig M, Gadow R (2014) Influence of feedstock preparation on ceramic injection molding and microstructural features of zirconia toughened alumina. J Eur Ceram 34(3):745–751
9. Imgrund P, Rota A, PetzoldtF SA (2007) Manufacturing of multi-functional micro parts by two-component metal injection moulding. Int J Adv Manuf Technol 33(1–2):176–186
10. Han JS, Gal CW, Park JM, Kim JH, Park SJ (2018) Experimental analysis for fabrication of high-aspect-ratio piezoelectric ceramic structure by micro-powder injection molding process. Mater Res Express 5(4):046303
11. Zhou PP, Wu GQ, Tao Y, Cheng X, Zhao JQ, Nan H (2018) Optimization of the injection molding process for development of high performance calcium oxide -based ceramic cores. Mater Res Express 5(2):025208
12. Liu ZY, Loh NH, Tor SB, Khor KA (2002) Characterization of powder injection molding feedstock. Mater. Charact. 49(4):313–320

13. Baek ER, Supriadi S, Choi CJ, Lee BT, Lee JW (2007) Effect of particle size in feedstock properties in micro powder injection molding. Mater Sci Forum 534–536:349–352
14. Islam ST, Samanta SK, Nagahanumaniah RH, Lohar AK, Das S, Bandyopadhyay A (2018) Rheological behavior of 316L stainless steel feedstock for μ-MIM. Mater Today Proc 5(2):8152–8158
15. Meng J, Loh NH, Fu G, Tor SB, Tay BY (2010) Replication and characterization of 316L stainless steel micro-mixer by micro powder injection molding. J Alloys Compd 496(1–2):293–299
16. Liu L, Loh NH, Tay BY, Tor SB, Murakoshi Y, Maeda R (2007) Effects of thermal debinding on surface roughness in micro powder injection molding. Mater Lett 61(3):809–812
17. Liu L, da Wang X, Li X, Qi XT, Qu XH (2017) Effects of size reduction on deformation, microstructure, and surface roughness of micro components for micro metal injection molding. Int J Miner Metall Mater 24(9):1021–1026
18. Meng J, Loh NH, Fu G, Tay BY, Tor SB (2011) Micro powder injection moulding of alumina micro-channel part. J Eur Ceram Soc 31(6):1049–1056
19. Song J, Gelin JC, Barrière T, Liu B (2006) Experiments and numerical modelling of solid state sintering for 316L stainless steel components. J Mater Process Technol 177(1–3):352–355
20. Choi JP, Lyu HG, Lee WS, Lee JS (2014) Densification and microstructural development during sintering of powder injection molded Fe micro-nanopowder. Powder Technol 253:596–601
21. Mohamad Nor NH, Muhamad N, Ismail MH, Jamaludin KR, Ahmad S, Ibrahim MHI (2009) Flow behaviour to determine the defects of green part in metal injection molding. Int J Mech Mater Eng 4(1):70–75
22. Chinesta F, Ausias G (2015) Rheology of non-spherical particle suspensions. Elsevier
23. Kwon TH, Park JB (1995) Finite element analysis modeling of powder injection molding filling process including yield stress and slip phenomena. Polym Eng Sci 35(9):741–753

Alumina Based Cutting Tools—A Review

Sourav Mondal, Rajashi Chatterjee, and Partha Haldar

1 Introduction

Ceramic materials used in cutting tools today are based either on alumina (Al_2O_3) or silicon nitride (Si_3N_4). Alumina based cutting tools (ACT) are extensively used as the benchmark for its abundance, cheapness and excellent structural properties [1]. ACT exhibits spectacular mechanical and structural properties, as these can provide long tool life and can carry out machining in hard and tough work pieces like stainless and hardened steel. Its physical properties can be enhanced by various toughening methods like fiber toughening or transformation toughening. Evidently, the machining of most of the complex and hard materials is done through alumina-ceramics and cubic boron nitrides which resembles high hardness at high temperature, chemical stability and its resistance to wearing. There are various advantages associated with using ACT, as it can work out with complex and hard shapes and giving quality surface finish even in tough situations. Various improvements can be made in its tool properties like resistivity to thermal shock and wearing, increased fracture strength and hardness etc. ACT has been found to substitute grinding operations in finishing part of steels, with the help of machining [2]. Machining is carried out between tool and work piece leading to intense abrasion, adhesion and diffusion

S. Mondal
Department of Ceramic Technology, Government College of Engineering and Ceramic Technology, Kolkata, India

R. Chatterjee
Department of Computer Science and Engineering, Government College of Engineering and Ceramic Technology, Kolkata, India

P. Haldar (✉)
Department of Mechanical Engineering, Government College of Engineering and Ceramic Technology, Kolkata, India

S. K. Ghosh et al. (eds.), *Advances in Thermal Engineering, Manufacturing, and Production Management*, Lecture Notes in Mechanical Engineering,
https://doi.org/10.1007/978-981-16-2347-9_22

over the specific section of the cutting tool. Various wear processing such as abrasion, adhesion and plastic deformation is significantly seen in alumina based cutting tools. The history of machining with alumina ceramic was found way back in 1905, in Germany [3]; whereas use of sintered form of alumina based ceramic for cutting tool was first found in Britian, in 1913 [4]. In 1950s, mixed Al-ceramic products were used for the very first time in metal cutting as these require specific mechanical property which a random ceramic product doesn't possess [5]. The wearing process is though regarded as a complex processes as this operates through several wear mechanisms. These mechanisms depends upon different factors such as the type of cutting tool chosen, type of work piece used and several other things [6]. It has been found over the years that, these alumina based cutting tools are more resistant to wearing than other products like cemented carbide and associated equipments.

In the next sections, some initial developments and improvements are discussed. Further, overall characteristics of ACTs in various kinds of work-pieces are discussed. Later on ACTs in machining of specific steel and alloy based work-pieces have been dealt, along with some important conclusions.

2 Some Initial Breakthroughs and Development

After the initial development in 1920s, when Britain and Germany laid the foundation of ACTs in the field of cutting tools as it slowly grew over the years. The earliest development occurred around 1950s when the concept of mixed ceramics came into existence [7], then in 1960s concept of fabrication was introduced for which carbides were used as additives [5]. In 1970s alumina–zirconia cutting tools became important because of its toughness and resistant to thermal shock, around in 1980s alumina reinforced silicon carbide whiskers entered the competition [8]. Since then, it took over the entire arena because of its better internal strength, shock resistance, toughness etc.

In order to make a more clear comparison between the earlier used ACT and today's ACT let's consider some parameters: cutting speed (V), depth of cut (d) and feed (f).

Jianxin et al. [9] in 1997, studied the wear mechanisms in machining of hardened steel and nickel oriented alloy; Inconel 718. V_{max} for steel was 200 m/min and V_{max} for Inconel was 250 m/min with $f = 0.15$ mm/rev and $d_{max} = 2.1$ mm for both. These are way better than the performance of ACT used in 70 s, when V_{max} of steel barely crossed 170 m/min. The feed value here represents the supremacy of today's ACT as compared to that of earlier (about 0.09). Hence, this cutting tool is recommended for finishing operation in today's world. Brandt [10] presented some data on the flank and crater wear of ACT in machining steel ($V_{max} = 450$ m/min, $f = 0.30$ mm/rev, $d = 2.0$ mm). It was observed that the addition of TiC to alumina increased the rate of rake wear. Today, technological advancements in the form of reinforcements and several coatings have been made which enhanced the internal strength, fracture toughness and other important parameters in ACTs. In a nutshell; ACT came a long way through its development over the years and is capable of more improvement in terms of its speed, feed and depth of cut in the coming days. The detailed study of several ACTs and its performance is discussed in the next points, including a vivid comparison of the performance of then and now.

2.1 Machining of Alumina-Boron Carbide (Composite) Cutting Tool

Jung et al. [11] studied performance of alumina-boron carbide cutting tool insert on S15C steel with $V_{max} = 205$ m/min, $f = 0.2$ mm/rev, $d = 0.5$ mm and diameter of work-piece was 120 mm. This composite cutting tool showed higher micro-hardness and toughness than pure Al_2O_3 because of the reduced grain growth due to B_4C particles. The fracture toughness was about 4.2 MPa $m^{1/2}$.

2.2 Machining Detail of Silicon-Carbide Whisker Reinforced Cutting Tool

Kumar et al. [12] studied cutting performance of whisker reinforced alumina cutting tool and Ti[C, N] mixed alumina cutting tool. The study was conducted in dry environment for about 15 min without the involvement of cutting fluid. The advantages of using these tools are that the hard materials with complicated shape and sizes are machined in a single step; also better quality of grinding with improved finishing of the surface can be gained by turning of much strong work pieces.

Schneider [13] stated that machining of hard materials is more often done by ceramics and cubic boron nitrides because of the hardness of their material, ability to resist wearing, superior stability and hot hardness at high temperature. From the various results obtained it is concluded that ceramic cutting tools can increase the metal removal at a much higher rate than those of the conventional tools. Most of these tools further got incorporated with alumina, and so born the concept of alumina based ceramic cutting tools which then spread all over; and got evolved over the years.

Narutaki et al. [14] concluded that while machining Inconel 718, SiC whisker reinforced ACTs showed good performance as compared to mixed ACTs during notch wear. After the experiments, it was found that these ACTs exhibits better performance in overall wear resistance than Ti[C, N] mixed ACTs on machining martensitic stainless steel.

2.3 Machining Detail of Zirconia Toughened ACT

Senthil et al. [12] conducted machining performance test on same work piece, but the tool was zirconia toughened ACT having particle and whisker reinforcement. They found that the performance of Ti[C, N] ACT is better than zirconia toughened ACT with respect to surface finish. Excellent strength and fracture toughness of zirconia is due to tetragonal to monoclinic phase transformation. Mondal et al. [15] examined the performance of the developed zirconia toughened alumina cutting tool, during machining of low and medium strength carbon steel and concluded that ceramic composite cutting tool is having low power consumption with good mechanical properties. Hard steel along with cast iron machined by ZTA based ACTs give result better than plain ceramics as it has improved the strength toughness. These features are responsible for its widespread application.

2.4 Machining Details of ACTs on Ultra High Strength Steel

Wang et al. [16] found important results on machining very high strength steel with mixed alumina-composite (Al_2O_3 + TiC + TiN). They reported ACTs are better choice for turning operation rather than other cutting tools, he also concluded turning at 300 m/min. or above gave comparatively longer tool life.

According to Barry and Brine [17] hard machining can offer several advantages over the traditionally carried out grinding or turning processes. In other cases it has been found that machining of steels which contains Ca mixed oxide can lead to a reaction between the oxide layer and alumina phase, resulting in tool wear in many cases. Recently, many experiments claimed that addition of the secondary phase multi-sized particles into the matrix of Al_2O_3 ceramic was a great method to improve both flexural strength and fracture toughness.

Blinov et al. [18] found that titanium carbide and TiN addition in alumina can restrict the growing up of alumina grain, and hence can improve the fracture toughness of the ACTs and hence chosen for machining of such high strength steel. These tools are mostly used to cut very high-strength steels at relatively low cutting speeds (less than 100 m/min).

3 Performance of Some Important ACTs

After particular review of the performances of the individual ACTs along with the respective work-pieces, let's now compare their performance as per the years they were performed (Please refer Table 1). Some conclusions drawn here are:

It can be seen from [11, 16, 19] that reinforcements and coating have effectively enhanced the strength of the cutting tool materials. This happened because B_4C

Table 1 List of some important alumina based cutting tools

Sl.No	Year	Source	Cutting tool	Work-piece	Cutting tool properties				
					Speed (V_{max}) (m/min)	Feed (mm/rev)	Depth of cut (mm)	Density (g/cm^3)	Fracture toughness
1	1976	[24]	Al_2O_3 + TiC + TiN	High carbon steel	200	0.40	2.0	4.40	7.60
2	1981	[25]	Al_2O_3 + TiC (mixed ceramics)	Incoloy 901	230	0.200	2.00	4.20	4.5
3			Al_2O_3 + SiC (mixed ceramics)	Inconel 718	150	0.125	1.80	4.30	7.5
4	1982	[20]	Lucas SYALON	Incoloy 901	310	0.180	2.40	5.40	6.2
5	1994	[26]	Al_2O_3 + TiC + SiC	ADI	300	0.100	1.00	4.20	7.5
6	1995	[21]	Whisker reinforced ACT	S.G. Iron	600	0.063	0.75	4.30	5.3
7	1997	[9]	Al_2O_3 + TiB_2	Inconel 718	100	0.150	1.50	4.40	5.1
8			Al_2O_3 + TiB_2 + SiC	Inconel 718	200	0.150	1.50	4.40	7.8
9	1997	[27]	Al_2O_3 + TiC	Inconel 718	500	0.190	0.50	3.97	4.5
10	2000	[28]	Al_2O_3 + TiC + WC	Mild carbon steel 0.45%C	160	0.200	0.30	4.30	5.6
11	2001	[17]	Al_2O_3 + TiC	BS817M40	250	0.100	0.20	4.20	5.6
12	2002	[29]	Aluminium carbide K 10	Carbon fibre reinforced plastics	20	0.150	1.00	1.6	4.4
13	2004	[30]	Al_2O_3 + CeO_2	Grey cast iron-170 BHN	170	0.120	0.50	4.00	4.4

(continued)

Table 1 (continued)

Sl.No	Year	Source	Cutting tool	Work-piece	Cutting tool properties				
					Speed (V_{max}) (m/min)	Feed (mm/rev)	Depth of cut (mm)	Density (g/cm^3)	Fracture toughness
14	2005	[11]	$Al_2O_3 + B_4C$	S15C steel	205	0.200	0.50	3.53	4.2
15	2005	[23]	$Al_2O_3 + WC$	Copper based alloy	360	0.100	1.50	15	4.3
16	2005	[19]	$Al_2O_3 + TiN$	Aerospace alloy	170	0.100	1.50	4.00	5.2
17	2006	[12]	Whisker reinforced alumina	Stainless steel 410	270	0.120	0.50	3.74	8.0
18	2007	[31]	Sintered YTCA	Grey cast iron	270	0.120	0.50	3.90	4.9
19	2007	[22]	$Al_2O_3 + SiN$	Inconel 718	300	0.150	1.82	4.40	5.0
20	2009	[32]	$Al_2O_3 + TiC$	Si alloys	400	0.100	0.50	4.20	5.6
21	2011	[33]	ZTA	AISI 4340 steel	420	0.240	2.00	9.80	14.0
22	2012	[34]	$Al_2O_3 + TiC + TiN$	AISI 52100 steel	300	0.140	0.50	4.40	7.6
23	2016	[16]	$Al_2O_3 + TiC + TiN$	High strength 5140 steel	300	0.100	0.10	4.50	7.8

particles enhanced the micro-hardness and decreased the grain size of the composites. From [9, 12, 20, 21]; it is observed that addition of TiN and TiC have enhanced the wear resistance and tool life of materials over the years. It is because of the reason that carbide and nitride, even after being used as low speed cutting tools; gave better surface finish with better tool life. In [22]; hot pressing of TiC based ACTs are being observed and found to have increased the flexural strength of tool. This is because of the chemical affinity between the work-piece and cutting tool; which gets activated at high temperatures. References [12, 23] shows, grain growth of alumina can be controlled by addition of SiC in the tool. This is because, at higher temperatures solubility of SiC on the metal increases which restricts the grain growth.

As, it can be seen from [11, 16, 19]; over the years whisker reinforced ACTs are found more promising as they can operate at very high speeds and can give a better surface finish. This is due to the reduced flank and notch wear, even at high speed; along with better tool life.

4 Performance on Specifically Steel Work-Pieces

Over the years, several types of steel work-pieces have been chosen to perform turning and drilling operation, the cutting details is shown in Table 2. Some conclusions drawn here are:

From [33, 35], it is observed that ZTA can perform operations in steel based work-pieces, both in high and low speed effectively. This is because of the physical parameters of the work piece, which leads to change in cutting speed according to the strength. Better surface finish shows, it is capable of carrying out operations at both low and high speed. References [36, 37] shows, molybdenum oxide as additives can increase strength and work-life of the tool; whereas Young's modulus is decreased with the high Mo-oxide content. This is because Mo addition acts as a second phase, which enhances the strength of the tool material. Mo-oxide is also responsible for the special lubricating properties of the tool. Mo-oxide at 15 wt.% condition shows good lubricating properties in dry conditions.

From [33, 38, 39], it can be seen that the wear rate decreases with the increase in concentration of alumina. This is due to alumina's chemical properties, which is responsible for crack and wear resistance of the tool. References [38–40] shows high cutting speed in machining of steels results; increase in flank wears but decrease in surface roughness. This is because flank wear is the most dominant of the other wears when cutting speed is higher than 250 m/min whereas; high speed cutting can give a better surface finishing, reducing the chances of roughness.

In general silver and molybdenum led ACTs are better than ZTA in terms of thermal conductivity. From [39] it can be seen, in turning of 42 HRC steel with coated ACTs; tool wear rate increases smoothly. This is because alumina powder was doped with MgO to prevent abnormal tool wear and grain growth of the material.

Table 2 Performance of alumina based cutting tools on steel based work-pieces

Sl.No	Year	Source	Cutting tool	Work-piece	Cutting tool properties				
					Speed (V_{max}) m/min	Feed mm/rev	Depth of cut (mm)	Density (g/cm^3)	Fracture toughness
1	1976	[24]	Al_2O_3 + TiC + TiN	High carbon steel	200	0.40	2.0	4.40	7.60
2	1997	[35]	ZTA	AISI 1040 steel	3.9 (slow)	0.25	2.0	9.80	14.00
3	2000	[28]	Al_2O_3 + TiC + WC	Mild carbon steel 0.45%C	160	0.200	0.30	4.30	5.6
4	2001	[17]	Al_2O_3 + TiC	BS 817M40 steel	250	0.10	0.2	4.20	5.60
5	2003	[41]	Al_2O_3 + TiC + TiN	100 CR6 steel	210	0.28	0.2	4.40	7.60
6	2005	[38]	Al_2O_3 + TiC + TiN	AISI 4340 steel	180	0.40	2.0	4.40	7.50
7			Al_2O_3 + TiC + TiN	AISI D2	220	0.15	0.2	4.40	7.50
8	2006	[39]	ZTA	C 45 steel	400	0.24	2.0	9.80	14.0
9	2007	[40]	Al_2O_3 + TiC	AISI D2	150	0.15	0.8	4.20	5.60
10	2011	[33]	ZTA	AISI 4340 steel	420	0.240	2.00	9.80	14.0
11	2013	[36]	Al_2O_3 + MoO_3	145 CR6 steel	200	0.17	0.5	4.70	4.24
12	2017	[37]	Al_2O_3 + MoS_2	AISI 304 steel	120	0.16	1.2	4.65	4.30

5 Performance on Specific Alloy Based Materials

Various turning operations have been carried out, in these vast range of alloy materials. The cutting details along with the year in which they were performed is shown here (Please refer Table 1). Some conclusions drawn here are:

From [25], the rate of tool wear can be determined by alloy carbide content of the work material and by the effect of temperature. This is because temperature is an important parameter for determining rate of wearing, and the wearing of tool is dependent on carbide content; as more carbide corresponds to less wearing and vice-versa. From [20], it was concluded that if the hardness of work-piece material made up to 70HRC, flank wear can be reduced. This is because flank wear is dependent on the hardness and density of the material, above 70 HRC the tool shows specific resistance to wearing; as more hardness corresponds to less flank wear and vice versa. From [27], it can be concluded that for aerospace materials and nickel-based alloys, cutting speeds usually ranges from 30 to 100 m/min. This is because the cutting speed depends on the material of the work-piece; some prominent machining parts involves cutting speed of 100 m/min. whereas some can be as less as 30 m/min, depending on the characteristic of the work piece material.

According to [19, 22, 31, 32]; ACTs are suitable for using at very high speed during machining. After the introduction of sialon, Inconel 718 can be machined using whisker reinforced ACT at very high speed. This is because, machining at high speed leads to maximum chances of notch and flank wear, this wearing tendency is resisted by whisker reinforcement of the tool. From [29]; it was observed that during machining of nickel based super alloys, notch wear is obtained at low cutting speed. It is because, in case of alloys; this wearing is usually seen at very high speed (300 m/min. or more); in case of nickel based super alloys, this wear was also seen at as low as 25 m/min.

According to [9], developments in ACTs have been made by refining of microstructures and improving manufacturing processes, toughening processes such as whisker toughening and transformation toughening. It is because of these toughening processes, the fractural strength of these ACTs, improved over the years, as the conventional ones having high brittleness possessed very low fractural strength. Hence, the quality of ACT in terms of strength was improved. From [22, 23] it is observed that in turning of hard steels $Al_2O_3/TiB_2/SiC$ tool exhibited lower flank wear resistance than that of the Al_2O_3/TiB_2. This is due to the presence of SiC in the former one, which restrained the flank wear even at high cutting speed.

6 Modern Fabrication and Coating

Coating is an important procedure to improve the surface strength; the materials used for coating are tungsten and titanium carbide powders having grain size the range of 0.65–1 μm. To obtain a good coating result the grain of the carbide powders should be clean and free from aggregation [42]. For this, the carbide powders are cleaned and then made finer and then added into the alumina sol. After that, when these are completely mixed with the sol, the powder needs to be coated with the sol of alumina. This will lead to a formation of uniform stagnant layer, and the whole process can be termed as fabrication. The coating can be removed by heating and drying inside a vacuum chamber, then crushed and screened to separate the powder. Interestingly, it has been found that low concentration of the alumina sol will result to thin layer of formation, and high concentration encountered the problem of aggregate formation [43]. The best thickness is achieved at a stage when maximum width can be managed without aggregation. The standard concentration used for this purpose is around 0.6 Molar with TiC powder. The fracture toughness, hardness and internal strength of these tool materials have been tested and compared with other tool materials and is found to be better than them.

The coating has improved several factors like Vicker's hardness, fracture toughness and tool life. In cases of machining ultra high strength steels; coated cemented carbide and ACTs are mainly used as they give high cutting temperature, low tool duration, improved cutting force even at very low speeds [44]. The wear resistances of these coated ACTs are found to be better than that of uncoated tools; the cutting speeds of coated tools for machining high strength materials can be kept even below 140 m/min, if required [45]. It is also regarded as one of the main working fields in future, as it does not affect the chemical strength and structure of the material but provides a physical strength of envelope to resist the surface deformation. Hence, it is concluded that alumina based cutting tools have shown better promising results in terms of strength and machining performance after coating, as compared to other cutting tools.

7 Conclusions

The development of alumina as a cutting tool material for last forty years is shown here, and it is still expanding. They are the most promising cutting materials in twenty-first Century. The researchers are trying to improve the machining performance of monolithic alumina with the introduction of coating, formation of nano-composites, altering the microstructures etc. Researchers are also trying to improve the fracture toughness of alumina since till date it is much lower than conventional materials. After going through all these prospects and developments of various alumina based cutting tools, it can be seen the progress took place both in forms of reinforcements as well as coating and surface oriented developments. These not only improved the surface strength but also the internal strength factors like fracture toughness and hardness. Hence, we conclude these cutting tools, specifically oriented with alumina have significantly went under various developments in the past; an extensive development and experimentation is going on in the present and these provides a promising future in the field of cutting tool and related industries. In near future, development and use of alumina based cutting tools will become increasingly important as high-speed cutting tools.

References

1. Rittidech A, Somrit R, Tunkasiri T (2013) Effect of adding Y_2O_3 on structural and mechanical properties of Al_2O_3 ZrO_2 ceramics. Ceram Int 39:S433–S436
2. Silva LR, Abrao AM (1999) Machining of the hardened AISI 4340 steel using ceramic and PCBN tooling. Sci Eng J 8:85–92
3. Ezugwu EO, Wallbank J (1987) Manufacture and properties of ceramic cutting tools: a review. Mat Sci Technol 3:881–887
4. Ai X, Li ZQ, Deng JX (1995) Development and perspectives of advanced ceramic tool materials. In: Key engineering materials, Trans Tech Pub Ltd., vol 108. pp 53–66
5. Furukawa M, Nakano O, Takashima Y (1985) Fracture toughness of the system Al_2O_3-TiC ceramics. Nippon Tungsten Rev 18:16–22
6. Hannink RH, Pascoe RT, Garvie RC (1975) Ceramic steel. Nature 258(703):704
7. Clausen N (1976) Fracture toughness of Al_2O_3 with an unstabilized ZrO_2 dispersed phase. J Am Ceram Soc 59:49–51
8. Tiegs TN, Becher PF (1987) Thermal shock behavior of an alumina SiC whisker composite. J Am Ceram Soc 70:C–109
9. Jianxin D, Xing A (1997) Wear behavior and mechanisms of alumina based ceramic tools in machining of ferrous and non-ferrous alloys. Tribol Int 30:807–813
10. Brandt G (1986) Flank and crater wear mechanisms of alumina-based cutting tools when machining steel. Wear 112(39):56
11. Jung CH, Lee SJ (2005) Machining of hot pressed alumina–boron carbide composite cutting tool. Int J Refract Met Hard Mater 23:171–173
12. Kumar AS, Durai AR, Sornakumar T (2003) Machinability of hardened steel using alumina based ceramic cutting tools. Int J Refract Met Hard Mater 21(109):117
13. Schneider J (1999) Ceramics and CBN. Manuf Eng 1(66):73
14. Narutaki N et al (1993) High-speed machining of Inconel 718 with ceramic tools. CIRP Ann 42:103–106
15. Mondal B et al (1992) Development and performance of zirconia-toughened alumina ceramic tools. Wear 156:365–383
16. Wang D et al (2017) Fabrication and cutting performance of an Al_2O_3 or TiC or TiN ceramic cutting tool in turning of an ultra-high-strength steel. Int J Adv Manuf Tech 91:1967–1976
17. Barry J, Byrne G (2001) Cutting tool wear in the machining of hardened steels: Part I: alumina or TiC cutting tool wear. Wear 247:139–151
18. Blinov EV et al (2008) Machinability of the high strength corrosion resistant high ductility austenitic steel 06Kh22AG15N8M2F. Russ Metall Met 2:128–132
19. Sánchez JM et al (2005) Microstructural characterisation of material adhered over cutting tool in the dry machining of aerospace aluminium alloys. J Mater Process Technol 164:911–918
20. Bhattacharyya SK, Wallbank J, Jawaid A (1982) Syalon proves its points in cutting ability and life research. Met Working Prod 126:104–108
21. Sornakumar T et al (1995) Development of alumina and Ce TTZ ceramic–ceramic composite cutting tool. Int J Refract Met Hard Mater 13:375–378
22. Altin A, Nalbant M, Taskesen A (2007) The effects of cutting speed on tool wear and tool life when machining Inconel 718 with ceramic tools. Mater Des 28:2518–2522
23. List G et al (2005) Wear behaviour of cemented carbide tools in dry machining of aluminium alloy. Wear 259:1177–1189

24. Funk R et al (1976) Coating of cemented carbide cutting tools with alumina by chemical vapor deposition. J Electrochem Soc 123:285
25. Baker RD (1981) Ceramic cutting tools: application guidelines. Carbide Tool, 28–35
26. Masuda M, Sato T, Chujo Y (1994) Cutting performance and wear mechanism of alumina-based ceramic tools when machining austempered ductile iron. Wear 174:147–153
27. Kitagawa T, Kubo A, Maekawa K (1997) Temperature and wear of cutting tools in high speed machining of Inconel 718 and Ti|6Al| 6V|2Sn. Wear 202:142–148
28. Huang CZ, Wang J, Ai X (2000) Development of new ceramic cutting tools with alumina coated carbide powders. Int J Mach Tools Manuf 40:823–832
29. Teti R (2002) Machining of composite materials. CIRP Ann 51(611):634
30. Kumar AS, Durai AR, Sornakumar T (2004) Development of alumina-ceria ceramic composite cutting tool. Int J Refract Met Hard Mater 22:17–20
31. Kumar AS, Durai AR, Sornakumar T (2007) Development of yttria and ceria toughened alumina composite for cutting tool application. Int J Refract Met Hard Mater 25:214–219
32. Roy P et al (2009) Machinability study of pure aluminium and Al 12% Si alloys against uncoated and coated carbide inserts. Int J Refract Met Hard Mater 27:535–544
33. Mandal N et al (2011) Optimization of flank wear using zirconia toughened alumina cutting tool: taguchi method and regression analysis. Measurement 44:2149–2155
34. Aslantas K, Ucun I, Cicek A (2012) Tool life and wear mechanism of coated and uncoated Al_2O_3/TiCN mixed ceramic tools in turning hardened alloy steel. Wear 274:442–451
35. Casto SL et al (1997) Wear rates and wear mechanisms of alumina-based tools cutting steel at a low cutting speed. Wear 208:67–72
36. Broniszewski K et al (2013) Al_2O_3-Mo cutting tools for machining hardened stainless steel. Wear 303:87–91
37. Sharma AK et al (2017) Novel uses of alumina-MoS_2 hybrid nanoparticle enriched cutting fluid in hard turning of AISI 304 steel. J Manuf Processes 30:467–482
38. Lima JG et al (2005) Hard turning: AISI 4340 high strength low alloy steel and AISI D2 cold work tool steel. J Mater Process Technol 169:388–395
39. Dutta AK, Chattopadhyaya AB, Ray KK (2006) Progressive flank wear and machining performance of silver toughened alumina cutting tool inserts. Wear 261:885–895
40. Davim JP, Figueira L (2007) Comparative evaluation of conventional and wiper ceramic tools on cutting forces, surface roughness, and tool wear in hard turning AISI D2 steel. Proc Inst Mech Eng Part B J Eng Manuf 221:625–633
41. Benga GC, Abrao AM (2003) Turning of hardened 100Cr6 bearing steel with ceramic and PCBN cutting tools. J Mater Process Technol 143:237–241
42. Bhattacharyya SK, Ezugwu EO, Jawaid A (1989) The performance of ceramic tool materials for the machining of cast iron. Wear 135:147–159
43. Wayne SF, Buljan ST (1990) Wear of ceramic cutting tools in Ni-based superalloy machining. Tribol Trans 33:618–626
44. Chandrasekaran H, Johansson JO (1994) Chip flow and notch wear mechanisms during the machining of high austenitic stainless steels. CIRP Ann 43(101):105
45. Klimenko SA, Mukovoz YA, Polonsky LG (1996) Cutting tools of superhard materials. In: Key engineering materials, Trans Tech Publications Ltd., vol 114. pp 1–66

Deposition of Duplex and Multilayer Electroless Ni–P/Ni–B Coating and Study of Their Wear Resistance

Palash Biswas, Suman Kalyan Das, and Prasanta Sahoo

1 Introduction

Electroless Nickel coating was initially developed by Brenner and Riddell [1]. Electroless deposition technique doesn't require electricity during deposition process. Electroless deposition method has experienced more popularity for industrial applications due to excellent tribological, electrical, mechanical and physical properties [2]. Producing homogeneous coatings with electroless deposition is possible which also exhibits high-quality adhesion on different substrate materials. Basically, electroless deposition method is carried out by a chemical reduction process. Here, reduction of metal ions occurs through autocatalytic way from aqueous solution containing source of metal ion, a reducing agent, stabilizers and operating in a specific pH value and temperature [3]. Electroless processes allow producing very uniform homogeneous coatings. Properties of the developed coating, deposition rate and structural nature predominantly depend on plating bath conditions and composition. During the deposition period, every parameter plays a specific role for the development of microstructure. Bath load and temperature is an effective parameter for higher deposition rate [4]. Electroless Ni–B-based coatings have better mechanical and tribological performance depending on the coating thickness while Ni–P coating provides the best corrosion resistance [5].

To increase corrosion resistance, alloying element like Zn, Mo, Cu, etc. [6–8] is added to electroless Ni–P and Ni–B deposits. However, researchers have also paid considerable attention to duplex Ni–P/Ni–B coating as well as multilayer coating deposition which combines the best of the mechanical and tribological properties of the individual deposits [9]. Wear rate is very low in the duplex coatings [10]. The present work involves with preparation of duplex Ni–B/Ni–P coating and multilayer

P. Biswas (✉) · S. K. Das · P. Sahoo
Department of Mechanical Engineering, Jadavpur University, Kolkata, India

S. K. Ghosh et al. (eds.), *Advances in Thermal Engineering, Manufacturing, and Production Management*, Lecture Notes in Mechanical Engineering,
https://doi.org/10.1007/978-981-16-2347-9_23

Ni–B/Ni–P coatings as given in Table 3, using dual bath technique. These coatings are also evaluated for their tribological (like wear resistance, friction) and mechanical performance (micro-hardness).

2 Experimental Procedure

2.1 Sample Preparation and Coating Deposition

Multilayer EN coatings deposited on mild steel substrate of 15 mm × 15 mm dimension and thickness of 2 mm (used for characterization) and cylindrical specimens of diameter of 6 mm and length 30 mm (for tribological test). All the surface of specimens was ground by 400 grade emery paper. To remove foreign materials and rust, the coatings are degreased using soap water and acetone. Deionized water is employed for rinsing the samples at each intermediate step. Finally, pickling treatment was performed for the samples using 50% HCl for 10 min. Ni–B coating bath is prepared with deposition conditions as given in Table 1 [11]. Before immersing into electroless bath, the specimens are activated by dipping into palladium chloride solution at 55 °C. For the duplex coatings, the specimens were first immersed into the Ni–B bath for inner layer of deposition for two-hour duration. For the source of nickel ions Nickel chloride is used and sodium borohydride ($NaBH_4$) is used as a reducing agent for the alkaline bath. As complexing agent, ethylenediamine ($C_2H_8N_2$) is used and lead nitrate Pb $(NO_3)_2$ acts as a stabilizer. NaoH is used as buffering agent. In the meantime, electroless Ni–P coating bath is made ready for further deposition as the outer layer, whose main constituents include deposition condition presented in Table 2 [12]. For the multilayer coatings, four successive layers of the Ni–B and Ni–P coatings (Coating C) are deposited for an hour and then altered [10]. Double bath technique is used to acquire desired coating thickness. An electronic heater cum stirrer is used to monitor the pH value of the solution, time by time with the help of a digital pH metre. After 4 hours of deposition, samples are again rinsed in deionized water.

Table 1 Electroless Ni–B bath and deposition condition [11]

S. No.	Component and condition	Quantity
1	Nickel chloride (g/l)	20
2	Sodium borohydride (g/l)	0.8
3	Ethylenediamine (g/l)	59
4	Lead nitrate (g/l)	0.0145
5	Sodium hydroxide (g/l)	40
6	pH	12.5
7	Temperature (°C)	90 ± 2

Table 2 Electroless Ni–P bath and deposition condition [12]

S. No.	Component and condition	Quantity
1	Nickel–chloride (g/l)	25
2	Nickel–sulphate (g/l)	25
3	Sodium–hypophosphite (g/l)	24
4	Sodium–succinate (g/l)	12
5	Temperature (°C)	90 ± 2
6	pH	12.5

Table 3 Coating developed and their nomenclature

Coating types	Coating composition	Nomenclature
Duplex	Ni–B/Ni–P (Ni–B inner layer and Ni–P outer layer) coating	Coating A
	Ni–P/Ni–B (Ni–P inner layer and Ni–B outer layer) coating	Coating B
Multilayer	Ni–B/Ni–P/Ni–B/Ni–P(4layers of alternate Ni–B and Ni–P) coating	Coating C
	Ni–P/Ni–B/Ni–P/Ni–B (4layers of alternate NiP and NiB) coating	Coating D

2.2 *Hardness Measurement of Multilayer EN Coating*

In the current study, micro-hardness test of samples is carried out in a Vickers tester (UHL-VMHT) with indentation load 100 g-f at 25 μm/s for dwell time of 15 s. A square-based diamond indenter is used for indentation of the samples. The indentation size is measured employing an optical microscope. Lower value of indentation load is chosen due to low coating thickness. Three repeated values are recorded and the average value is reported.

2.3 *Tribological Test of Multilayer EN Coating*

The friction and wear tests are performed on the coated samples in dry condition with 40% humidity and at an ambient temperature 32 °C in tribo-tester apparatus (pin-on-disc TR-20LE-CHM-400, Ducom, India). ASTM standard G99 provides the guidelines regarding the tests. With the help of holder, a developed duplex electroless Ni–P/Ni–B coated pin having a length of 30 mm and diameter of 6 mm is placed stationary on a rotating disc. The counter face disc is prepared of hardened steel so that it suffers negligible wear. The track diameter and the speed are fixed at 60 mm and 50 RPM, respectively [13]. The sliding distance is fixed at 94,200 mm. Winducom 2010 is the software that records and displays the COF. $K = V/(S \times F)$ is the relation

Table 4 Deposition coating types and hardness values

Sl No	Deposition coating types	Hardness values
1	Coating A	797 (HV0.1)
2	Coating B	852 (HV0.1)
3	Coating C	620 (HV0.1)
4	Coating D	1007 (HV0.1)

which is used to calculate the wear rate where K, S, F and V are wear rate, sliding distance (m), normal load (N), and wear volume (mm^3), respectively.

2.4 *Microstructural Study*

SEM (Scanning Electron Microscope—JSM 6360, JEOL, Japan) is employed to study the surface characteristics of the developed coatings.

For the preparation of cross-section, first the samples are cut by hacksaw and ground with different grades of emery paper up to 2400 grit. Finally, for the mirror finish, polishing with diamond paste was done in a polishing machine and etched with 10% nitral solution.

3 Results and Discussion

3.1 *Microstructure*

Surface morphology of multilayer Electroless nickel coating reveal nodular structure as shown in Fig. 1. The coating sample is free from any impurity. Appeared nodule size in the deposited sample has radii between 10 μm and 15 μm. The surface of the substrate appears to be homogeneous and also free from pores.

3.2 *Hardness Characterization of Multilayer Electroless Nickel Coating*

Electroless nickel coatings are renowned for high hardness and the multilayer nickel coating is found to have improved hardness values. Coating B is found to have better hardness than coating A in as-deposited condition as shown in Fig. 2. Coating B hardness values have been raised by 7% from coating A due to microcrystalline deposit with lower phosphorous content. Coating C less hard than coating A. Hardness value has been decreased by nearly 28%.

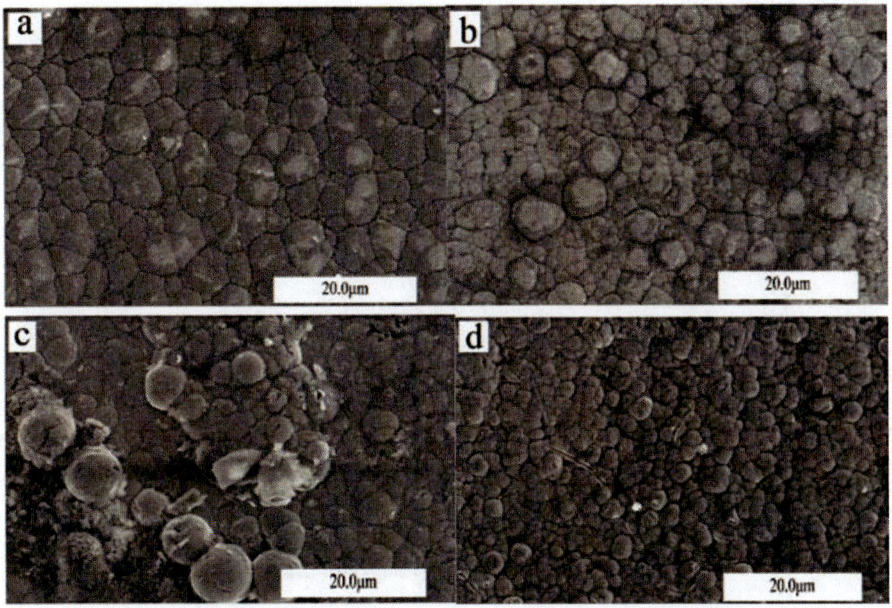

Fig. 1 **a** Coating types A. **b** Coating types B. **c** Coating types C. **d** Coating types D

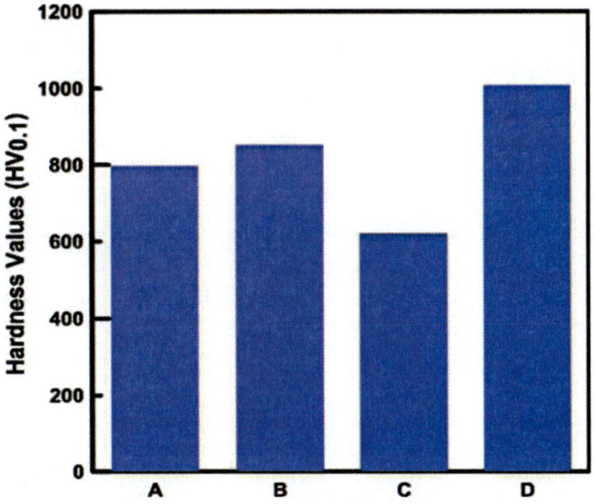

Fig. 2 Hardness values with various types of coatings

3.3 Friction Performance

Multilayer electroless nickel coatings friction performance is inherently smooth. Coefficient of friction (COF) plot for various types of coating is given in Fig. 3. COF of coating A is higher compared to all other coatings due to smoother surface. For

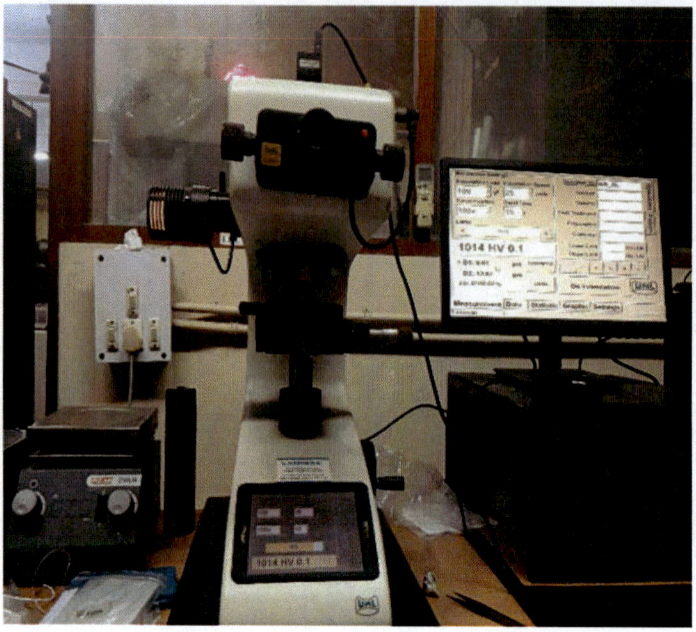

Fig. 3 Micro-hardness tester

coating B, COF reduces significantly by 24% with respect to coating A. COF for multilayer (Coating C) is also reduced by 10% with respect to A. Coefficient of friction for Coating D (multilayer Ni–P/Ni–B) is lesser than duplex coating B. The reduction rate of COF is almost around 26%. The maximum COF is observed for duplex Ni–B/Ni–P for polycrystalline materials. The grain size affects the performance of friction.

The minimum COF of Coating D (multilayer Ni–P/Ni–B) has been displayed in Fig. 4 which was near about (0.484 grain size). Other factors, i.e. development of different phases, distribution of phases, etc. have influence over the friction behaviour of the coatings.

3.4 Wear Behaviour

Wear tests carried out for different types of coated samples are displayed in Fig. 5. The maximum wear rate is found for coating A due to the presence of Ni–B coating in the outer layer. Wear rate is minorly low for coating B where the outer layer is Ni–P. The minimum wear and higher hardness were observed in coating D. Figure 2 and Fig. 5 shows higher hardness and lower wear rate for multilayer Ni–P/Ni–B coating and shows good correlation.

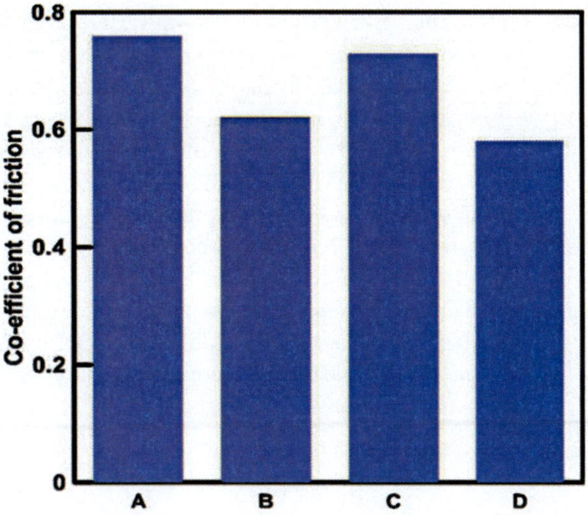

Fig. 4 Coefficient of friction with various types of coatings

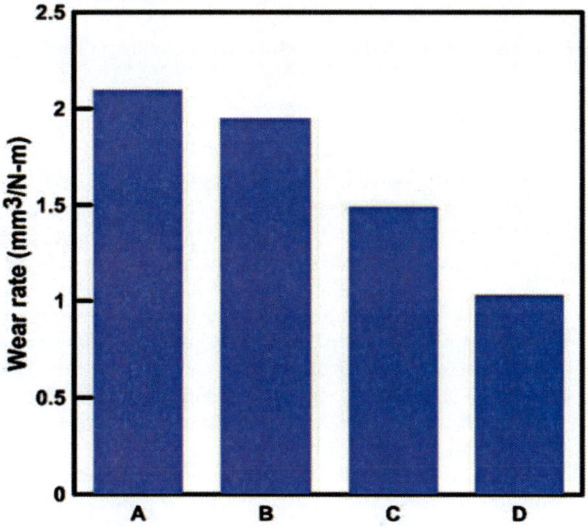

Fig. 5 Wear rate with different types of coating

4 Conclusion

(i) The present study reveals that it is possible to prepare electroless duplex Ni–P/Ni–B and multilayer Ni–P/Ni–B coatings by sequentially dipping into the previously prepared two coating baths. Ni–P solution seems to decompose minorly due to fluctuation of temperature. Ni–B bath is found to be much more stable without bath decomposition for a period of 2 h.

(ii) From microstructural view it was observed that nickel duplex coatings reveal uniform coatings and good compatibility is found between the layers with respect to nodular formation and grain boundary.

(iii) High wear resistance is achieved in coating D in comparison with others. Comparing the two types of multilayer coatings, the higher resistance to wear was observed for Ni–B as the outer layer and lower wear resistance for Ni–P as an outer layer. For the duplex coating Ni–B/Ni–P, wear resistance is higher than the duplex Ni–P/Ni–B coating.

(iv) The hardness of coating C is lowest compared to all types of coatings. It also reveals that hardness values are lower for coating A. The maximum value of hardness was found for multilayer Ni–P/Ni–B. The multilayer coating D is the most effective coatings among all the types of coatings on the performance of hardness and wear resistance.

References

1. Brenner A, Riddell GE (1946) Nickel plating on steel by chemical reduction. J Res Natl Bur Stand 37:31–34
2. Riedel W (1991) Electroless nickel plating. ASM International, Ohio
3. Kundu S, Das SK, Sahoo P (2018) Tribological behaviour of electroless Ni-P deposits under elevated temperature. Silicon 10:329–342
4. Oraon B, Majumder G, Ghosh B (2006) Application of response surface method for predicting electroless nickel plating. Mater Des 27:1035–1045
5. Vitry V, Delaunois F, Dumortier C (2008) Mechanical properties and scratch test resistance of nickel-boron coated aluminium alloy after heat treatments. Surf Coatings Technol 202:3316
6. Wang YW, Xiao CG, Deng ZG (1992) Plat Surf Finish 79(3):57
7. Schlesinger M, Meng X, Snyder DD (1990) J Electrochem Soc 137:1858
8. Palaniappa M, Seshadri SK (2007) Hardness and structural correlation for electroless Ni alloy deposits. Mater Sci 42:6600–6606
9. Sankara Narayanan TSN, Krishnaveni K, Seshadri SK (2003) Electroless Ni-P/Ni-B duplex coatings: preparation and evaluation of microhardness, wear and corrosion resistance. Mater Chem Phy 82:771–779
10. Vitry V, Bonin L et al (2017) Formation and characterization of multilayers borohydride and hypophosphite reduced electroless nickel deposits. Electrochim Acta 243:7–17
11. Mukhopadhyay A, Duari S, Barman TK et al (2016) Tribological performance optimization of electroless Ni–B coating under lubricated condition using hybrid grey fuzzy logic. J Inst Eng India Ser D 97:215–231
12. Biswas A, Das SK, Sahoo P (2017) Correlating tribological performance with phase transformation behavior for electroless Ni-(high)P coating. Surf Coat Technol 328:102–114
13. Sahoo P (2009) Wear behavior of electroless Ni-P coatings and optimization of process parameters using taguchi method. Mater Des 30:1341–1349

Effect of Process Parameters on Force Trends During Friction Stir Welding of AISI-316L Pipes

Suresh Gain, Suman Kalyan Das, Dipankar Sanyal, and Sanjib Acharyya

1 Introduction

Friction stir welding (FSW) is a solid-state joining process developed by Thomas et al. in the year 1991 at The Welding Institute (TWI, UK). FSW process enables the advantages of solid-state joining for fabrication of continuous linear and circular weld. This process leads to lower residual stress [1] in comparison to fusion welding since the process is done at solid state. FSW is considered to be the most significant development in metal joining process in the last two decades. This is a green technology due to its energy efficiency and environment friendliness.

Stainless steel is the most useful structural material used in marine industry, nuclear, and aviation [2] industries due to its superior corrosion resistance. Austenitic stainless steels are mostly used because of their higher tensile strength, corrosion resistance, and higher creep strength [3, 4]. AISI-316L is the second most used grade of austenitic stainless steel. The microstructure and mechanical properties of friction stir welded AISI-316L steel [5] joints were studied, and it was observed that the weld zones were free from macro defects. Friction stir welding was performed on 4 mm thick 316 austenitic stainless steel plate with 1100 rpm tool rotational speed and 8 mm/min transverse speed. It was seen that tensile strength increases 104% compared to base metal with 37% elongation [6].

Aydin [7] successfully carried out FSW of X80 API-grade pipe-line steel with a thickness of 11 mm by PCBN tool with rotating and traverse speeds of 350 rpm and 127 mm/min, respectively, and partial penetration mode using a depth-controlled process. Meran et al. [8] produced defect-free weld by FSW technique on 304

S. Gain (✉) · S. K. Das · D. Sanyal · S. Acharyya
Department of Mechanical Engineering, Jadavpur University, Kolkata 700 032, India

S. K. Das
e-mail: sumankalyan.das@jadavpuruniversity.in

S. K. Ghosh et al. (eds.), *Advances in Thermal Engineering, Manufacturing, and Production Management*, Lecture Notes in Mechanical Engineering,
https://doi.org/10.1007/978-981-16-2347-9_24

austenitic stainless steel with 40–100 mm/min traveling speed and constant rotational speed of 1000 rpm. Li et al. [9] conducted their FSW study on 2.4 mm thick sheet at a fixed transverse speed of 100 mm/min and two rotational speeds of 300 and 400 rpm for super austenitic stainless steel S32654. They found that lower rpm yielded better results. FSW experiment was carried out by Kumar et al. [10] on 316L grade steel by *W* tool doped with 1% La_2O_3 and 11 kN vertical force. Liu and Nelson [11] used PCBN tool for joining 12.5 mm thick plates of 304L stainless steel with 8 kN axial force. A number of studies [12–14] have been published which focused on identifying the process parameters for welding austenitic stainless steel welded joint by friction stir welding technique. But pipe welding by friction stir welding has remained mostly unaddressed due to the special arrangement required for the process. This provides the motivation for the present study.

2 Experimental Procedure

The as-received machined AISI-316L austenitic stainless steel pipe (ID 259 mm and OD 272 mm) of 6.5 mm thickness has been welded by friction stir welding technique. The presence of 17.4% chromium provides corrosion resistance which makes the steel sometimes referred to as marine-grade stainless steel. The chemical composition of the as-received AISI-316L austenitic stainless steel pipe is given in Table 1.

A Research Machine 1A-0.7 with a programmable rotary turntable from Bond Technologies, USA was used for joining of AISI-316L pipe by friction stir welding. The machine is capable of applying a maximum torque of 186 N m about the spindle axis *Z*, which is considered positive vertically downward. The maximum *Z*-force of 67 kN and maximum *X*-force (lateral direction) of 22.25 kN can also be applied by the machine. Following the null setting, the welding is initiated by moving the PCBN coated rotating tool vertically downward, so that the tool pin plunges slowly into the pipes symmetrically about the welding line for the pipe pair until reasonable contact between tool shoulder and the pipes is established. During engagement of the tool, the downward feeds were given in air for 3 mm at 20 mm/min feed rate, into the material for 6 mm at 10 mm/min feed rate and 1 s dwell. For extracting the tool, an upward feed was provided for 9 mm at 50 mm/min feed rate. The pin and the shoulder rotation produce friction heat that together with the forging pressure from the tool accomplish the welding. The pipes held over a mandrel are given rotation about *X* direction so as to set a desired linear welding speed along *Y* direction or the welding line. Different process parameters like, welding speed, spindle speed,

Table 1 Chemical composition of AISI-316L austenitic stainless steel

Element	C	Mn	Si	S	P	Cr	Ni	Mo
Wt%	0.023	1.85	0.48	0.018	0.024	17.4	10.42	2.14

Fig. 1 Friction stir welding of AISI316L pipe **a** pipe welding in progress, **b** 360° pipe weldment

and plunge depth have been set before start of the welding. The machine does not have any provision to provide any tilt to the tool. Friction stir welding of AISI-316L pipe under position control mode is shown in Fig. 1. Position control mode was used to maintain constant depth of penetration for the tool. A cooling arrangement for circulating coolants to prevent overheating of the tool and the mandrel has been employed. Friction stir welding has been done at different set of welding parameters as given in Table 2.

3 Results and Discussion

Welding temperature is the most significant parameter during friction stir welding. The temperature increases mainly with friction heating between the tool shoulder and the workpiece as well as on the residence time of the tool near the welding zone dictated by the spindle rpm. Hence, the friction heating increases with increase in

Table 2 Values of FSW parameters used in the present study

Set no	Tool rotational speed/spindle speed (RPM)	Welding speed (mm/min)	Tool penetration (mm)
Set 1	300	100	6
Set 2	325	100	6
Set 3	350	100	6
Set 4	400	100	6
Set 5	400	50	6

the spindle rpm and the Z-force, whereas decreases with the welding speed. The tool coating may fail at higher rotational speed, as high RPM generates high amount of heat. Hence, it has been decided to carry out some trial experiments depicted in Figs. 2, 3, 4, and 5 with spindle speed of 300 rpm, 325 rpm, 350 rpm, and 400 rpm and a welding speed of 100 mm/min. A result of 400 rpm and 50 mm/min has also been included in Fig. 6. In Figs. 2, 3, 4, 5, and 6, the zones of tool engagement, welding and extraction are shown separated by vertical dashed lines.

Figures 2, 3, 4, 5, and 6 show that the average engagement length achieved in all the cases is seen to be very near to 5.65 mm. Of course, the welding depth is expected to be greater than the engagement depth. The vertical force (*Z*-force) and the spindle torque can be seen to approach their peak values, as the tool engagement increases, reflected in the figures through increase in *Z*-depth. Thereafter, the force and the torque become nearly steady as the process moves over to the welding zone. Clearly, the horizontal forces (*X*-force and *Y*-force) start developing as soon as the welding begins. These appear to be much lower than the *Z*-force. The lateral force along *X* direction could be a scribed due to the advancing and retreating nature of the pipe motion across the welding line with respect to the tool rotation.

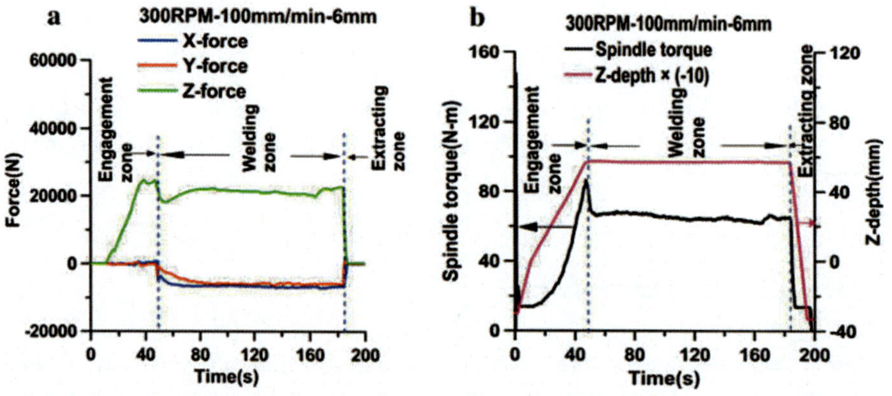

Fig. 2 Welding of SS316L pipe under position control at 300 rpm, 100 mm/min, 6 mm depth. **a** *X*, *Y*, and *Z*-forces. **b** spindle torque, *Z*-depth

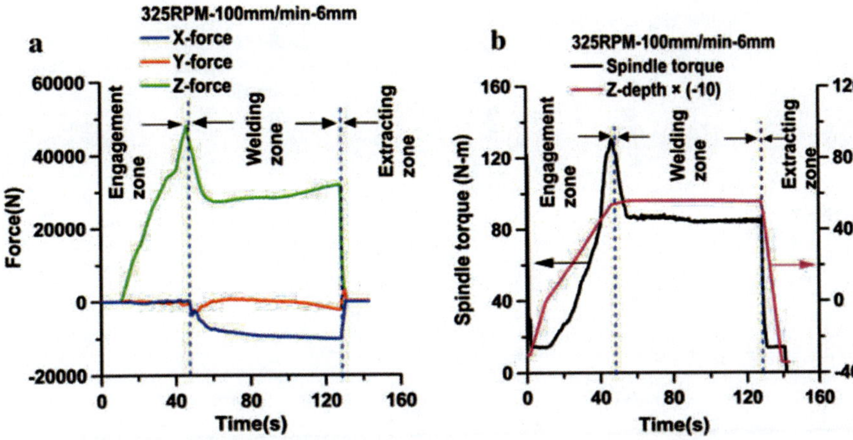

Fig. 3 Welding of SS316L pipe under position control at 325 rpm, 100 mm/min, 6 mm depth. **a** *X*, *Y*, and *Z*-forces. **b** spindle torque, *Z*-depth

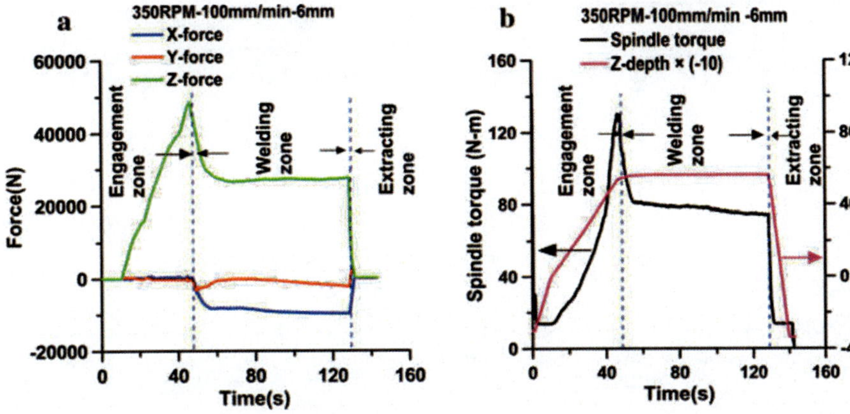

Fig. 4 Welding of SS316L pipe under position control at 350 rpm, 100 mm/min, 6 mm depth. **a** *X*, *Y*, and *Z*-forces. **b** spindle torque, *Z*-depth

For the combination of 300 rpm spindle speed and 100 mm/min welding speed, the *Z*-force achieved is 22.46 kN at the start of welding zone and 22.75 kN observed at the end of welding (Fig. 2). Spindle torques of 82.51 N m and 65.16 N m are noticed at the start and end of the welding zone, respectively (Fig. 2b). The *X*-force values perceive −0.51 kN, −6.69 kN, respectively, at start and end of welding zone. The *Y*-force varies from −2.00 kN to −5.96 kN within welding zone (Fig. 2). A good surface quality welding achieved visually at these parameter combinations. Here, observed two peaks of the vertical force in Fig. 2a could be due to well separated events of attaining the maximum *Z*-depth and the onset of the mandrel rotation overcoming the static friction. The first one may be due to force acting on the tool pin on reaching

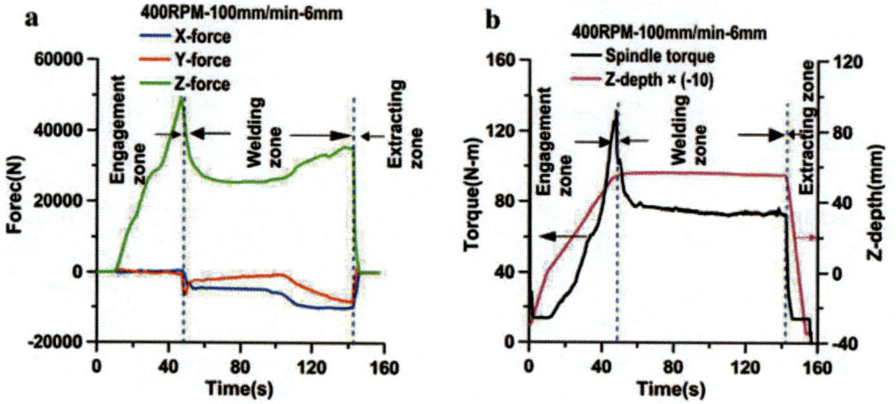

Fig. 5 Welding of SS316L pipe under position control at 400 rpm, 100 mm/min, 6 mm depth. **a** *X*, *Y*, and Z-forces. **b** Spindle torque, Z-depth

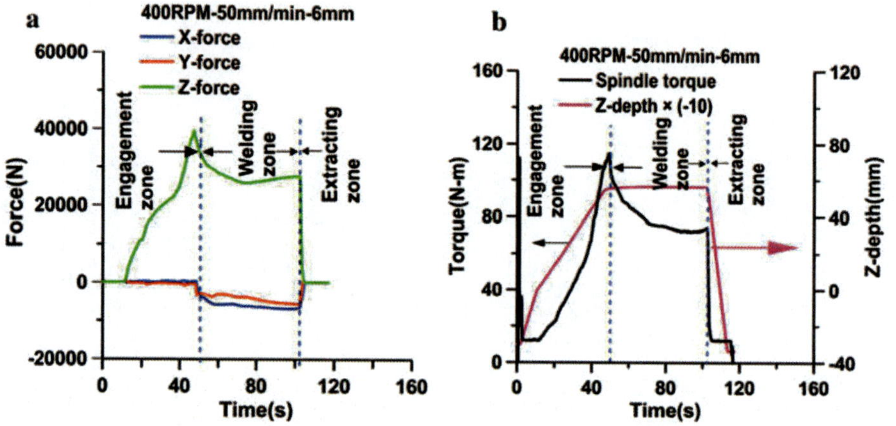

Fig. 6 Welding of SS316L pipe under position control at 400 rpm, 50 mm/min, 6 mm depth. **a** X, Y, and Z-forces. **b** spindle torque, Z-depth

depth of penetration, and the second peak may be for shoulder contact with the workpiece.

For 325 rpm and 100 mm/min combination the vertical force lies between 31.9 kN to 44.35 kN in the welding zone (Fig. 3). The spindle torque varies from 127.2 N m to 84.96 N m, which is significantly higher than the previous set of parameters (Set1 in Table 2). At the start and end of the welding zone, the *X*-forces are, respectively, −0.20 kN and −10.10 kN, and the *Y*-forces are, respectively, −0.99 kN and −2.3 4kN.

The vertical force is approximately similar (43.5 kN at welding start) for the combination of parameters 350 rpm, 100 mm/min and 400 rpm, 100 mm/min. It has been noticed that *X*-force, *Y*-force, *Z*-force, and spindle torque plots are almost identical for these two set of parameters (Figs. 4 and 5).

For welding of 400 rpm, 50 mm/min the vertical force recorded is 36.9 kN at the start of welding zone. This is greater than first set of parameters (Fig. 2), but lowers than the following three sets (Set 2, Set 3 and Set 4) of parameters (Figs. 3, 4 and 5). No welding defect has been observed during the welding of 400 rpm and 50 mm/min.

The forces and the torque recorded during the welding process are compared with the help of bar charts in Fig. 7. Nearly, steady values of forces and torque are reported in the charts. The lateral force *X* is found to increase with increase in spindle speed upto 350 RPM and then falls (Fig. 7a). The force *Y* along the welding line does not show any prominent trend (Fig. 7b). However, the vertical *Z*-force (Fig. 7c) and spindle torque (Fig. 7d) show to possess a definite correlation with spindle speed as long as the welding speed do not vary. Under the same welding speed, both of force and the torques increase with increase in spindle speed from 300 to 325 rpm beyond which it falls marginally with increase in the spindle speed. This correlated decrement could be attributed to the softening of the material due to high heat generation from

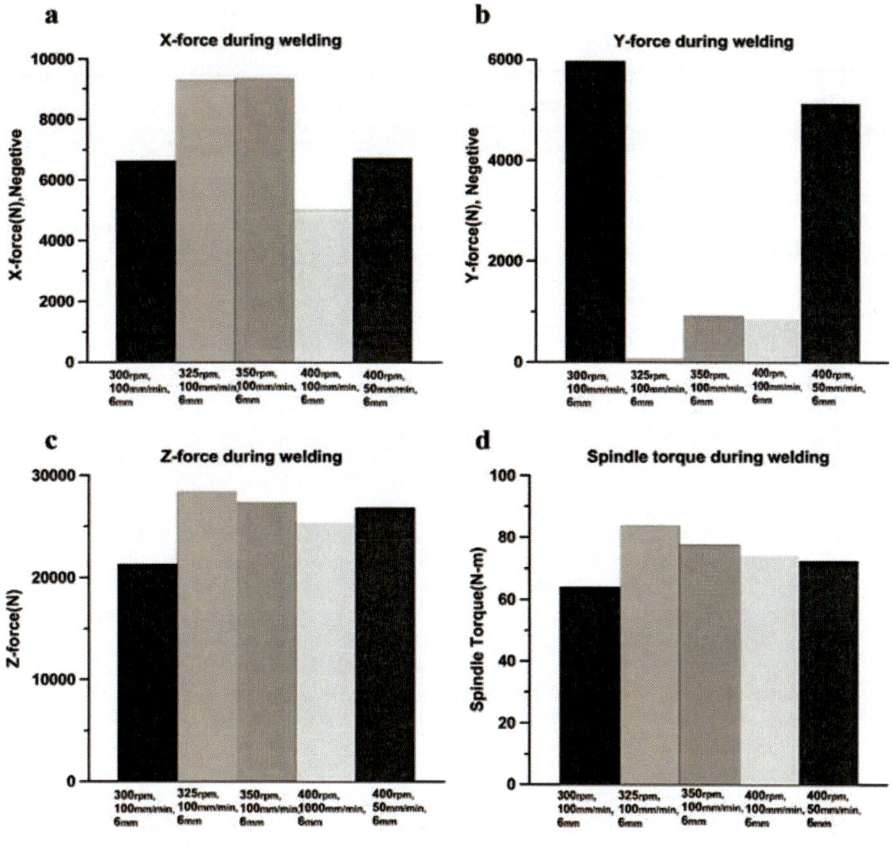

Fig. 7 Bar chart comparing the forces at position control mode. **a** X-force. **b**Y-force. **c** Z-force. **d**spindle torque

increased friction (due to increased Z-force). However, more investigation in this regard is planned in the future work based on microstructural analysis. Based on the outcomes of the present work, it can be suggested that carrying out welding under a force of about 30 kN would yield visually good weldments.

4 Conclusions

In the present study, the variations of the tool forces and spindle torque with welding parameters during friction stir welding of SS316L pipe under position control mode have been investigated. The forces and the torque recorded for different parameter combinations display similar trends in different zones, namely the tool engagement, welding process, and tool extraction. However, it is seen that the forces and torque vary with different parameter combinations. For all the parameter combinations, the steady value of vertical force in the welding zone is observed to be between 23 and 29 kN and the steady value of the spindle torque between 64 and 84 N m. Vertical Z-force and spindle speed increases with increase in the spindle speed. However, above 325 rpm, both decreases which may be due to softening of the material due to increased friction heating.

Acknowledgements The authors thank Department of Metallurgical and Material Engineering of Jadavpur University for providing the laboratory facilities and Ministry of Steel, GoI for providing necessary funding.

References

1. Brewer LN, Bennett MS, Baker BW, Payzant EA et al (2015) Characterization of residual stress as a function of friction stir welding parameters in oxide dispersion strengthened (ODS) steel MA956. Mater Sci Eng A 647:313–321
2. Davis JR (1994) ASM specialty handbook, stainless steel. Materials Park, OH
3. Karjalainen LP, Taulavuori T, Sellman M et al (2008) Some strengthening methods for austenitic stainless steels. Steel Res Int 79:404–412
4. Mironov S, Sato YS, Kokawa H et al (2011) Structural response of superaustenitic stainless steel to friction stir welding. Acta Mater 59:5472–5481
5. Kumar SS, Murugan N, Ramachandran KK (2018) Microstructure and mechanical properties of friction stir welded AISI 316L austenitic stainless steel joints. J Mater 254:79–90
6. Meshram MP, Kodli BK, Dey SR (2014) Friction stir welding of austenitic stainless steel by PCBN tool and its joint analyses. Procedia Mater Sci 6:135–139
7. Aydin H (2014) Relationship between a bainitic structure and the hardness in the weld zone of the friction-stir welded X80 API-grade pipe-line steel. Mater Technol 48:15–22
8. Meran C, Kovan V, Alptekin A (2007) Friction stir welding of AISI 304 austenitic stainless steel. Mater Sci Eng Technol 38:829–835
9. Li H, Yang S, Zhang S et al (2017) Microstructure evolution and mechanical properties of friction stir welding super-austenitic stainless steel S32654. Mater Des 118:207–217

10. Kumar SS, Murugan N, Ramachandran KK (2016) Influence of tool material on mechanical and microstructural properties of friction stir welded 316L austenitic stainless steel butt joints. Int J Refract Met H 58:196–205
11. Liu FC, Nelson TW (2016) In-situ material flow pattern around probe during friction stir welding of austenitic stainless steel. Mater Des 110:354–364
12. Sabooni S, Karimzadeh F, Enayati MH et al (2015) Friction-stir welding of ultrafine grained austenitic 304L stainless steel produced by martensitic thermomechanical processing. Mater Des 76:130–140
13. Saeid T, Abdollah-zadeh A, Assadi H et al (2008) Effect of friction stir welding speed on the microstructure and mechanical properties of a duplex stainless steel. Mater Sci Eng 496:262–268
14. Nathan SR, Balasubramanian V, Malarvizhi S et al (2015) Effect of welding processes on mechanical and microstructural characteristics of high strength low alloy naval grade steel joints. Def Technol 11:308–317

Finite Element Modeling of TIG Arcing Process and Its Validation on AISI 4340 Structural Steel Plate

Sudhir Kumar, Deepak Sharma, and Kaushal Kumar

1 Introduction

Considering the rapidly growing world, a faster and better approach is required in every sector of the industry including welding and surface modification. TIG welding is a widely used industrial process that is used for joining the materials as well as for surface treatment to enhance the surface properties [1–7]. To incorporate the correct use of TIG welding for such industrial applications, the selection of parameters plays a very critical role because the primary failure starts from the welded joint. This failure is because of the improper fusion or unwanted residual stresses caused due to the selection of inappropriate parameters. While in the case of surface treatment, the formation of the desired phase on the surface is the major concern which depends on the cooling rate and this itself can be controlled by choosing the correct parameters [8, 9]. Therefore, a better approach is required to predict the welding parameters for which Finite Element (FE) modeling is an effective technique to predict and analyze the welding residual stresses, weld bead geometry, and thermal cycle [10–14].

Mathematical modeling is generally used to explain the effect of thermal cycle on the residual stress, weld pool geometry, and distortion within the structure [15, 16]. However, these mathematical models require some constants which can be calculated through experiments. Therefore, to simplify and avoid experimental processes, FEM

S. Kumar (✉)
School of Engineering and Technology, Central University of Haryana, Mahendragarh 123031, India

S. Kumar · D. Sharma
Department of Metallurgical and Materials Engineering, Indian Institute of Technology Roorkee, Roorkee, Uttarakhand 247667, India

K. Kumar
Department of Physics, J. C. Bose University of Science and Technology, YMCA Faridabad, Faridabad, Haryana 121006, India

S. K. Ghosh et al. (eds.), *Advances in Thermal Engineering, Manufacturing, and Production Management*, Lecture Notes in Mechanical Engineering,
https://doi.org/10.1007/978-981-16-2347-9_25

models using different tools like ANSYS, ABAQUS, SWS WELD, etc. are used to find the bead geometry along with thermal cycle and residual stresses.

The welding simulation requires two analyses: thermal and structural analysis. For welding simulation, the mesh should be fine enough to accommodate the high thermal gradients developed (in time and space) during the transient heat analysis phase and at the same time, it should be able to solve the plastic-structural phase of the problem. The use of computers reduces the time for welding simulation and makes it possible to perform the 3D simulation of welding by using the fine mesh required for accurate analyses [17, 18]. The present study is motivated towards the understanding and implementation of the heat flux mechanism of the TIG welding process on ANSYS to find the temperature and residual stresses at various locations and also to calculate the thermal cycle and weld isotherm.

2 Analytical Modelling

A model named as double ellipsoidal moving heat source was considered by Rosenthal in 1941, the mathematical representation and solution will be proposed by Nguyen et al. (1999) as shown in Fig. 1, the temperature profile was measured at a point by using the following equation, and get closer results [19]:

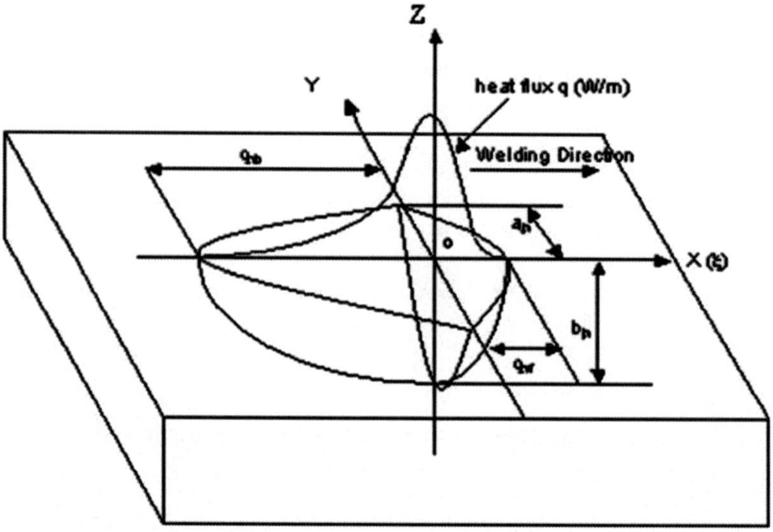

Fig. 1 The double ellipsoidal heat source for an arc

$$T_d = \frac{3\sqrt{3}.Q_{AW}}{\rho.c.\pi\sqrt{\pi}} \int_0^t \left[\frac{dt}{\sqrt{(12a(t-t')+a_h^2)}.\sqrt{(12a(t-t')+b_h^2)}} . \left(\frac{A'}{\sqrt{(12a(t-t')+c_{hf}^2)}} + \frac{B'}{\sqrt{(12a(t-t')+c_{hb}^2)}} \right) \right] + T_0 \quad (1)$$

$$A' = r_f.\exp\left(-\frac{3(x-v.t')^2}{12a(t-t')+c_{hf}^2} - \frac{3y^2}{12a(t-t')+a_h^2} - \frac{3z^2}{12a(t-t')+b_h^2} \right) \quad (2)$$

$$B' = r_b.\exp\left(-\frac{3(x-v.t')^2}{12a(t-t')+c_{hb}^2} - \frac{3y^2}{12a(t-t')+a_h^2} - \frac{3z^2}{12a(t-t')+b_h^2} \right) \quad (3)$$

where a = thermal diffusivity ($a = k/\rho\, c$); c = specific heat; k = thermal conductivity; ρ = mass density; t, t' = time. The Parameters for Double-ellipsoidal 3-D moving power source has been taken using calculated weld pool geometry (a_h, b_h, C_{hf}, C_{hb} and $\eta = 0.7$) and the material properties were assumed to be uniform with Temperature. A computer program is written to plot transient temperature curves with variation in welding speed and current.

3 Experimental

The AISI 4340 steel plate with dimensions 150 mm × 70 mm × 10 mm was considered as base material with composition as shown in Table 1. Autogenous (without filler) Arcing was done on the AISI 4340 steel plate by using different arcing parameters, as mentioned in Table 2. A 3.2 mm electrode was used for TIG arcing with a Argon flow rate of 15 l/min.

A macroscopic view (width and the depth of penetration) of the welded cross-section of the sample has been shown in Fig. 2. The thermal profile of the weld bead was measured by 0.25 mm diameter of the thermocouple. The placement of

Table 1 Chemical Composition of AISI 4340 steel

C	Ni	Cr	Mo	Mn	Si	P	S	Fe
0.39	1.68	0.64	0.15	0.67	0.22	0.017	0.014	Rest

Table 2 Input parameters for TIG arcing

Arc velocity (m/min)	Current (A)	Voltage (V)
0.1	70	10.5 ± 1.0
	100	
	130	
	160	
	190	

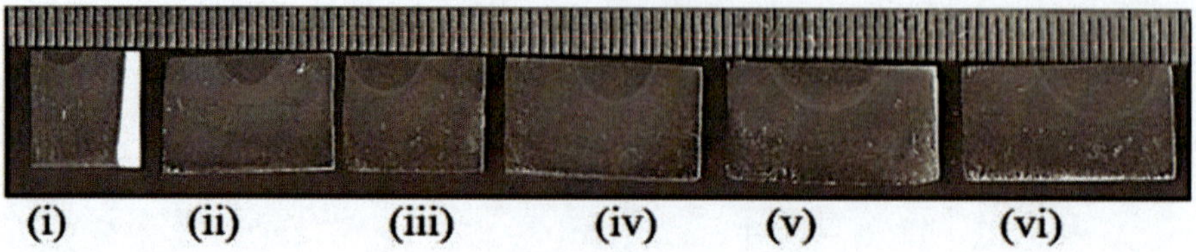

Fig. 2 Typical appearance of cross section of bead on plate at different welding current (i) 70 A, (ii) 100 A, (iii) 130 A, (iv) 160 A, (v) 190 A, (vi) 220 A

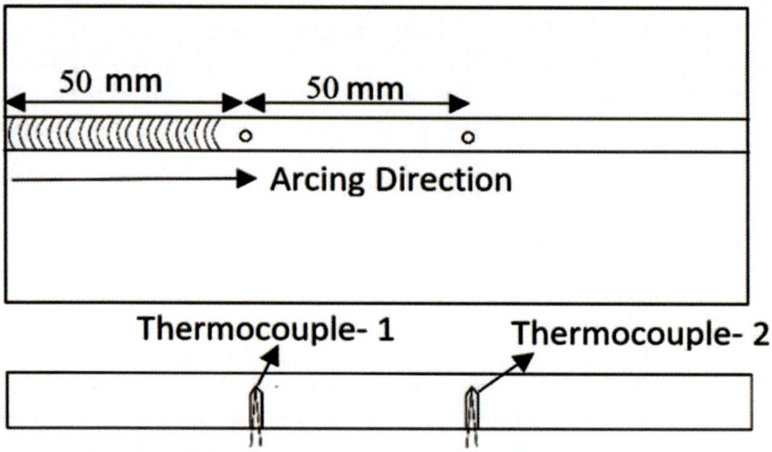

Fig. 3 Typical view of thermocouple position to measure thermal cycle

the thermocouple as shown in Fig. 3 to measure the thermal cycle in weld bead. The Hole Drill Method [20] was used to assess the residual stresses of the sample.

For the FE model, the material properties (temperature-dependent) i.e. density, specific heat, thermal conductivity, and enthalpy are shown in Table 3, which was taken from ASM Handbook [21] and a handbook [22].

4 Results and Discussion

4.1 Process of Modeling and Analyzing Problem

The thermo-mechanical analysis of TIG arcing was performed using two different software/tools. Firstly, a 3D model with dimensions 150 mm × 70 mm × 10 mm was created using CATIA, as shown in Fig. 4. This model was subsequently imported into the Ansys software for generating and optimizing the number of nodes and elements

Table 3 AISI 4340 steel thermal material properties

Temperature (K)	Density (kg/m^3)	Specific heat (J/kg K)	Thermal conductivity (W/mK)	Enthalpy (J/m^3)
273	7870	345	44.5	9.76E+08
373	7868	369.6	44.1	1.05E+09
473	7862	396.1	43.6	1.35E+09
573	7859	422.2	43.0	1.75E+09
673	7855	444.9	42.3	2.16E+09
773	7854	471.3	41.7	2.59E+09
873	7851	494.8	40.4	3.03E+09
973	7850	521.7	36.9	3.48E+09
1073	7850	544.3	34.5	3.94E+09
1173	7850	570.2	32.0	4.41E+09
1273	7850	595.8	32.0	4.90E+09
1373	7847	620.1	32.0	5.40E+09
1473	7846	565.0	32.0	5.93E+09
1573	7844	525.9	32.0	6.49E+09
1673	7842	519.8	32.0	7.09E+09
1774	7841	519.8	32.0	7.21E+09

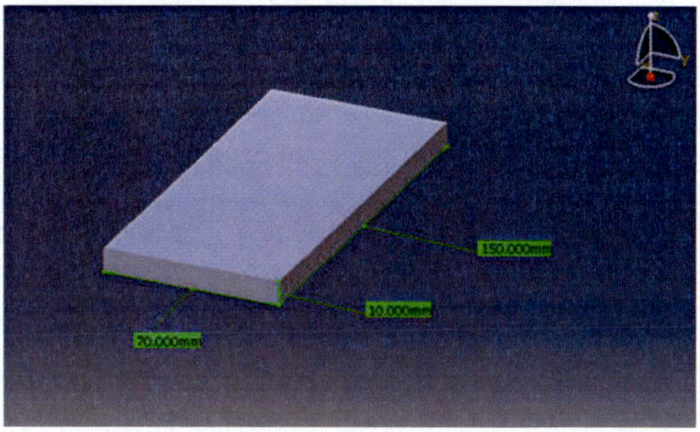

Fig. 4 Model of the plate in CATIA

on the substrate model, also called the meshing of the model. During meshing, the primary consideration was to generate a finer mesh near the welding region. The minimum size of the element was kept as 1.5 mm × 0.5833 mm × 0.653 mm at the center of the plate. The size of these elements was decided by making different mesh size element programs and results were correlated with the experimental results such that it will be optimized based on the accuracy of results and should be less time-consuming. The final finite elemental model was associated with 43,032 nodes and 51,146 elements.

For thermal—structure couple field analysis, we used a SOLID 5 element type. This type of element is generally used to solve the three-dimensional problem of magnetic, thermal, electric, piezoelectric, and structure. Besides, the SOLID5 element supports a couple field analyses. The ANSYS program where used non-linear equations for the model. Following assumptions were taken into consideration while solving the equations:

- There is no effect on the thermal distribution of parts during welding due to the displacement of parts.
- Material properties, which are dependent on temperature, were taken till liquid phase temperature.
- Conduction and convection effects are considered. Radiation effects are neglected.
- Autogenous TIG methodologies for a bead on the plate are modeled.
- Element Birth and death procedure is used.
- A coupled field analysis (Thermal + structural) model is used (Fig. 5).

4.2 *Validation of Model*

The validation of a welding process can be possible by two considerations: (i) Thermal validation (ii) weld isotherm (bead width and depth of penetration) validation. In view of that following validation is done.

4.2.1 Thermal Validation

At a given arcing parameter such as arcing current, voltage, and travel speed as 130 A, 10.9 V, and 0.1 m/min, respectively. The temperature profile obtained from the FE model and experiment in the fusion zone (FZ) and heat affected zone (HAZ) is shown in Fig. 6. Figure shows the approximate similar profile in the analytical and experimental result. It was also analyzed that the FZ and HAZ temperature reached in the order of 1700 K and 1300 K respectively. Hence validation from Fig. 6 gives a clear indication that the thermal profile obtained from the FE model matches with the thermal profile obtained experimentally.

The thermal profile at a different location in the transverse direction of the welding direction was determined by the FE model as shown in Fig. 7. The distance is calculated from the weld centre line which is considered as reference line. The temperature at point 1.1667 mm away from the centerline was found to be the order of 2300 K that means, it lies in the fusion zone and the temperature at point 1.75 mm was found to be 1610 K so it is adjacent to the fusion line followed by the thermal cycle in the base material.

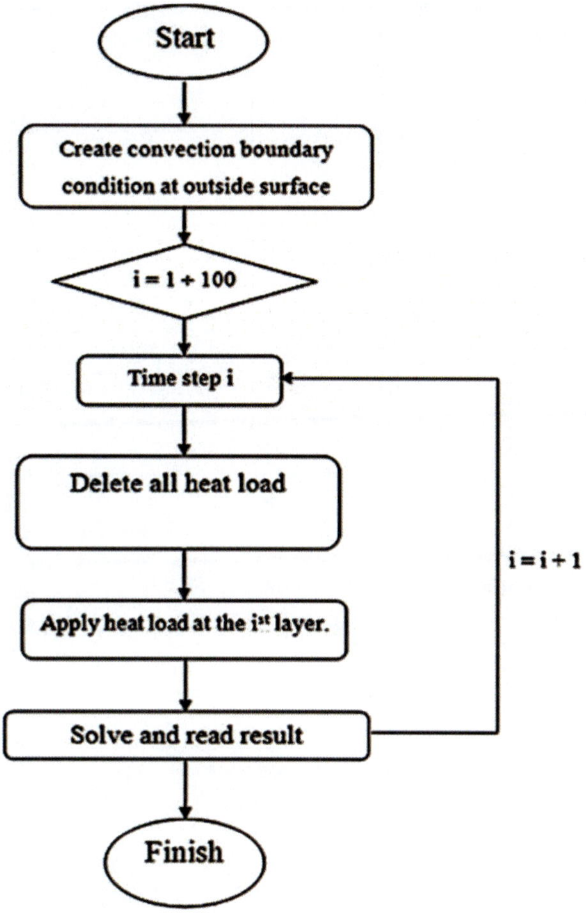

Fig. 5 Simulation procedure of welding in ANSYS

4.2.2 Weld Isotherm Validation

This validation was done by considering of weld bead profile with the thermal profile of the model. Figure 8 shows the weld bead profile obtained from experimentally and from FE model at a voltage of 10.5 ± 1.0 V and speed of 0.1 m/min, and welding current varies from 70 to 220 A with an interval of 30 A. From Fig. 8, it can be said that the weld bead profiles are approximately the same in the FE model as compare with the experimental result.

Further movement of three-dimensional heat source has been studied as shown in Fig. 9. Figure shows the temperatures at different steps during welding such as temperature distribution at 1.8, 9, 18, 18.72, 27, 36, 45, 54, 63, 72, 81, and 90 s of arc welding. Temperature is varying from 1774 K to approximately 3000 K in the FZ and HAZ temperature was 1000–1774 K. It was given a quite similar movement as it happened in the actual welding process.

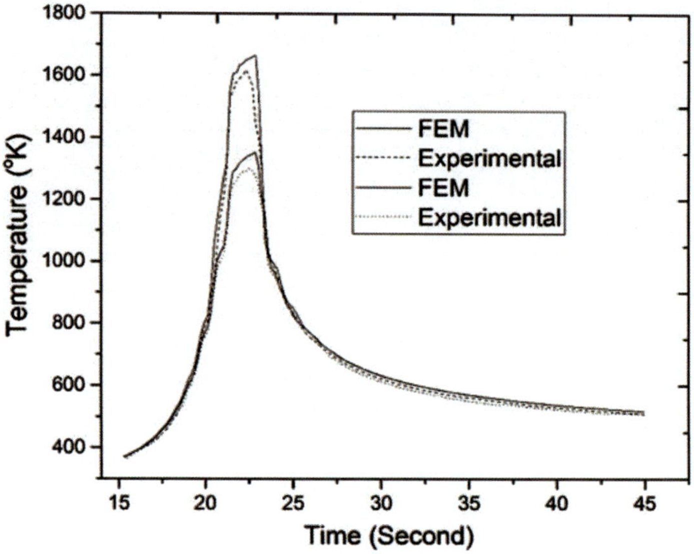

Fig. 6 Thermal cycle obtained from the FE model and experimental

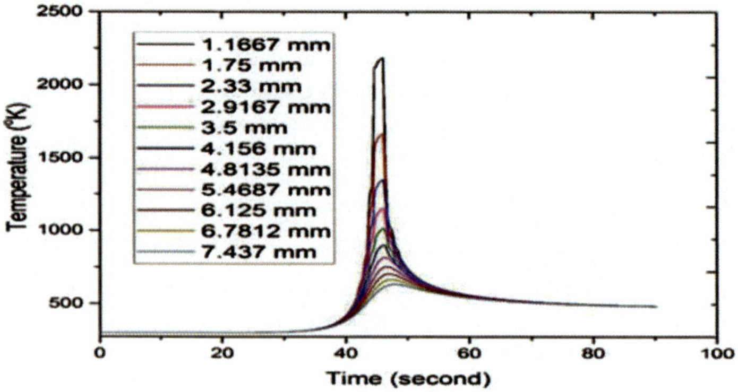

Fig. 7 Temperature profile at a different point along the transverse direction of the weld

4.3 Residual Stresses

The residual stress in the longitudinal and transverse direction of the weld was calculated by the hole drill method at the center of the plate which is lying in the center of the weld bead as shown in Fig. 10a. At a given arcing current, voltage, and speed as 130 A, 10.9 V, and 0.1 m/min respectively, the residual stresses are shown in Fig. 10b, which also includes the residual stress obtained by the FE method. Figure 10b depicts that the transverse stresses are tensile stress of 200 MPa at the

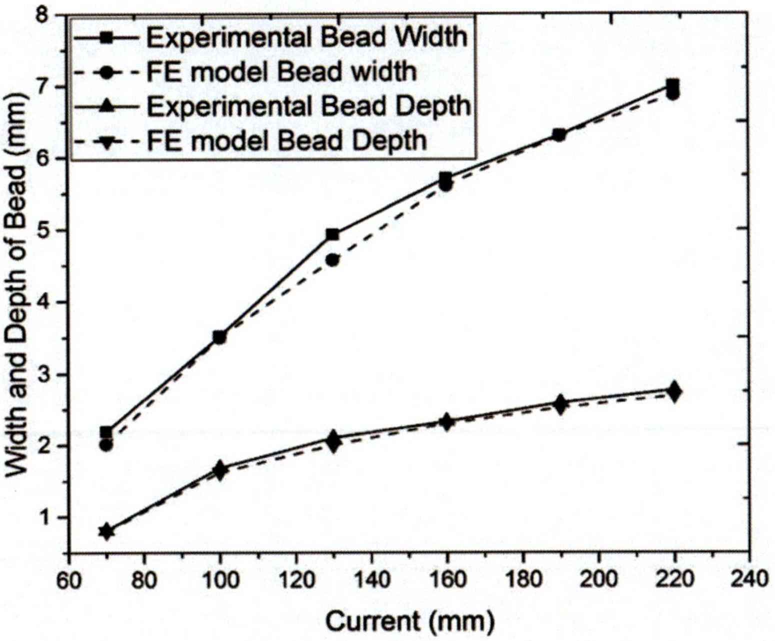

Fig. 8 Comparison of bead width and depth of penetration by using experimental and FE model

top surface and followed by the compressive stress (from 250 to 50 MPa) but the transverse stresses are complete tensile nature up to 1.4 mm depth.

To predict the amount and nature of internal stresses, the model consists of a non-linear mechanical analysis along with thermal analysis. The output of the thermal profile will be considered as the input for mechanical properties. In the model, it was assumed that the edge of the plate in the longitudinal direction will be clamped and fixed.

At given parameters of welding current, voltage, and speed as 130 A, 10.9 V, and 0.1 m/min respectively for FE model, the residual stresses were developed on the top and bottom surfacing of the sample after TIG welding process is showing in Figs. 11, 12, and 13. Residual stress of x-direction (σ_x) as shown in Fig. 11a, b on the top and bottom surface where stable zone stresses vary in the range of 76 MPa (compressive) to 273 MPa (tensile). Figure 12a, b shows the y-direction stresses (σ_y) of the top and bottom surface where stresses vary in the range of 258.8 MPa (compressive) to 175.5 MPa (tensile). Similarly, the z-direction of residual stresses varies in the range of 206 MPa (compressive) to 64.2 MPa (compressive) as shown in Fig. 13a, b.

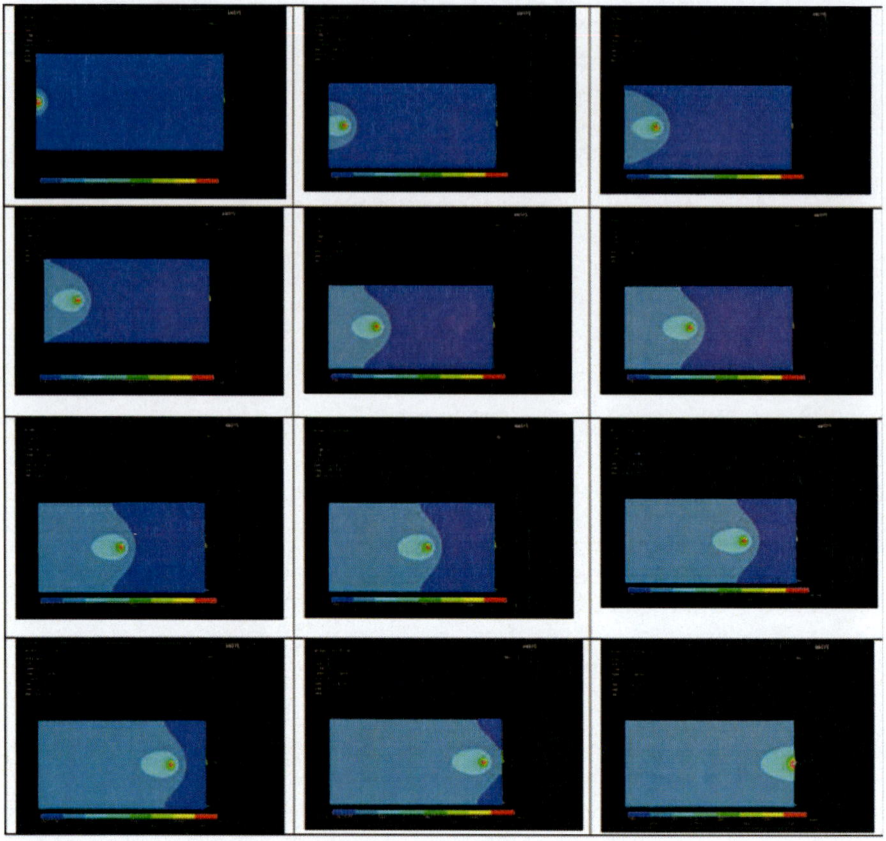

Fig. 9 Movement of TIG arc in FE model after every 9 s during the process

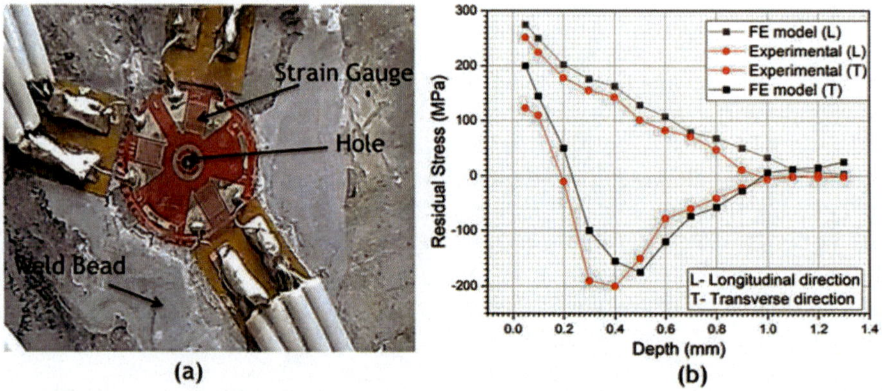

Fig. 10 Residual stress analysis **a** strain gauge, **b** residual stresses at given arcing parameter

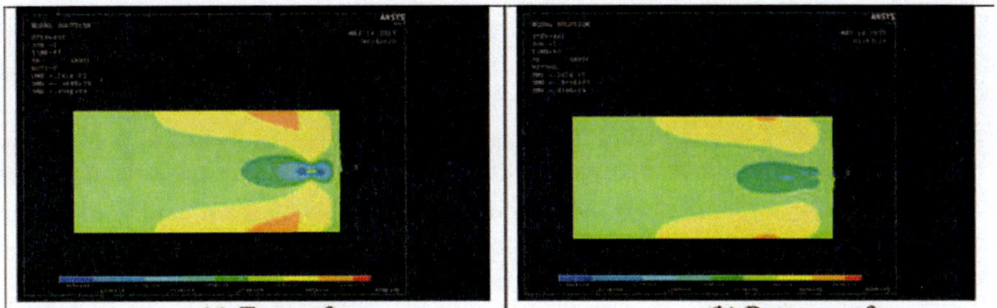

(a) Top surface (b) Bottom surface

Fig. 11 Stresses in X-direction (σ_x)

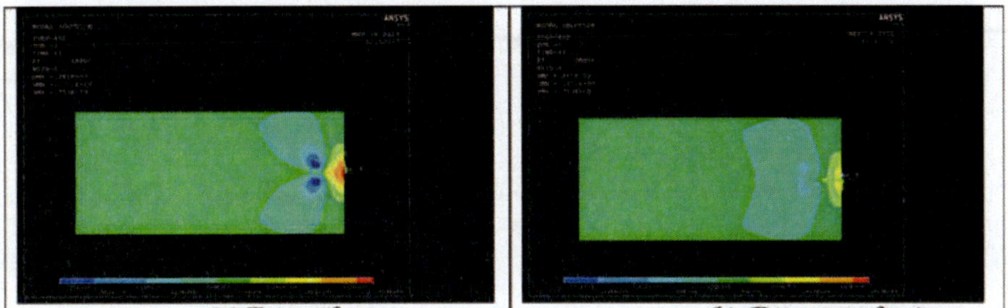

(a) Top surface (b) (Bottom surface)

Fig. 12 Stresses in Y-direction (σ_y)

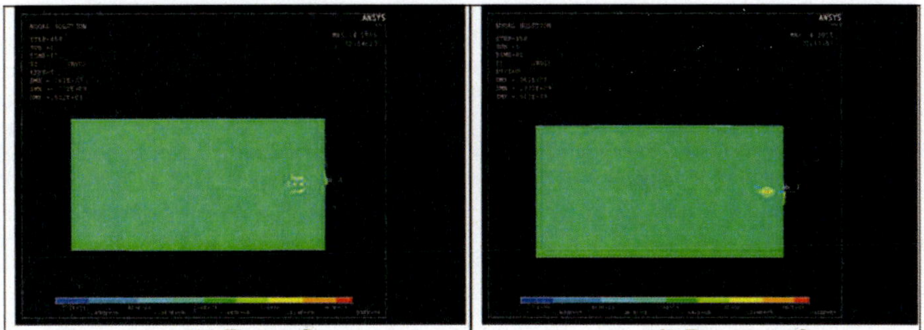

(a) Top surface (b) Bottom surface

Fig. 13 Stresses in Z-direction (σ_z)

5 Conclusion

The three-dimensional FE model using birth and death element procedure in ANSYS has been created for autogenous TIG welding process on AISI 4340 steel. The thermal cycle obtained by the FE model has the same nature as an experimentally measured thermal cycle with an accuracy of average variation 5 ± 1% which means that the FE model has the correct representation of the TIG welding process. Further model is again validated in terms of weld bead profile with the experimental result. The main feature of the model which makes it unique is flexibility and independence from any empirical parameter. The model can be widely applicable to other materials. Residual stresses analysis can be done by this model to find stress in X, Y, and Z direction (σ_x, σ_y and σ_z). A hole drill method has been analyzed and it confirms the stress is in a similar range with the FE model in the weld zone.

References

1. Ragavendran M, Chandrasekhar N, Ravikumar R, Saxena R, Vasudevan M, Bhaduri AK (2017) Optimization of hybrid laser—TIG welding of 316LN steel using response surface methodology (RSM). Opt Lasers Eng 94:27–36. https://doi.org/10.1016/j.optlaseng.2017.02.015
2. Madadi F, Shamanian M, Ashrafizadeh F (2011) Effect of pulse current on microstructure and wear resistance of Stellite6/tungsten carbide claddings produced by tungsten inert gas process. Surf Coatings Technol 205:4320–4328. https://doi.org/10.1016/j.surfcoat.2011.03.076
3. Stegemann R, Cabeza S, Lyamkin V, Bruno G, Pittner A, Wimpory R, Boin M, Kreutzbruck M (2016) Residual stress characterization of steel TIG welds by neutron diffraction and by residual magnetic stray field mappings. J Magn Magn Mater 426:580–587. https://doi.org/10.1016/j.jmmm.2016.11.102
4. Sahoo CK, Masanta M (2017) Microstructure and mechanical properties of TiC-Ni coating on AISI304 steel produced by TIG cladding process. J Mater Process Technol 240:126–137. https://doi.org/10.1016/j.jmatprotec.2016.09.018
5. Okano S, Mochizuki M (2017) Transient distortion behavior during TIG welding of thin steel plate. J Mater Process Technol 241:103–111. https://doi.org/10.1016/j.jmatprotec.2016.11.006
6. Sharma D, Ghosh PK, Kumar S, Das S, Anant R, Kumar N (2018) Surface hardening by in-situ grown composite layer on microalloyed steel employing TIG arcing process. Surf Coatings Technol 352:144–158. https://doi.org/10.1016/j.surfcoat.2018.08.009
7. Sharma D, Ghosh PK, Anant R, Kumar S (2019) Surface modification of microalloyed steel by silicon carbide reinforcement using tungsten inert gas arcing Surface modi fi cation of microalloyed steel by silicon carbide reinforcement using tungsten inert gas arcing
8. Ghosh PK, Kumar R (2015) Surface modification of micro-alloyed high-strength low-alloy steel by controlled TIG arcing process. Metall Mater Trans A 46:831–842. https://doi.org/10.1007/s11661-014-2670-x
9. Kumar R, Ghosh PK, Kumar S (2017) Thermal and metallurgical characteristics of surface modification of AISI 8620 steel produced by TIG arcing process. J Mater Process Technol 240:420–431. https://doi.org/10.1016/j.jmatprotec.2016.10.020
10. Yang LJ, Xiao ZM (1995) Elastic-plastic modelling of the residual stress caused by welding. J Mater Process Tech 48:589–601. https://doi.org/10.1016/0924-0136(94)01698-Z
11. Fujii S, Takahashi N, Sakai S, Nakabayashi T, Muro M (2000) Develoment of 2D simulation model for laser welding. In: Chen X, Fujioka T, Matsunawa A (eds), p 218. https://doi.org/10.1117/12.377023
12. Scott DA (1999) Finite element model of pulsed laser welding. Weld J 78:15s-22s. https://aws.org/wj/supplement/FREWIN/ARTICLE3.pdf.
13. Lankalapalli KN, Tu JF, Gartner M (1996) A model for estimating penetration depth of laser welding processes. J Phys D Appl Phys 29:1831–1841. https://doi.org/10.1088/0022-3727/29/7/018

14. Tsirkas S, Papanikos P, Kermanidis T (2003) Numerical simulation of the laser welding process in butt-joint specimens. J Mater Process Technol 134:59–69. https://doi.org/10.1016/S0924-0136(02)00921-4
15. Goldak J, Chakravarti A, Bibby M (1984) A new finite element model for welding heat sources. Metall Trans B 15:299–305. https://doi.org/10.1007/BF02667333
16. Deng D, Murakawa H (2008) Prediction of welding distortion and residual stress in a thin plate butt-welded joint. Comput Mater Sci 43:353–365. https://doi.org/10.1016/j.commatsci.2007.12.006
17. Lindgren LE (2006) Numerical modelling of welding. Comput Methods Appl Mech Eng 195:6710–6736. https://doi.org/10.1016/j.cma.2005.08.018
18. Moraitis GA, Labeas GN (2008) Residual stress and distortion calculation of laser beam welding for aluminum lap joints. J Mater Process Technol 198:260–269. https://doi.org/10.1016/j.jmatprotec.2007.07.013
19. Nguyen NT, Ohta A, Matsuoka K, Suzuki N, Maeda Y (1999) Analytical solutions for transient temperature of semi-infinite body subjected to 3-D moving heat sources. Weld Res Suppl I 265–274. e:%5CBIBLIO%5CArticles%5CSolution_transient-temp-3D-moving-heat-source_Nguyen_1999.pdf
20. Kumar S, Ghosh PK (2018) TIG arc processing improves tensile and fatigue properties of surface modified of AISI 4340 steel. Int J Fatigue 116:306–316. https://doi.org/10.1016/j.ijfatigue.2018.06.036
21. A.I.H. Committee (2000) ASM handbook: heat treating. ASM International
22. Barnes S, Nash MJ, Kwok YK, Bag S, De A, Madhusudhan Reddy G, Srinivasa Rao K (2003) Surface modification of powder metallurgy components with a direct diode laser. Def Technol 11:2698–2710. https://doi.org/10.1007/s11661-008-9607-1

Studies of Lip Profiles for Micro-Indentation of a Frictionless Rigid Wedge into a Semi-Infinite Plate by Numerical Slip-Line Field Techniques and Finite Element Analysis

Arup K. Biswas, Santanu Das, and Sanjoy Das

1 Introduction

Slip-lines are the network of hypothetical lines along the direction of flow of maximum shear stress. This network consists of mainly two families of orthogonal lines viz. α-lines and β-lines. Slip-line field constitutes an important part in plastic deformation analysis of ductile materials as the ductile materials fail by shear along the direction of maximum shear stress. With this type of analysis, the pressures, forces, strains and the velocity of flow of material can be determined for any type of plastic deformation process. The present problem of wedge indentation to ductile material with the help of slip-line field was first initiated by Hill et al. in 1947 [1]. Later, it finds many developments over the decades.

The previous researchers in this field considered that the profile of the lip to be linear. The works of Grunzweiz et al. [2], 1954, Dugdale, 1954 [3], Locket, 1963 [6] and Johnson et al. [4], 1964 are noteworthy in this field. The first numerical implementation of this problem was reported by Dodd et al. [5] in 1973. Later, Chitkara and Butt, 1992 [8] had given a detailed formulation of producing slip-line field network numerically. Multhearn [7] was the first researcher to show first experimentally that during wedge indentation, the lip might be of nonlinear nature. Biswas et al. [9–12] worked extensively on this problem and showed that the nature of lip profile is nonlinear and is function of semi-apex angle of the cone and the friction parameter of the mating surfaces. Biswas et al. also proposed the new numerical

A. K. Biswas (✉) · S. Das
Department of Mechanical Engineering, Kalyani Government Engineering College, Kalyani 741235, West Bengal, India

S. Das
University of Kalyani, Department of Engineering and Technological Studies, Kalyani 741235, West Bengal, Kalyani, India

S. K. Ghosh et al. (eds.), *Advances in Thermal Engineering, Manufacturing, and Production Management*, Lecture Notes in Mechanical Engineering,
https://doi.org/10.1007/978-981-16-2347-9_26

scheme of this problem considering parabolic profile of the lip with the help of slip-line field theory.

The present work mainly concentrates on the profile of lip that is formed during wedge indentation around the wedge.

2 Finite Element Analyses (FEA) of the Problem

Finite element analysis (FEA) is a mathematical tool by which varieties of complex problems related to engineering can be solved. When differential equations and the boundary conditions are known but the exact solution is difficult to find out, and the FEA approach is preferred as it solves differential equations numerically. The present problem is nonlinear FEA problem as it involves two types of nonlinearity one material, and the other is contact. For this problem, semi-infinite block is taken. Semi-infinite means the block length should be such that the deformation process may not affect the overall length of the block. The block and the indenter are meshed with an average mesh size of 0.5 μ as shown in Fig. 1. The block material is mild steel with elastic modulus value 200 GPa and Poisson's ratio 0.33. The wedge material is diamond with elastic modulus value of 1400 GPa and Poisson's ratio of 0.33. The present problem is solved by using commercial package ANSYS Release-19 version. The wedge is given a 1.5 μ downward displacement for each semi-apex angle. The semi-apex angles are 10°, 20°, 30°, 40°, 45°, 50°, 60°, 70° and 80°. Some typical FEA results for deformation are shown in Figs. 2, 3 and 4.

Figures 2, 3 and 4 show the deformation plots when the semi-apex angle of the wedge are 10°, 45° and 80°, respectively. Figures clearly show the bulged out lip around the wedge. Figures also ensure that with the increase of the wedge angle, the nonlinearity of the lip gradually decreases. When the semi-apex angle of the wedge is small, the ease of indentation is more. The Y-component of velocity is more than X-component of velocity. With the increase of semi-apex angle, the X-component of velocity is more as for the same depth of indentation, the deformed material gets more surface area to push the material in rightward direction. The uneven distribution of velocity component gives rise to formation of parabolic lip. In later section, the

Fig. 1 Typical meshing of semi-infinite block and the rigid wedge

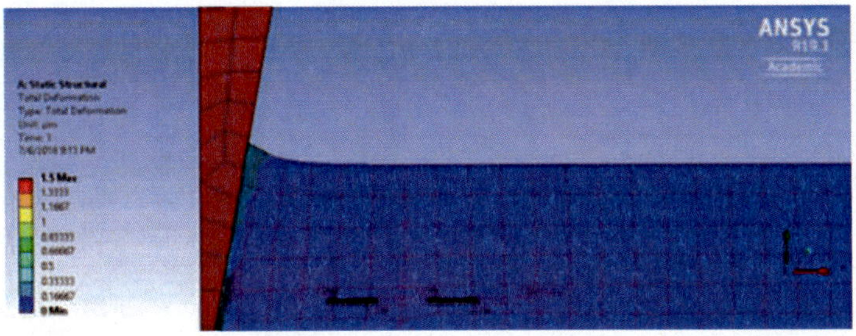

Fig. 2 Total deformation for $\mu = 0.0$ for rigid indentation with a cone of 10° semi-apex angle into mild steel block

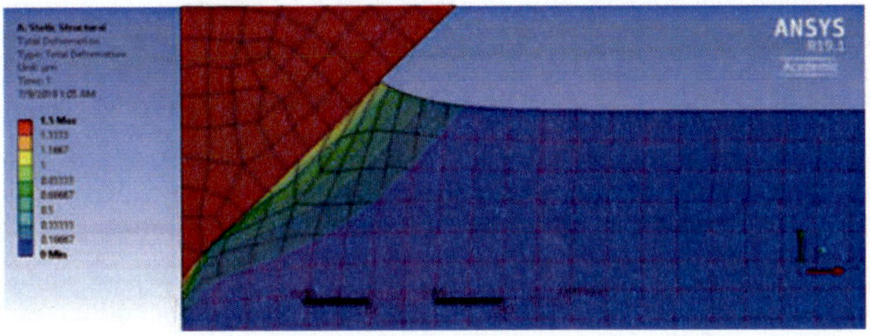

Fig. 3 Total deformation plots for $\mu = 0.0$ for rigid indentation with a cone of 45° semi-apex angle into mild steel block

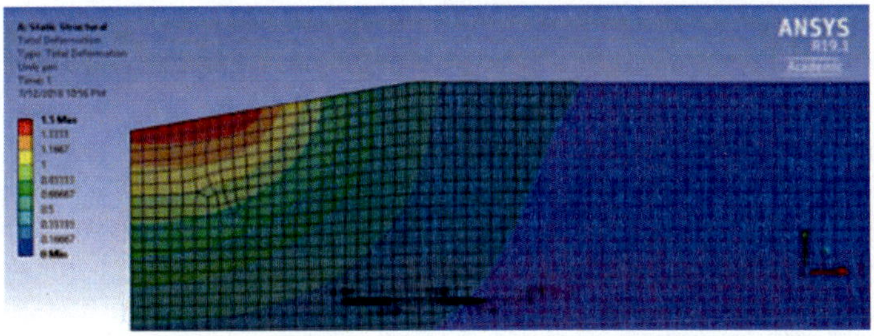

Fig. 4 Total deformation for $\mu = 0.0$ for rigid indentation with a cone of 80° semi-apex angle into mild steel block

lip profiles for different angles are analysed with the help of slip-line field theory. Table 1 shows the normal force required for indenting wedge up to depth of 1.5 μ for varying wedge angles. The force values in table show that with the increase of semi-apex angle of wedge, the force is more, and it is exponential in the zone of high wedge angle. For the same depth of indentation, with increasing semi-apex angle, more volume is displaced. Hence, the plastic deformation work is more, and the force values are more.

Table 1 Values of normal and tangential forces in wedge indentation by FEA

Cone angle (°)	Coefficient of friction	Maximum normal force (μN)	Minimum normal force (μN)	Average normal force values (μN)
10	0.0	82.0	705.0	390
20	0.0	341	2416	1980
30	0.0	780	6753	1500
40	0.0	1328	13,045	7200
45	0.0	2867	22,823	12,850
50	0.0	3429	33,929	18,680
60	0.0	7145	67,388	37,266
70	0.0	40,125	255,440	147,700

3 Slip-Line Field (SLF) Analysis of the Problem

As it is discussed in introduction section, the slip-line field network is composed of two families of orthogonal lines viz. α-lines and β-lines. Figure 5 shows a typical slip-line field network for wedge indentation problem considering lip is linear. Figure shows that it has three distinctive zones. The mathematical procedure and numerical schemes for producing this slip-line field have been discussed in detail by the work of Biswas et al. [9–12]. The present problem considers the lip to be nonlinear. But, the generation of slip-line becomes difficult for small values of semi-apex angles when nonlinear lip is considered. To overcome the difficulty, Biswas et al. [10] proposed a new hybrid approach. Figure 6 shows the approach of producing hybrid parabolic SLF network. The most outer SLF is pure parabolic profile, and the most interior SLF is pure linear profiles. The intermediate SLF is hybrid parabolic SLF which is actually the interpolated data of two extreme SLFs with weghted parameters.

Figure 7 depicts the hybrid parabolic SLF for 30° semi-apex angle of the wedge. Figure 8 shows the velocity distribution associated with the SLF shown in Fig. 7. As it was discussed in previous section that uneven distribution of velocity field gives rise to formation of parabolic lip generation. Table 2 shows the normal force value found from SLF theory for indentation up to a depth of 1.5 μ. The force values obtained

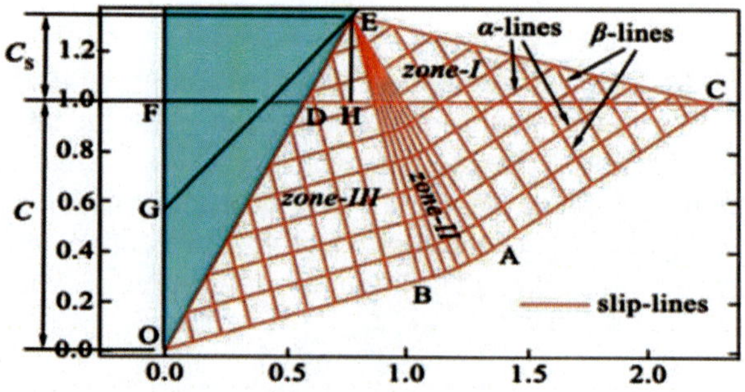

Fig. 5 Typically generated slip-line field for straight lip for 30° wedge angle

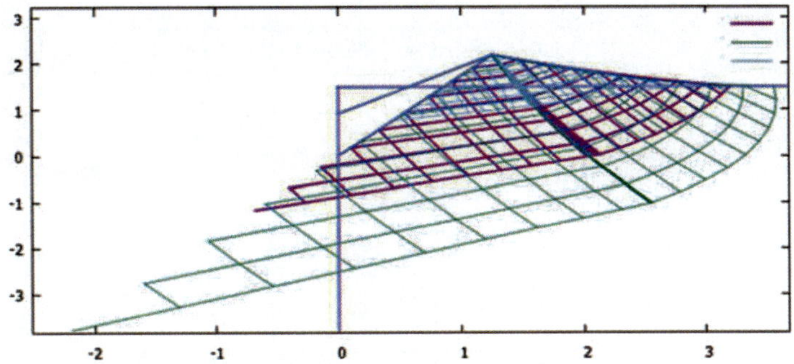

Fig. 6 Typical formation of hybrid parabolic slip-line field for straight lip for 30° wedge angle

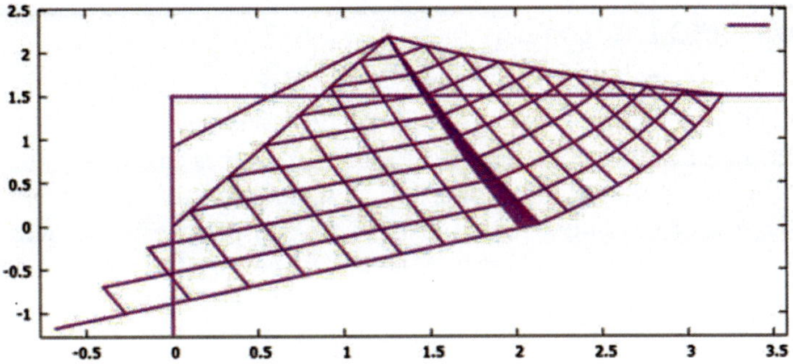

Fig. 7 Hybrid parabolic slip-line field for straight lip for 30° wedge angle

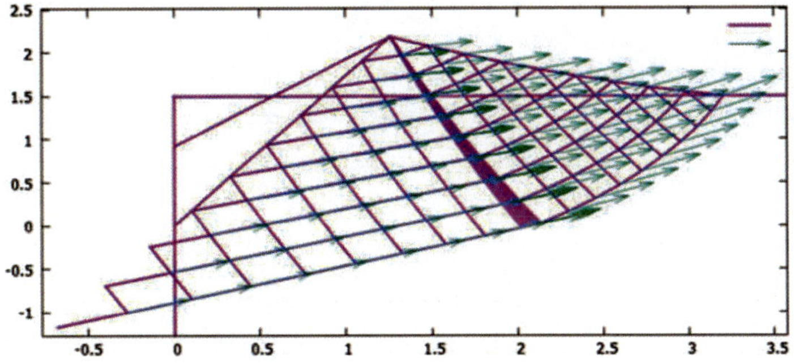

Fig. 8 Velocity distribution of deformed material for 30° wedge angle

from FEA are closer to that with SLF theory, though at high semi-apex angle, the force value found from FEA is more. The trend of force values is the same for SLF and FEA, both are increasing with the increase of semi-apex angle of the wedge.

Table 2 Values of normal forces in wedge indentation by SLF theory

Cone angle (°)	Coefficient of friction	Normal force (μN)
10	0.0	348
20	0.0	910
30	0.0	1870
40	0.0	2850
45	0.0	12,760
50	0.0	17,800
60	0.0	28,400
70	0.0	54,000
80	0.0	88,400

4 Results and Discussion

Figures 9 and 10 show the variation of lip profiles obtained from FEA analysis for different values of semi-apex angles when the mating surfaces are smooth that means coefficient of friction, $\mu = 0.0$. The associated equations of the lips are shown in figures. The degree of nonlinearity is actually given by value of coefficient of x^2 in equation. From analysis, it is clear that the coefficient of x^2 is the maximum at 20° value of semi-apex angle. The values gradually decrease with increase of semi-apex angle values. It is nearly negligible at high values of semi-apex angle which means that at those values, the lips are nearly linear.

Figures 11 and 12 show the variation of lip profiles obtained from SLF analysis. λ is the angle made by zone-III with wedge as shown in Fig. 5. This angular value actually determines the coefficient of friction in between mating surface. When the

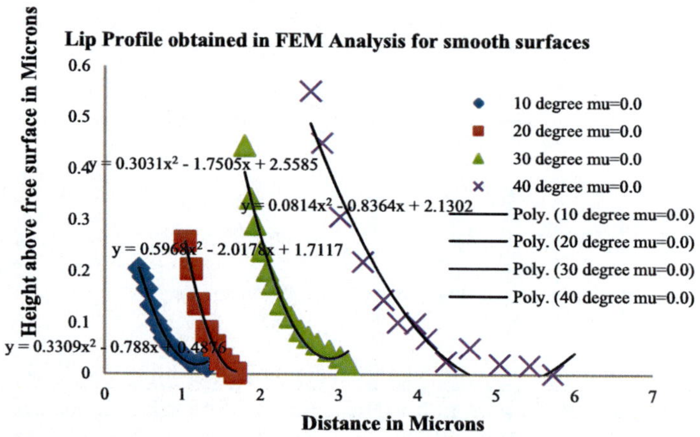

Fig. 9 Nature of lip profiles for smooth surfaces ($\mu = 0.0$) for indentation of a rigid cone with 10, 20, 30 and 40° to a mild steel block

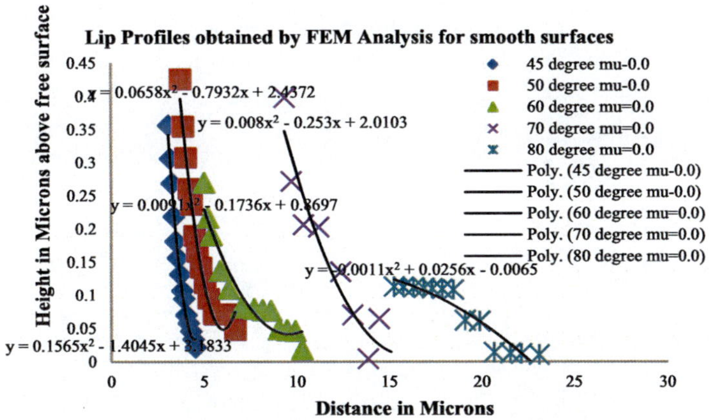

Fig. 10 Nature of lip profiles for smooth surfaces ($\mu = 0.0$) for indentation of a rigid cone with 45, 50, 60, 70 and 80° to a mild steel block

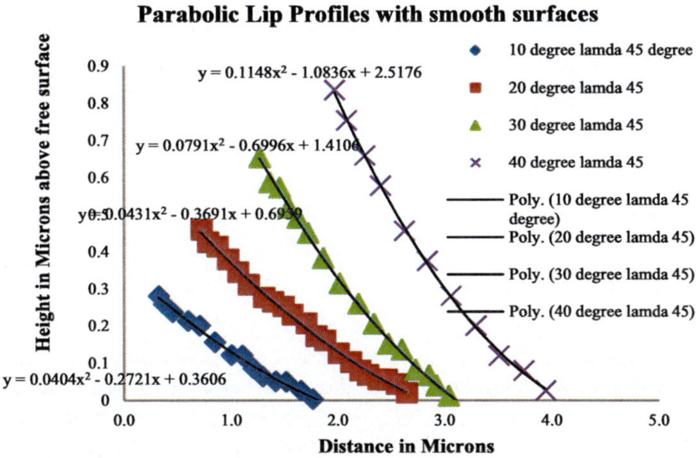

Fig. 11 Variation of lip for different semi-apex angle of wedge for $\lambda = 45°$

surfaces are smooth, the value of λ is 45°. The value of coefficient of x^2 is maximum at 40° value of semi-apex angle. This difference is due to the formation of hybrid parabolic lip below the semi-apex angle value of 45°. Above this, pure parabolic SLF formation is possible. The trends of decrement of values of coefficient of x^2 remain the same as that with FEA analysis.

The difference between the FEA and SLF analysis is that the nonlinear lip is automatically generated in case of FEA analysis, and in case of SLF, the lips are considered to be parabolic, and then SLFs are generated. In SLF analysis, the apex of the parabolic profile is set such that it always lies on the free surface or on the X-axis. But, analysis of curves obtained using FEA depicts that the apex of the parabola

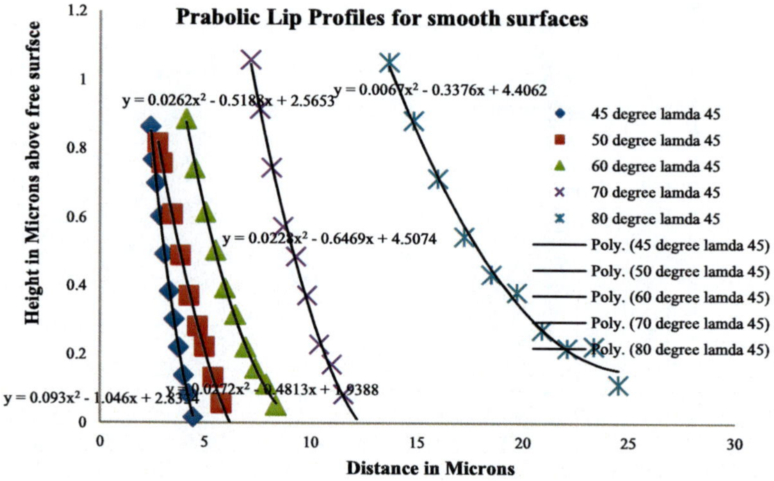

Fig. 12 Variation of lip for different semi-apex angle of wedge for $\lambda = 45^{\circ}$

may not positioned at X-axis. Hence, there are some differences which appeared in analyses of profiles with FEA and SLF results.

5 Conclusion

From finite element results, it is clear that the lip produced during wedge indentation is nonlinear in nature. The degree of nonlinearity gradually decreases with the increase of wedge angle. Taking this consideration of formation of parabolic lip, the slip-line filed network for wedge indentation is developed accordingly. The normal force values tend to show good matching. Therefore, the proposition of considering parabolic lip instead of linear lip is established in case of wedge indentation to ductile materials.

References

1. Hill R, Lee EH, Trupper SJ (1947) The theory of wedge indentation of ductile materials. Proc R Soc Lond A 188:273–289
2. Grunzweiz M, Longman IM and Petch NJ (1954) Calculation and measurement of wedge-indentation. J Mech Phys Solids 2:81–86
3. Dugdale DS (1954) Cone-indentation experiments. J Mech Phys Solids 2:265–277
4. Johnson W, Mathab FU, Haddow JB (1964) The indentation of a semi-infinite block by a wedge of comparable hardness-I-theoretical. Int J Mech Sci 6:329–336
5. Dodd B, Osakada K (1976) A note on the type of slip-line field for wedge indentation determined by computer. Int J Mech Sci 16:931–938

6. Locket FJ (1963) Indentation of a rigid/plastic material by a conical indenter. J Mech Phys Solids 11:345–355
7. Multhearn TO (1959) The deformation of metals by Vickers type pyramidal indenter. J Mech Phys Solids 7:85–96
8. Chitkara NR, Butt MA (1992) Numerical construction of axisymmetric slip-line field for indentation of a thick blocks by rigid conical indenters and friction at tool-metal interface. Int J Mech Sci 34:849–862
9. Biswas AK, Das S, Das B, Das B (2015) A slip-line field solution for micro-indentation of a rigid conical wedge by numerical technique. Int J Innov Res Sci Eng Technol 4:90–94
10. Biswas AK, Das S, Das S (2017) Micro-indentation of conical rigid wedge by numerical slip-line field theory: a hybrid approach. Indian Sci Cruiser 31(3):36–42
11. Biswas AK, Das S, Das S (2018) Estimating conical indentation load by numerical slip-line field technique. J Struct Eng 45(1):118–124
12. Biswas AK, Das S, Das S (2017) Determination of hardness for cone indentation to a semi-infinite block for various cone angles and friction parameters by numerical slip-line field technique. In: Proceedings of ICTACEM-2017, IIT KGP, India

Numerical Simulation of Welding for Residual Stress Prediction of Low Carbon Steel

Sagnik Choudhury, Debasish Biswas, and Santanu Das

1 Introduction

The origin of welding can be traced back to the Bronze Age. Methods of welding continued to evolve and by the end of the nineteenth century, gas welding, arc welding, and resistance welding became prevalent means of permanently joining two metallic materials. The history of simulation of heat and mechanical behavior by finite element during welding dates back to 1970s. Residual stresses and distortions in welding structures can render detrimental impacts on them.

Teng et al. [1] presumed thermo-mechanical behaviour for welded objects. Many of the researchers published works on transient heat transfer analysis [2–6]. And within few years, a new tool of coupled thermo-mechanical analysis brought the revolution in numerical analysis of welding [3–6]. There are many published works on the design of heat source model [7–10].

Uncoupled thermal and structural analysis has been simulated in the present work. The simulation is carried out in ANSYS 17 software using 3-D finite element method. A 3-D model has been developed using CAD software. The whole 3-D object has been meshed and divided into small elements or nodes. Different meshing techniques and sizes have been used to check mess independency. And the model is first solved in transient structural to get thermal profile and then the data is fed into static structural model to analyze the structural behavior of the weld. Element birth and death technique has been implemented to simulate the addition of filler material during welding.

S. Choudhury · D. Biswas (✉) · S. Das
Mechanical Engineering Department, Kalyani Government Engineering College, Kalyani, India

S. K. Ghosh et al. (eds.), *Advances in Thermal Engineering, Manufacturing, and Production Management*, Lecture Notes in Mechanical Engineering,
https://doi.org/10.1007/978-981-16-2347-9_27

2 Modeling of Transient Heat Flow

Formulation of finite elements has the greatest capacity to analyze and deal with complicated geometry in a nonlinear way.

$$T(x, y, z, t) = N_i(x, y, z) * T_1(t) \tag{1}$$

where, N_i are basic units that depend only on the type of components and their size and shape, T_i are the nodal temperature values.

The temperature profile can be evaluated directly from the following equation:

$$\left[\frac{\partial T}{\partial x}, \frac{\partial T}{\partial y}, \frac{\partial T}{\partial z}\right] = \left[\frac{\partial N_i}{\partial x}T_i, \frac{\partial N_i}{\partial y}T_i, \frac{\partial N_i}{\partial z}T_i\right] \tag{2}$$

2.1 *Heat Sources Used in Simulation*

Rykalins [11] did extensive research on welding heat sources. His works are the authentic source for the study of heat source. The flux of heat source ranges from 10^5 to 2×10^5 kJ/mm^2. The sources which are very concentrated like electron plasma, the energy density can reach up to 10^{10} kJ/mm^2 or even higher values. From different measurement techniques, it can be confirmed that mostly all stationary heat sources follow Gaussian power distribution. Lawson and Kerr [12] did extensive research on the effectiveness and accuracy of different heat sources in welding simulation. They established the fact that all heat sources were not being same. Rosenthal [13] suggested various practical heat source models. These included few one-dimensional models like a very small point line heating source and few two-dimensional heat sources like rectangular heat source and circular heat source. The model is shown in Fig. 1 was used in earlier models to predict temperature profile in a weld but they were not accurate in doing so. Boundary conditions and thermal properties were taken as fixed values, the effects of convective heat transfer and radiative heat transfer were not accounted for.

The effectiveness of the model used for welding simulation is directly dependent on external heat source geometry. The most important factor for choosing heat source or designing it is to accurately predict the temperature distributions [14].

The time consumed during the simulation can be reduced by using 3d solid and shell elements. The elements are concentrated along the weld centre line and it gets less dense as we move away from the centreline [15].

Here,

$$T(\varepsilon, y, z) = T_0 + \frac{Q}{2\varepsilon k}\exp\left(\frac{v\varepsilon}{2\lambda}\right)\left\{\sum_{n=-\infty}^{\infty}\frac{1}{R_n}\exp\left(\frac{vR_n}{2\lambda}\right)\right\} \tag{3.1}$$

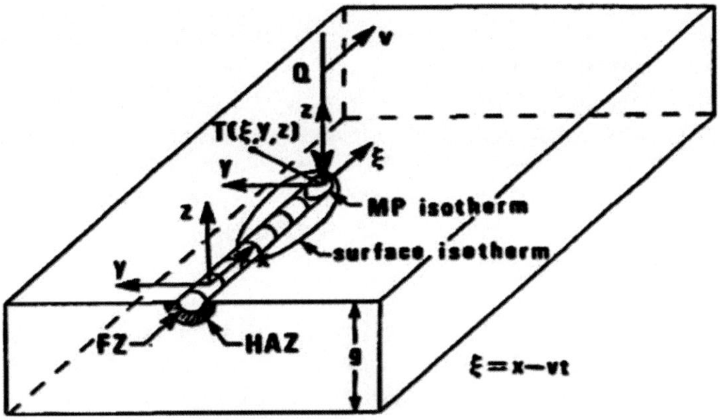

Fig. 1 A heat source in transit along the x-axis [13]

$$R_n = \sqrt{\varepsilon^2 + y^2 + (z + 2ng)^2} \tag{3.2}$$

$$\varepsilon = x - vt \tag{3.3}$$

3 Problem Description and Simulating Details

The model is first solved in transient structural system to get thermal profile and then the data is fed into static structural model to get the desired results.

The present simulation deals with certain assumptions, they are:

1. The movement of the plates during the procedure does not affect the temperature distributions of the plates themselves.
2. The material properties of metals are described till the liquid phase.
3. Heat loss by convection and radiation are considered.

A 3-D solid 90 type element has been implemented for the simulation of thermal model. The solid 90 is a tetrahedral element having 20 nodes, each having a single degree of freedom. In this analysis, surface heat flux model has been considered. The moving surface heat source along the weld centerline has been considered to describe the heat source. The speed of the moving heat source depends on welding speed. The model has about 28,000 elements, 5136 nodes, and 300 load steps. The welding part has been completed in the first 20 load steps and the rest of the remaining load steps have been used to complete the weld thermal cycle. The effects of convection, radiation, and conduction have been incorporated into the simulation. Two different values of film coefficient have been considered to account for forced convection effects due to the flow of inert gas and to incorporate cooling due to combined effects of radiation and convection.

The temperature-dependent properties of material have been used here. The values of all the properties of material are taken from standard sources and research papers. The filler material properties for the plates, thermal properties, mechanical properties, and the properties of Heat Affected Zone (HAZ) are assumed to be the same. The plasticity model to simulate the residual stress is von Misses rate-independent isotropic bilinear hardening system.

The heat input has been modeled through surface flux following Gaussian heat distribution profile. The reason for using Gaussian distribution is that it is very accurate in predicting the temperature profile during thermal analysis. The heat source is moved from one face to another face during each load step, emulating the movement of a welding torch during welding. The amount of heat input is given as follows:

$$Q = \eta \frac{UI}{v} \tag{4}$$

Here, U and I are arc voltage and current respectively, η is the efficiency of the welding arc and v is the traverse speed of welding heat source. The current used in this analysis is 180A, the voltage 24 V. Welding heat source speed has been considered as 5 mm $(\text{sec})^{-1}$, while the arc efficiency of the process as 85%.

3.1 Element Birth and Death Technique

In this analysis, element birth and death technique has been implemented to simulate the addition of filler material during welding. In this method, the element is not killed or deleted; rather it multiplies the stiffness with very small value in the order of 10^{-6}. During simulation of thermal model, the elements along the weld centreline are killed or deactivated at the beginning. They are activated or born later according to the movement of heat source, at the desired time.

3.2 Meshing

Meshing is an important tool for solving any problems using FEM. The use of meshing divides the model into small elements and nodes. The governing equations are solved for each nodes and elements. The final result has been given by summing up all the individuals results from the nodes. The index (density, method, distortion, jacobian) of meshing governs the accuracy of the solution. The type of elements formed in the meshing, i.e., hexahedral, tetrahedral, etc. also affects the accuracy of the solution. Due to the movement of heat flux along the weld centre line, high temperature and stress gradients are concentrated near the centre line. Adaptive

Fig. 2 Meshed model used in the analysis

meshing techniques have been implemented in the FEM model to increase the density of mesh in both sides near the weld centre line. The moving heat source is simulated by applying heat flux as surface flux at each time step. The meshed model is shown in Fig. 2.

The meshing method used here is multi-zone. It is a patch-independent meshing method which increases the accuracy of solution and also reduces the CPU time needed to solve the simulation. The elements formed in this meshing are a tetrahedral solid with twenty nodes. The elements are so chosen to support the use of Element Birth and Death technique.

4 Results and Discussion

In the present analysis, a moving heat source has been developed and its effect on temperature distribution, residual stress distribution, and angular distortion of steel specimen (ASTM 36) have been studied along with the effect of different values of convective film coefficient on stress distribution. 2-D Gaussian surface flux has been used for moving heat sources. The Gaussian heat flux model has been chosen for its simplicity in analysis and also provides accurate results in predicting thermal profiles in welding simulations. Four different models are solved by varying the convective film coefficient. The values of film coefficient considered are 35 , 100 , 150 , and 500 W/m^2 °C.

4.1 Temperature Distributions

The movement of the heat source causes temperature distribution in the weldment. The temperature profile generated in welding is directly proportional to parameters like welding speed, arc efficiency, weld current, weld voltage, and thickness of the plate. The factors like conduction, convection, radiation also play a significant part in determining the temperature profile in welding. Temperature profile for different time steps in transient model have been figurized in Fig. 3a–c.

4.2 Cooling Curve

The cooling curves by varying the value of convective film coefficient *h* have been shown in Fig. 4a-d. The cooling curve is generated by probing the centre of the plate. The probing tool was selected and put on a surface in the middle of the plate. The moving heat source was active for 20 s and then the weld was allowed to cool to ambient temperature; the thermal cycle ends at about 300 s. The cooling curve is observed and plotted for 1000 s. The rate of cooling is directly proportional to the conductivity of the material used, coefficient of radiation, and the convective film coefficient. Both the conduction and the convection effect have been considered here.

Figure 3a-c show the temperature profile in the centre of the plate. As the heat source moves towards the centre of the plate, temperature rises due to conduction heat transfer effect. It reaches the maximum when the heat source is directly at the centre. Then as the heat source moves away, heat is lost due to convection and radiation effect. The plate cools down rapidly first, and then the process becomes gradual. This can be explained as the heat lost due to convection is proportional to the temperature gradient, the temperature gradient at the beginning of the welding is quite low, and then it increases very quickly and reaches a maximum value as the heat source starts moving and imparting heat energy to the work piece. The heat quickly spreads throughout the model due to the conductivity of the material, and after the heat source is withdrawn i.e., as the welding process is finished, the temperature gradient starts falling and the heat starts dissipating and the work-piece starts cooling down.

4.3 Residual Stress

The non-uniform cooling and contraction of weld material introduced residual stress in the weldment. The failure of certain welds can be attributed to residual stress. In the present analysis, residual stresses are produced due to the movement of the heat flux, which causes non-uniform expansion and contraction of the weld material. The workpiece is clamped in eight outer vertices, and the clamping of workpiece also

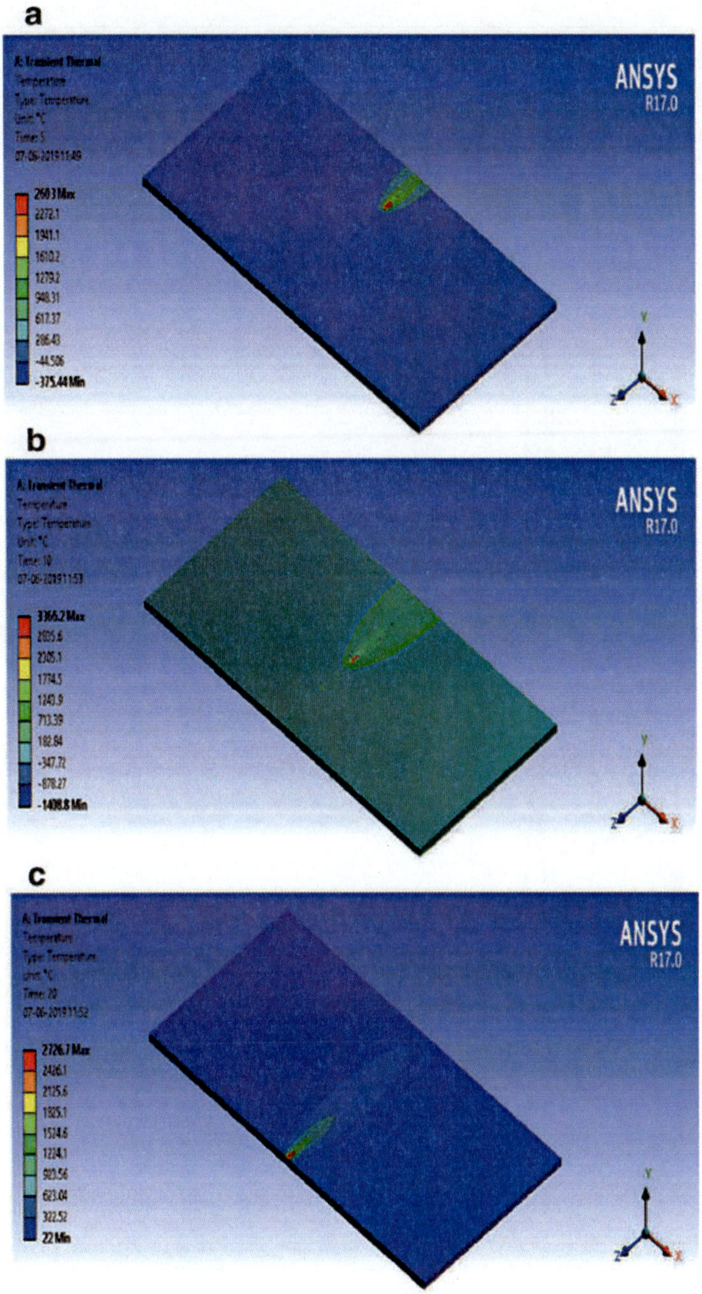

Fig. 3 **a** Temperature distributions after $t = 5$ s, **b** Temperature distribution after $t = 10$ s, **c** Temperature distributions after $t = 20$ s

has significant effect on the distribution of residual stress in the weldment. Residual stresses are calculated and plotted by solving the thermal model. The data from the thermal profile is taken and fed into the mechanical model. The mechanical model is solved using modified Newton–Raphson method to eliminate non-linerities.

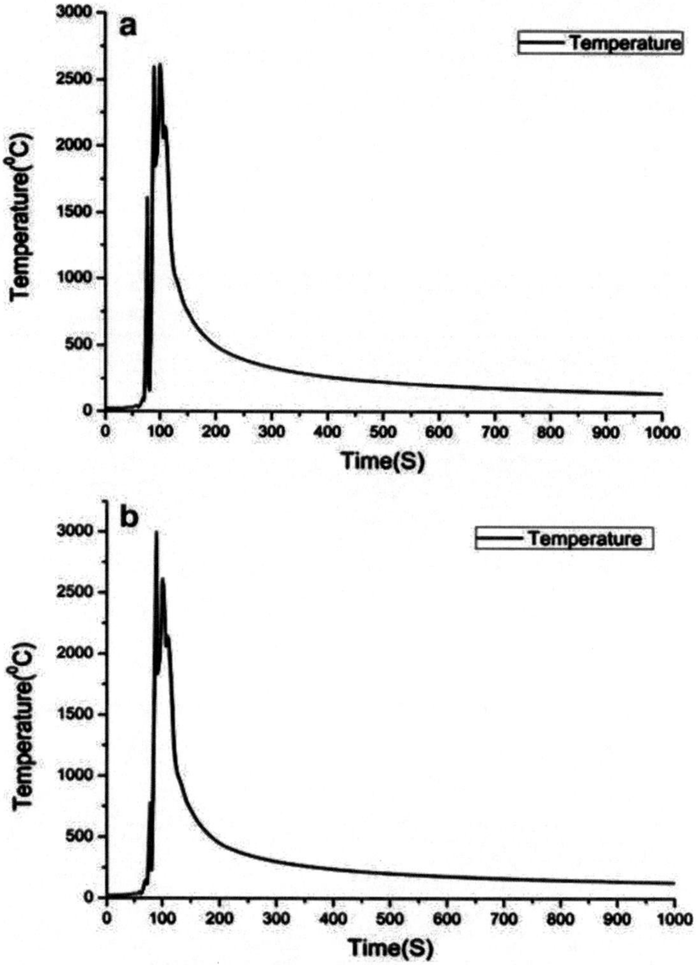

Fig. 4 **a** Cooling Curve for $h = 35$ W/m^2 °C, **b** Cooling Curve for $h = 100$ W/m^2 °C, **c** Cooling curve for $h = 150$ W/m^2 °C, **d** Cooling Curve for $h = 500$ W/m^2 °C

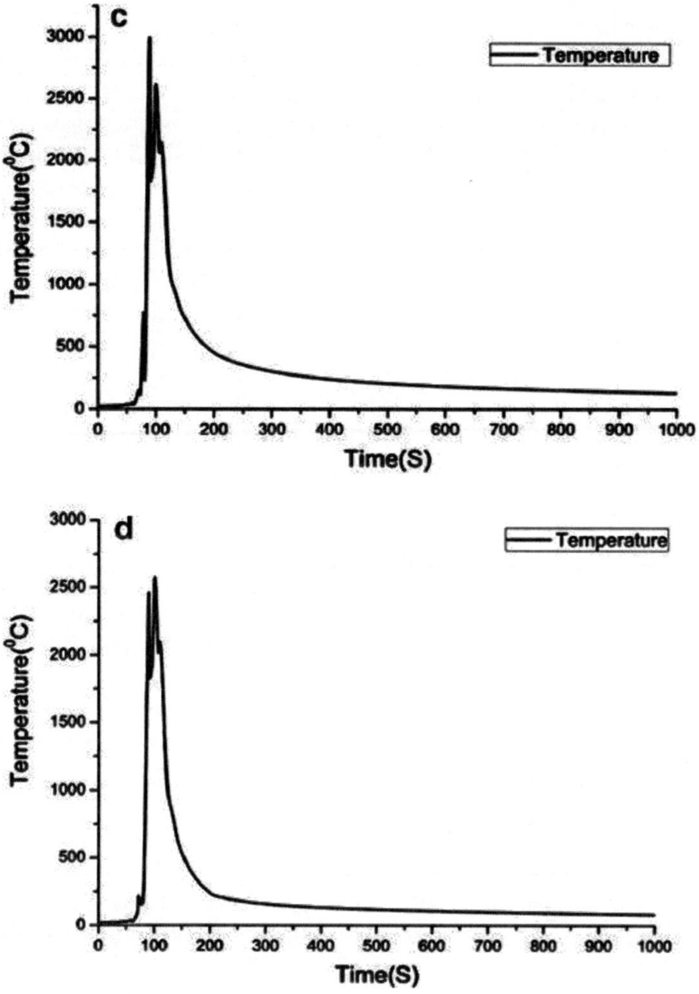

Fig. 4 (continued)

4.4 *Equivalent Stress*

The equivalent (Von Mises) stresses are predicted after structural analyses of the welded plates. The meshing of the mechanical model is done differently other than that in the thermal model. The element type used in meshing of the mechanical model is solid 185. The solid 185 has plasticity, large strain, and deflection capabilities. The solid 185 is compatible with coupling solution and supports isotropic hardening. The accuracy of the results was increased by taking large number of substeps in the modeling.

It can be clearly seen that the magnitude of stress decreases with the distance away from the weld centerline. The reason for this can be attributed to the fact that the weld centerline is the region with most heat concentration resulting in strong yielding and is the last region to cool to ambient temperature, giving other regions sufficient time to cool and shrink early. This results in mismatching strains and plastic deformations. Hence, as one moves towards the weld centerline, the stress keeps on increasing and reaches its maximum value along the centerline.

From Fig. 5b, c it can be shown that for $h = 100$ and $h = 150$, there is no significant difference between the stress profile. It can be attributed to the fact that for small change in the value of convective heat transfer coefficient there is no significant change in cooling rate of the weldment.

5 Conclusions

Uncoupled thermal and structural analysis has been performed in the present work. From the analysis, it can be concluded that

1. Conduction heat transfer effect rises with temperature as heat source moves towards the centre of the plate and becomes maximum when the heat source is directly at the centre.
2. A maximum value of 684.79 MPa has been observed for $h = 150\ \text{W/m}^2\ °\text{C}$. This can be explained as the plate cools it tries to retain its original shape, but due to clamping of the weldment cannot do so and as a result, the region which cools earlier exerts pressure on the region which cools slowly and causes mismatching strains.
3. Relatively low value of convective heat transfer coefficient implies that heat rejection rate due to convection and radiation is very low and hence, some time has been taken by the plate to get cooled and as a result, large amount of stresses are developed.
4. The least value of equivalent stress has been obtained for $h = 500\ \text{W/m}^2\ °\text{C}$. The reason for this can be explained as due to the high value of convective heat transfer coefficient, heat rejection rate of the plate increases and hence, the plate cools quickly not giving enough time for the development of stress. It can be viewed as the fact of a higher flow rate of gas helps in faster cooling of the weld zone and hence, reduces the equivalent stress in welding.
5. The use of element birth and death technique does not significantly affect the temperature profile because of the low thickness of the plate compared to the length. The heat transfer by conduction occurs fast and the temperature gradient decreases quickly.

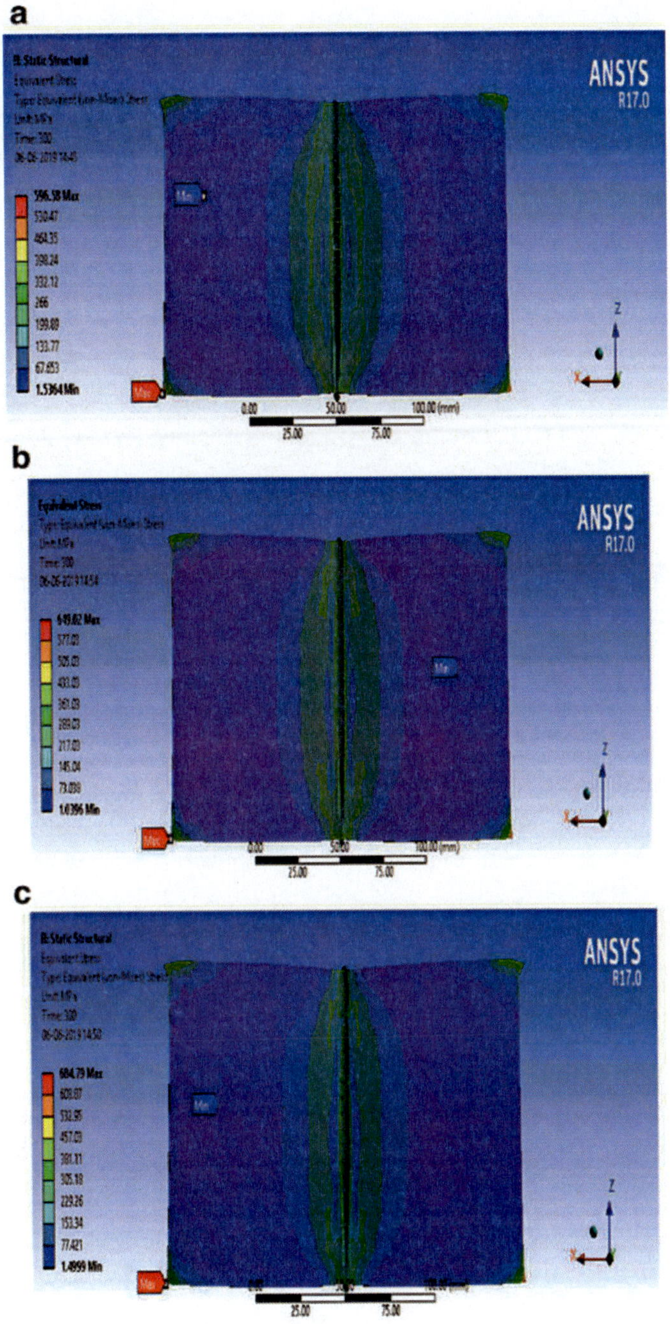

Fig. 5 **a** Equivalent stress for $h = 35$ W/m^2 °C, **b** Equivalent stress for $h = 100$ W/m^2 °C, **c** Equivalent stress for $h = 150$ W/m^2 °C, **d** Equivalent stress for $h = 500$ W/m^2 °C

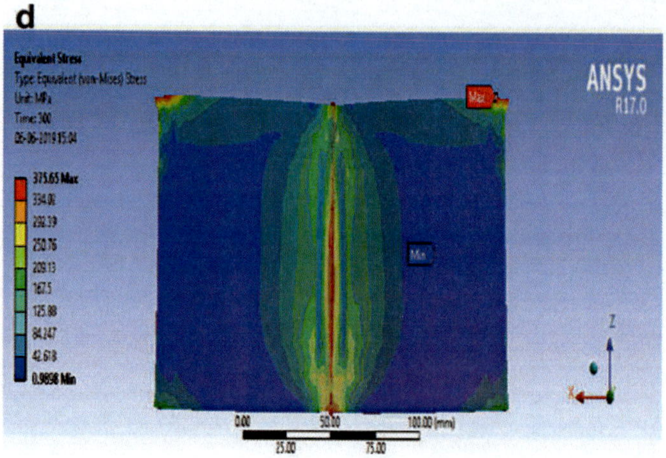

Fig. 5 (continued)

References

1. Teng T, Chang P, Tseng W (2003) Effect of welding sequences on residual stresses. Comput Struct 81(5):273–286
2. Andersson B (1978) Thermal Stresses in a Submerged-Arc Welded Joint Considering Phase Transformations. J Eng Mater Technol 100(4):356–360
3. Goldak J, Bibby M, Moore J, House R, Patel B (1986) Computer modeling of heat flow in welds. Metall Trans B 17(3):587–600
4. Giedt WH, Wei XC, Wei SR (1984) Effect of surface convection on stationary GTA weld zone temperatures. Welding Journal 63(12):376–380
5. Rosenthal D (1941) Mathematical Theory of Heat Distribution during Welding and Cutting. Welding Journal 20:220–234
6. Goldak J, Chakravarti A, Bibby M (1984) A new finite element model for welding heat sources. Metall Trans B 15(2):299–305
7. Nandan R, Roy G, Debroy T (2006) Numerical simulation of three-dimensional heat transfer and plastic flow during friction stir welding. Metallurgical and Materials Transactions A 37(4):1247–1259
8. Colegrove P, Shercliff H, Zettler R (2007) Model for predicting heat generation and temperature in friction stir welding from the material properties. Sci Technol Weld Joining 12(4):284–297
9. Zhang W, Roy G, Elmer J, Deb Roy T (2003) Modelling of heat transfer and fluid flow during gas tungsten arc spot welding of low carbon steel. J Appl Phys 93(5):3022–3033
10. Farzadi A, Serajzadeh S, Kokabi A (2007) Modelling of heat transfer and fluid flow during gas tungsten arc welding of commercial pure aluminium. The International Journal of Advanced Manufacturing Technology 38(3–4):258–267
11. Rykalin RR (1974) Welding in the World. 12(9/10):227–248
12. Lawson S, Kerr H (1976) Welding Research International 6(5–6):23–28
13. Rosenthal D (1946) The theory of moving sources of heat and its application to metal treatments. Transactions of the ASME 68(8):849–866
14. Yue J, Dong X, Guo R, Liu W, Li L (2019) Numerical simulation of equivalent heat source temperature field of asymmetrical fillet root welds. Inter J Heat Mass Transfer 130:42–49
15. Peric M, Tonkovic Z, Rodic A, Surjak M, Garasic I, Boras I, Svaic S (2014) Numerical analysis and experimental investigation of welding residual stresses and distortions in a T-joint fillet weld. Mater Des 53:1052–1063

Numerical Simulation and Study of Solidification Process During Casting Using FLUENT

Dipanjan Dey

1 Introduction

In order to simulate the filling up process, the various complex factors such as computational fluid dynamics (CFD), methods of numerical simulation and heat transfer need to be considered and the feasibility of using FEM to model casting process is already studied [1–4]. FEM also provides a better solution for solving complicated manufacturing problems such as these and is also useful for research at a lower cost, also giving consistent results [5]. The estimation of solidification time which depends upon the solidification rate of the material is one of the very important factors and is studied here. While the long pouring time that is slow pouring rate can lead to problems such as cold shut, air entrapment, reoxidation of the metal, lump formation or starting of solidification prior to filling up the mould cavity, etc.; short pouring time is normally advised for casting. However, if the pouring rate is too fast, there are risks of sand inclusions, erosion of the mould wall or even breaking of the mould structure altogether. Therefore, there is a necessity to form models so as to determine the range at which the pouring must be done apart from other design considerations, prior to the actual casting.

2 Methodology

In industry, the pouring time for casting processes is recorded and can be generally expressed as

D. Dey (✉)
Department of Mechanical Engineering, Jalpaiguri Government Engineering College, Jalpaiguri, West Bengal 735102, India

S. K. Ghosh et al. (eds.), *Advances in Thermal Engineering, Manufacturing, and Production Management*, Lecture Notes in Mechanical Engineering,
https://doi.org/10.1007/978-981-16-2347-9_28

$$Tp = k\sqrt{W} \tag{1}$$

where

W is the weight of the casting,
k is a constant dependent on the material and
Tp is the pouring time.

The standard transport (model) equations are used for getting the turbulence kinetic energy

$$\frac{\partial}{\partial t}(\rho k) + \frac{\partial}{\partial x_i}(\rho k u_i) = \frac{\partial}{\partial x_j}\left[\left(\mu + \frac{\mu_t}{\sigma_k}\right)\frac{\partial k}{\partial x_j}\right] + G_k + G_b - \rho\varepsilon - Y_M + S_k \tag{2}$$

$$\frac{\partial}{\partial t}(\rho \varepsilon) + \frac{\partial}{\partial x_i}(\rho \varepsilon u_i) = \frac{\partial}{\partial x_j}\left[\left(\mu + \frac{\mu_t}{\sigma_\varepsilon}\right)\frac{\partial \varepsilon}{\partial x_j}\right] + C_{1\varepsilon}\frac{\varepsilon}{k}(G_k + C_{3\varepsilon}G_b) - C_{2\varepsilon}\rho\frac{\varepsilon^2}{k} + S_\varepsilon \tag{3}$$

where G_k and G_b denote the production of turbulent kinetic energy due to the mean velocity gradients and buoyancy, respectively,

$C_{1\varepsilon}$, $C_{2\varepsilon}$, and $C_{3\varepsilon}$ are constants,

σ_k and σ_ε are the turbulent Prandtl numbers for k and ε, respectively. S_k and S_ε are user-defined source terms.

The turbulent (eddy) viscosity,

$$\mu_t = \rho C_\mu \frac{k^2}{\varepsilon} \tag{4}$$

here, Cμ is constant (Table 1).

Energy equations used are shown in the following equations:

Since enthalpy,

$$H = h + \Delta H, \tag{5}$$

Table 1 Values of model constant

$C_{1\varepsilon}$	**1.44**
$C_{2\varepsilon}$	1.92
C_μ	0.99
σ_k	1.0
σ_ε	1.3
coef	0.1

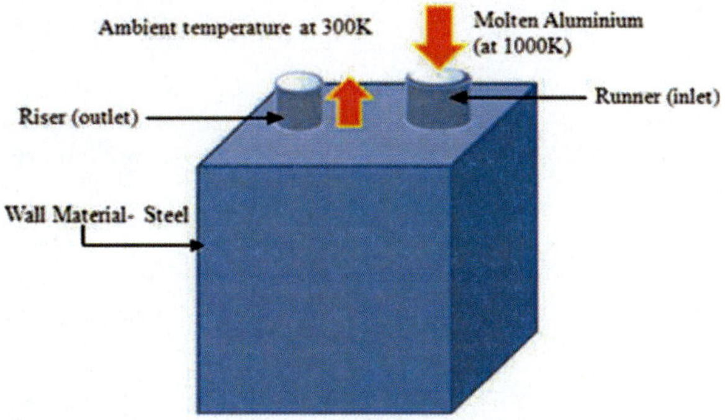

Fig. 1 Schematic diagram of the casting process

where ΔH is latent heat and sensible enthalpy *h*,

$$h = h_{\text{ref}} + \int_{T\text{ref}}^{T} CpdT \tag{6}$$

where h_{ref} = Reference enthalpy,
T_{ref} Reference Temperature,
c_p = Specific heat at constant pressure
For solidification and melting problems, the energy equation used is [5],

$$\frac{\partial}{\partial T}(\rho H) + \nabla.\left(\rho\vec{\mathbf{v}}H\right) = \nabla.(k\nabla T) + S \tag{7}$$

where
H is enthalpy,
p is density,
$\vec{\mathbf{v}}$ is the fluid velocity
S = source term
Here, we have used a permanent mould made of steel, and pure aluminium is selected as the casting material. Figure 1 shows the schematic diagram of the casting process. However, in order to reduce the complexity that comes with analysis of a 3D model, a 2D cross-section is considered to reduce the size of the model. The solidification rate, strain rate and temperature variation were studied.

3 Geometry of the Model and Mesh Parameters

The wall material is steel and the thickness of the wall is 0.01 m. The pouring of the molten aluminium at the inlet or runner was done under both the conditions of velocity inlet at 0.01 m/s and also made to fill up the mould using gravity. The model geometry used for the study was pouring in the direction of gravity, unlike a model

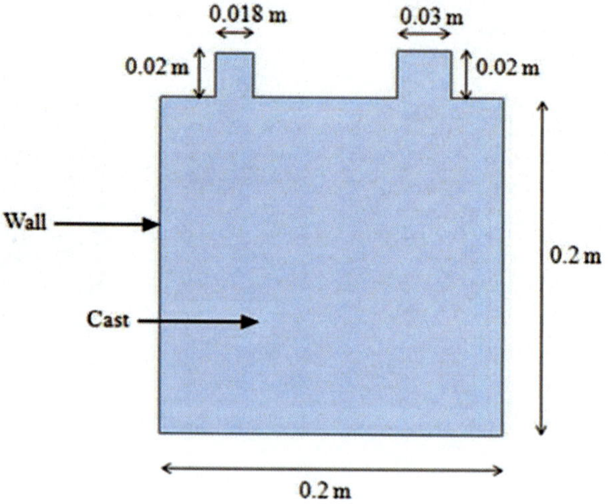

Fig. 2 2D geometry of the casting

where anti-gravity filling [1, 4] or side filling was explored [2] in previous works. The mesh was refined continuously until an element size of 0.1 mm^2 was used, which gave approximately correct results (Fig. 2).

4 Simulation

Here, the simulation of the casting process using a 3D model was done using FLUENT solver available in ANSYS package. The heat transfer model was carefully studied and the boundary conditions were selected so as to replicate the actual environment in which the casting takes place. Solidification model, energy equation and volume of fluid models were used. The turbulent model was used in this study and the viscosity model selected was k epsilon-2. For the casting material, phase change material was defined under the fluid material, and accordingly, the properties of liquid aluminium were provided. The initial temperature of the liquid at the inlet was 1000 K and the ambient temperature was given 300 K.

5 Results

The total solidification time was found to be 148.56 s and the total time taken by the solid to reach the ambient temp was found to be 1049.56 s.

Figure 3 shows the variation of solidification of the metal with time. The contours of solidification are shown in time scales of 10 s, 20 s, 60 s, 90 s and 150 s. Figure 4 and Fig. 5 show, respectively, the strain rates and temperature contours within the same time frames.

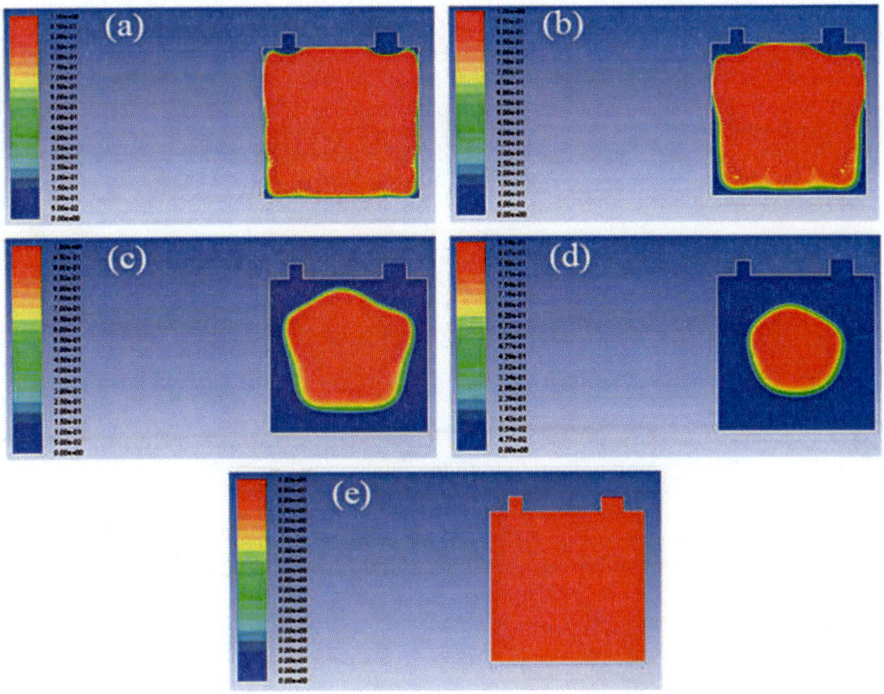

Fig. 3 Contours of solidification (liquid fraction) at time **a** 10 s, **b** 20 s, **c** 60 s, **d** 90 s and **e** 150 s

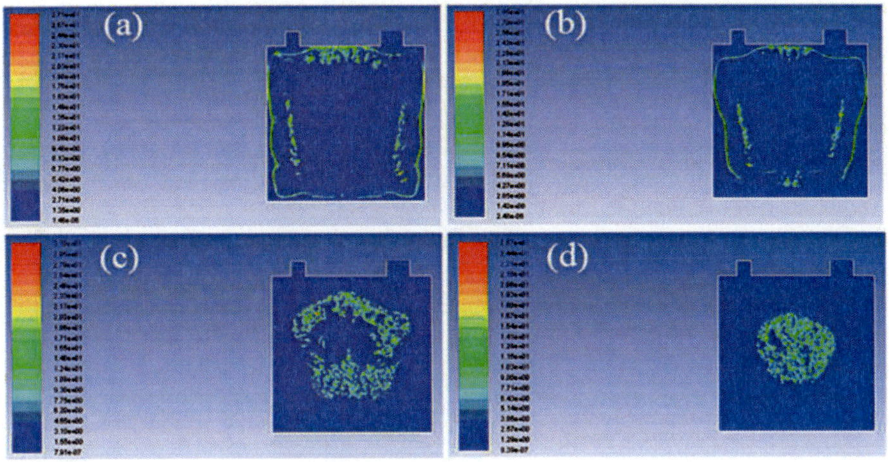

Fig. 4 Strain rate at time **a** 10 s, **b** 20 s, **c** 60 s and **d** 90 s

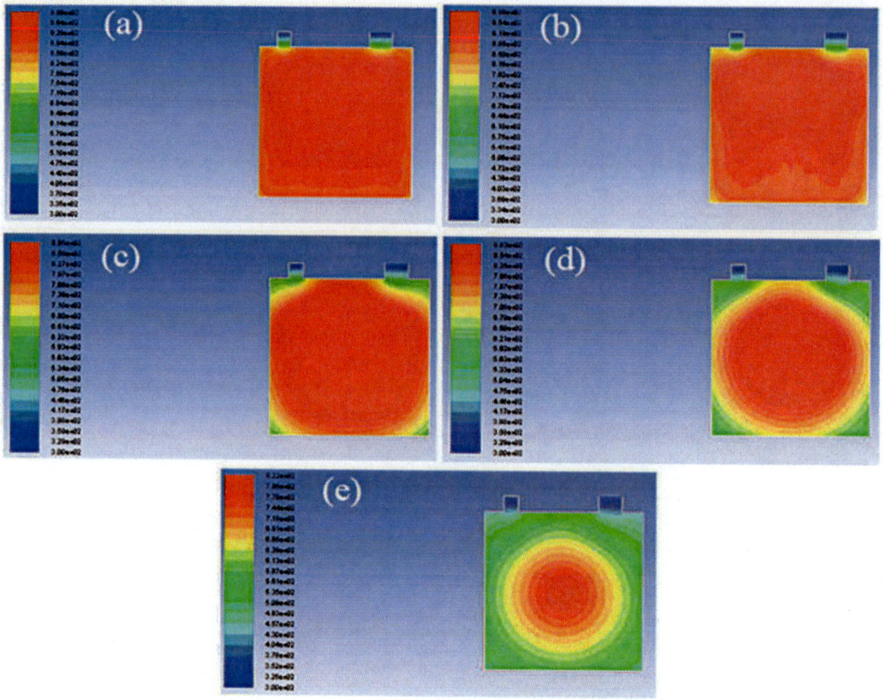

Fig. 5 Temperature at time **a** 10 s, **b** 20 s, **c** 60 s, **d** 90 s and **e** 150 s

6 Conclusion

In this study, we arrive at the following conclusions:

- The solidification rates can be successfully predicted using the FEM method in the FLUENT solver.
- The material property needs to be a phase change material for the simulation of the casting process to take place successfully.
- The standard model and settings used can be refined to predict further the pouring time, the grain structure with a finer mesh, riser designs, etc.
- Further exploration and scope for research can be the simulation of casting with different mould properties and cast materials along with an analysis of the different types of casting methods in industry practice.

References

1. Yuwen XX, Chen L, Han YJ (2012) Numerical simulation of casting filling process based on FLUENT. Energy Procedia 17:1864–1871
2. Choudhari CM, Narkhede BE, Mahajan SK (2013) Modeling and simulation with experimental validation of temperature distribution during solidification process in Sand casting. International journal of computer applications 78(16):23–29
3. Venkatesan A, Gopinath VM, Rajadurai A (2005) Simulation of casting solidification and its grain structure prediction using FEM. J Mater Process Technol 168(1):10–15
4. Changzhong J. and Ling C. (2010) Coupling Simulation of Flow Field and Temperature Field in Mold Filling Process Based on ANSYS Software. In: Third International Conference on Information and Computing, June 2010. Vol. 4. IEEE, p 246–249
5. Fluent, A.N.S.Y.S. (2011) 14.0 User's Manual. ANSYS Inc., Canonsburg, PA

Numerical Simulation and Study of the Effects of Machining Parameters Under Dry Turning Conditions

Dipanjan Dey and Nripen Mondal

1 Introduction

The turning operation is regarded as one of the most basic machining operations in industry practice. The tool life, tool wear rate, cutting temperature and power consumption by the machine are the major factors which are looked upon in order to increase the efficiency of the machining process. For reduction and optimisation of the cutting temperature and tool wear rates, different types of methodology such as the use of various types of cooling systems like flood cooling, minimum quantity lubrication (MQL) and liquid CO_2 are used to reduce the cutting temperature [1–3]. However, due to various reasons such as degradation of the machining environment quality, cost and affordability and unable to be reused again, dry turning condition is preferable in most circumstances [4]. The emerging field of research in this respect is to develop the tool geometry such that the temperature generated during machining operation is optimised [5, 6]. It can be shown that with the development of the carbide cutting tool insert geometry, the machining process can be made more efficient [7, 8]. In this study, we have developed a finite element model and analysed the cutting temperature and load in the turning process under dry condition.

D. Dey · N. Mondal (✉)
Department of Mechanical Engineering, Jalpaiguri Government Engineering College, Jalpaiguri, West Bengal 735102, India

S. K. Ghosh et al. (eds.), *Advances in Thermal Engineering, Manufacturing, and Production Management*, Lecture Notes in Mechanical Engineering,
https://doi.org/10.1007/978-981-16-2347-9_29

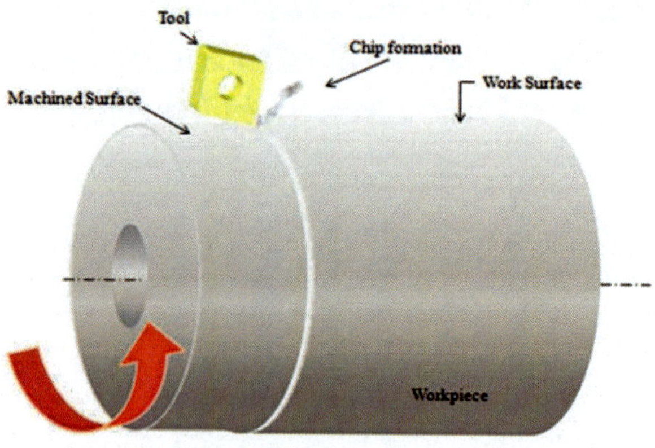

Fig. 1 Schematic diagram of turning operation using a cutting insert

Table 1 Process conditions

Ambient temperature (°C)	20
Convection coefficient (N/s/mm/°C)	0.02
Shear friction factor	0.6
Heat transfer coefficient (N/s/mm/°C)	45

2 Finite Element Modelling

2.1 *Process Setup and Controls*

Finite element models have become very popular recently for its use in turning processes, especially for orthogonal turning process [9–13]. Here, the simulation of the turning process is carried out using Deform 3D software. The process involves the input of the machining parameters, initial environmental conditions, positioning of the tool, assigning the material of the tool and workpiece, mesh formation, setting up of the boundary conditions and defining the equations. The tool material used for the purpose is uncoated tungsten carbide (WC), and workpiece material is selected as AISI-1013 for machining. The software allows the user to import the required tool as CAD files, its material properties and define workpiece properties or to use the tool library. Figure 1 shows the schematic diagram of the turning operation where a cutting insert is used.

The initial process conditions are shown in Table 1.

2.2 *Tool and Workpiece Setup*

The tool insert can be positioned as per the user's choice. The tool holder can be loaded from the library or imported externally. The properties of tool holder required

Table 2 Tool geometry

IC (mm)	12.7
T (mm)	4.7625
R (mm)	0.79375
B (mm)	3.08864
H (mm)	5.1562

are as follows—side cutting angle, back rake angle and side rake angle which in this case are −3, −5 and −5, respectively. The tool geometry is shown in Table 2.

2.3 *Mesh Generation*

The mesh generation needs to be determined carefully as finer mesh will result in more number of elements and thus greater accuracy. During the actual running of the simulation, there is the deformation of the material, and thus, every time remeshing is done by the software. Eulerian meshing or Lagrangian meshing is employed for this purpose. At this step, the number of elements was increased gradually so that the mesh can be made more suitable for greater accuracy of the results. The number of mesh elements for the tool insert, after which there was consistency found in the results was 45,000. For the workpiece sample used, the number of elements in the mesh was 30,000. Figure 2 shows the mesh generated for tool and workpiece before and after chip formation.

Table 3 shows the tool and workpiece meshing conditions used for this study. The boundary conditions include velocity at the x, y and z directions and heat exchange with the environment.

2.4 *Simulation Controls and Database Generation*

The initial number of simulation steps given was 10,000, and step increment to save was given 25.

Models that are commonly used for instantaneous tool wear calculation are the Archard model (Eq. 1) or the Usui model (Eq. 2).

$$w = \int K \frac{p^a v^b}{H^c} dt \tag{1}$$

$$w = \int apVe^{-b/T} dt \tag{2}$$

where w = wear volume

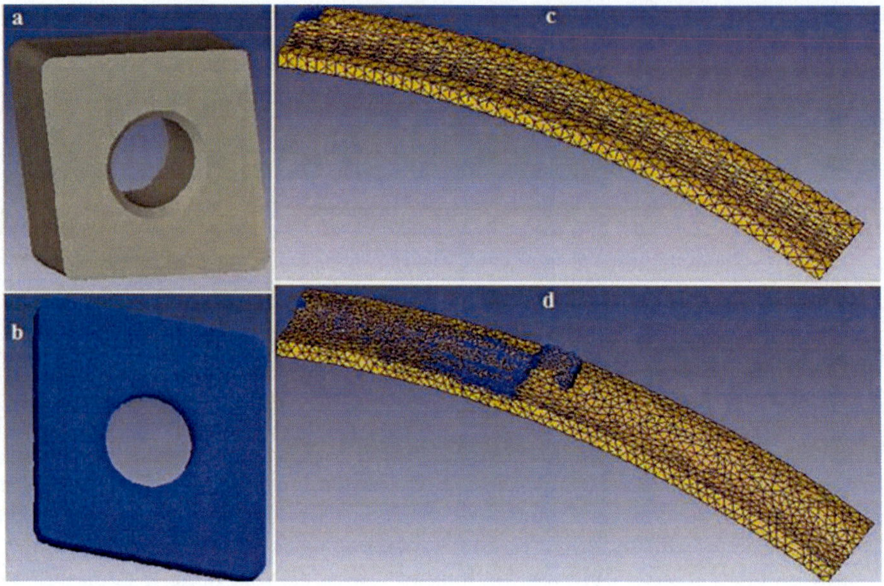

Fig. 2 **a** Simulated model of tool, **b** mesh generated for tool, mesh for workpiece **c** before machining and **d** after chip formation

Table 3 Tool and workpiece conditions

Tool material	Tungsten carbide (WC)
Workpiece material	AISI1013
Tool temperature (°C)	20
Number of elements for tool mesh generation	45,000
Size ratio	4
Workpiece type	Plastic
Workpiece temperature (°C)	20
The diameter of the workpiece (mm)	20
Number of elements for workpiece mesh generation	30,000
Size ratio	7

P = interface pressure;
v = sliding velocity;
H = hardness of tool material;
T = interface temperature (in degrees absolute);
a, b, c and K are experimentally calibrated coefficients.

Here, we have used the Usui model (Eq. 2) for tool wear calculation with the tool wear parameters a = 1e-05 and b = 1000.

The last step of the pre-processor step is where the database is generated followed by the simulation.

3 Results and Data Interpretation

The results for the temperature generated at the workpiece, tool and interface regions with the variation of the cutting velocity were noted. The temperature increases as the cutting velocity increases. Figures 3 and 4 show the effects obtained for cutting velocities 40, 50 and 60 m/min in maximum temperature versus time plots.

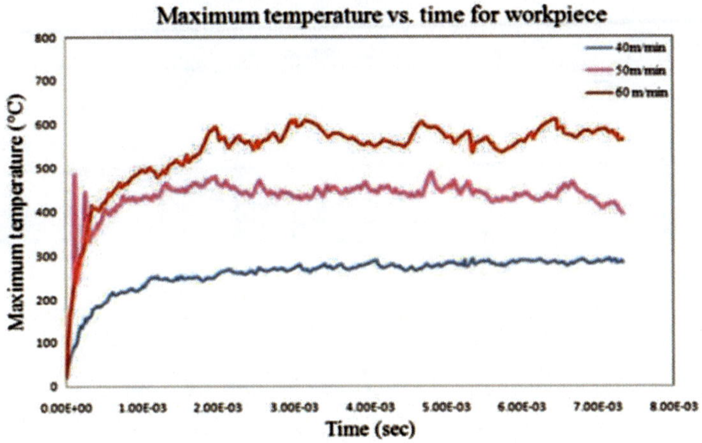

Fig. 3 Temperature generated in workpiece versus time graph for different values of cutting velocity

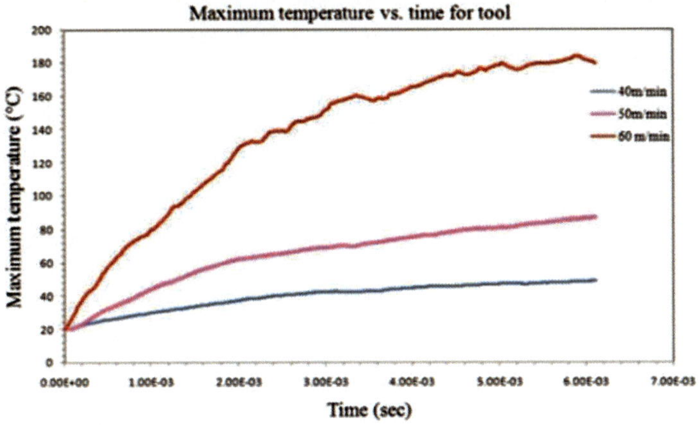

Fig. 4 Temperature generated in tool versus time graph for different values of cutting velocity

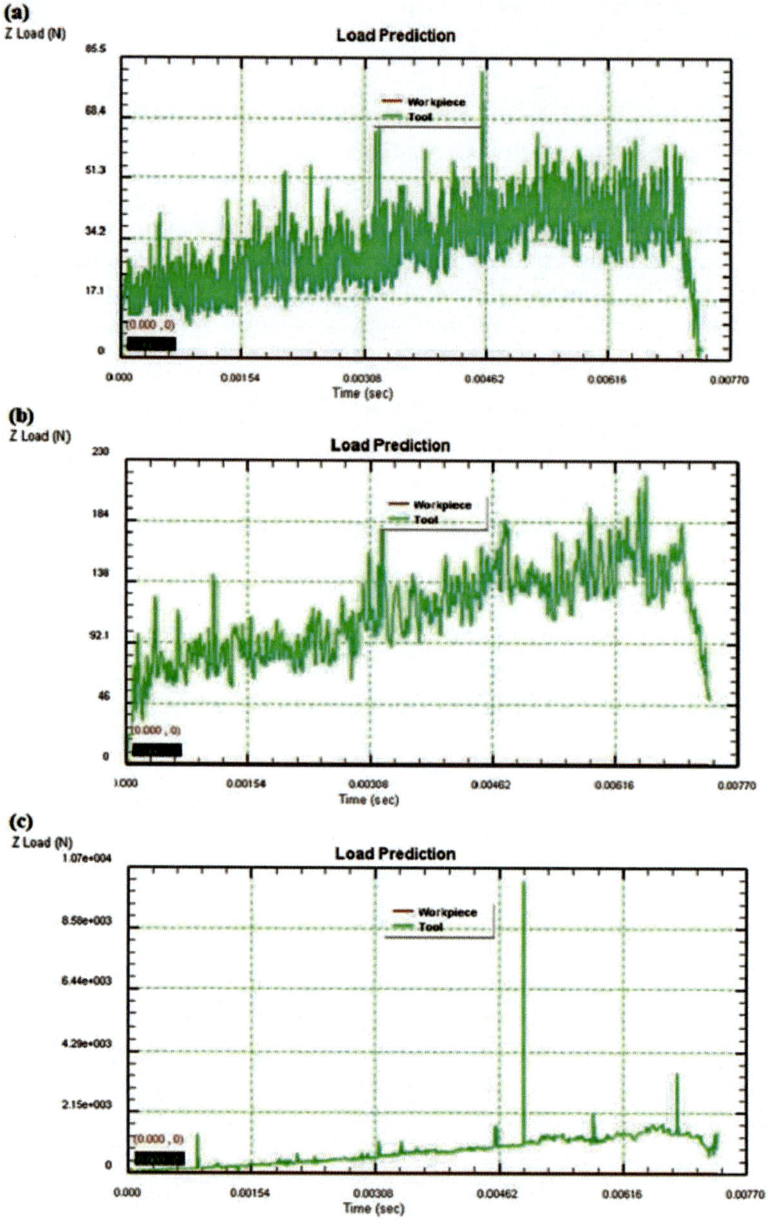

Fig. 5 Load prediction curves for the z load with the depth of cut **a** 0.12 mm, **b** 0.3 mm, **c** 4.8 mm

Also as the simulation results were noted, this time with variation of depth of cut, the tool and workpiece temperature values show increment with increasing values of depth of cut which matches with the expected theory. Figure 6 shows the material deformation, chip formation temperature generation and heat distribution at the interface.

Different studies show that the prediction of tool wear [14, 15], analysis of the residual stress [16], surface roughness and tool life [17], etc. can be successfully done using similar FEM methods.

Now, the load prediction curves for the z load are shown in Fig. 5.

4 Conclusions

The results that were obtained during this study show that the FEM analysis can prove beneficial as it can successfully give the following information based on calculations and user-defined models:

- Show the different relations and relative dependence between the machining and response parameters during machining,
- Help define tool geometry and its development according to user-defined models so that the efficiency of the cutting tool and machining operation can be increased,
- Calculate the temperature generated at the tool and workpiece and show the possible heat distribution,
- Calculate the load, stress, strain and damage caused,
- Show the possible tool wear, tool life and also other response parameters,
- The basic model developed can be used for the optimisation of the temperature and tool wear and thus reduce power consumption,
- Find out the difference in the efficiency of turning operations under different cooling conditions, i.e. in dry turning or application of different types of coolants.

Thus, modelling and simulation models can help in determining the machining conditions, and based upon the results obtained, the desired optimum condition can be determined prior to the actual machining process.

Also, numerical simulation provides a further scope of research and development in areas such as

(a) Design and modelling of tool to reduce tool wear and power consumption,
(b) Determine new numerical models for tool wear,
(c) Study of chip formation and its geometry,
(d) Determine models to calculate heat generation and its effects that agree with the experimental results,
(e) Study the effects of machining under various cooling environments.

Selection and analysis of the proper set of tool and workpiece material suitable for maximum efficiency in turning.

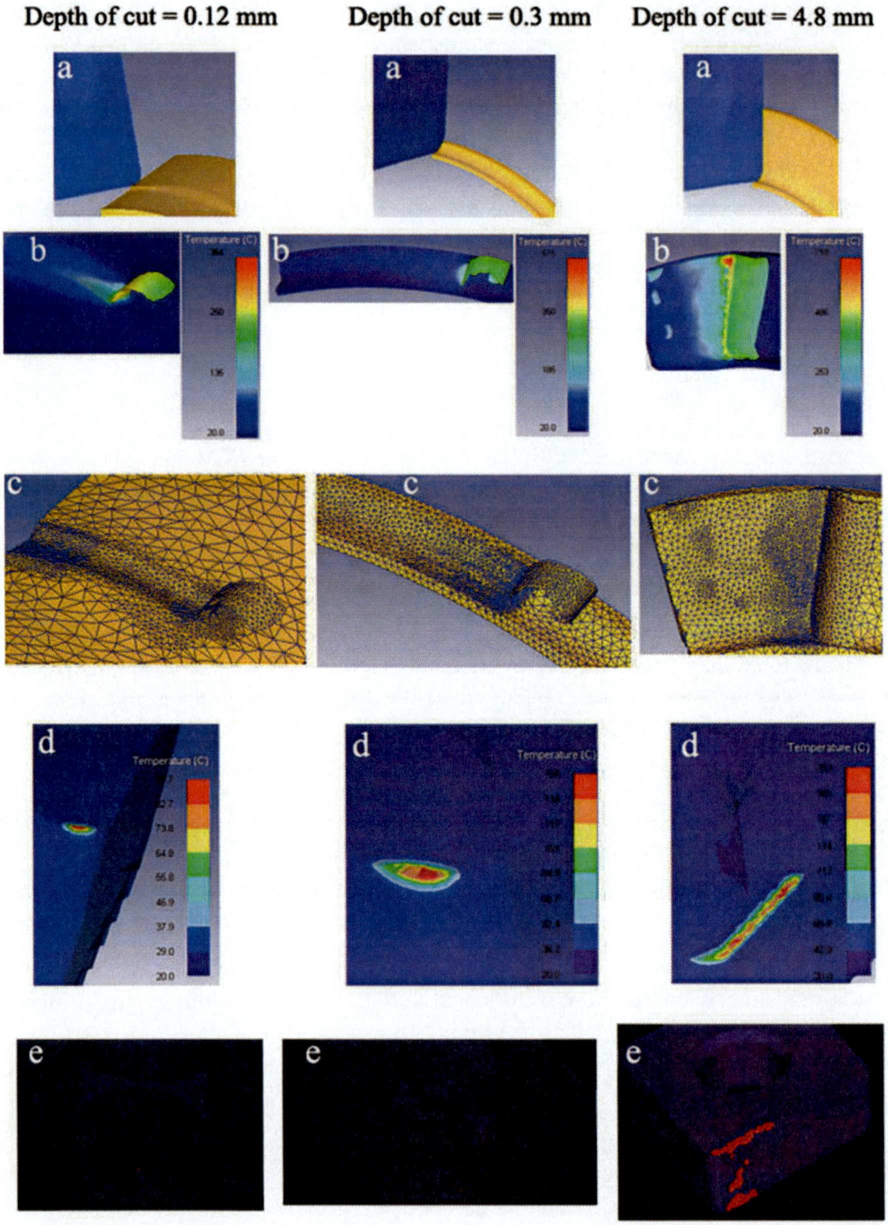

Fig. 6 **a** Settings of the turning operation in FEA, **b** temperature generated at the workpiece during machining, **c** material deformation and chip formation, **d** tool temperature and **e** heat distribution due to wear at the interface

References

1. Mia M, Gupta MK, Singh G, Królczyk G, Pimenov DY (2018) An approach to cleaner production for machining hardened steel using different cooling-lubrication conditions. J Clean Prod 187:1069–1081
2. Khanna N, Shah P (2020) Comparative analysis of dry, flood, MQL and cryogenic CO2 techniques during the machining of 15-5-PH SS alloy. Tribol Int 146:
3. Kaynak Y, Gharibi A, Yılmaz U, Köklü U, Aslantaş K (2018) A comparison of flood cooling, minimum quantity lubrication and high pressure coolant on machining and surface integrity of titanium Ti-5553 alloy. Journal of Manufacturing Processes 34:503–512
4. Sreejith PS, Ngoi BKA (2000) Dry machining: machining of the future. J Mater Process Technol 101(1–3):287–291
5. Neşeli S, Yaldız S, Türkeş E (2011) Optimization of tool geometry parameters for turning operations based on the response surface methodology. Measurement 44(3):580–587
6. Singh D, Venkateswara Rao P (2007) Optimization of tool geometry and cutting parameters for hard turning. Mater Manuf Processes 22(1):15–21
7. Özel T, Sima M, Srivastava AK, Kaftanoglu B (2010) Investigations on the effects of multi-layered coated inserts in machining Ti–6Al–4 V alloy with experiments and finite element simulations. CIRP Ann 59(1):77–82
8. Ee KC, Dillon OW Jr, Jawahir IS (2005) Finite element modeling of residual stresses in machining induced by cutting using a tool with finite edge radius. Int J Mech Sci 47(10):1611–1628
9. Bil H, Kılıç SE, Tekkaya AE (2004) A comparison of orthogonal cutting data from experiments with three different finite element models. Int J Mach Tools Manuf 44(9):933–944
10. Carroll JT III, Strenkowski JS (1988) Finite element models of orthogonal cutting with application to single point diamond turning. Int J Mech Sci 30(12):899–920
11. Mamalis AG, Kundrák J, Markopoulos A, Manolakos DE (2008) On the finite element modelling of high speed hard turning. The International Journal of Advanced Manufacturing Technology 38(5–6):441–446
12. Movahhedy M, Gadala MS, Altintas Y (2000) Simulation of the orthogonal metal cutting process using an arbitrary Lagrangian-Eulerian finite-element method. J Mater Process Technol 103(2):267–275
13. Grzesik W, Bartoszuk M, Nieslony P (2005) Finite element modelling of temperature distribution in the cutting zone in turning processes with differently coated tools. J Mater Process Technol 164:1204–1211
14. Attanasio A, Ceretti E, Fiorentino A, Cappellini CRISTIAN, Giardini C (2010) Investigation and FEM-based simulation of tool wear in turning operations with uncoated carbide tools. Wear 269(5–6):344–350
15. Lotfi M, Jahanbakhsh M, Farid AA (2016) Wear estimation of ceramic and coated carbide tools in turning of Inconel 625: 3D FE analysis. Tribol Int 99:107–116
16. Outeiro J.C., Pina J.C., M'saoubi R., Pusavec F. and Jawahir I.S. (2008) Analysis of residual stresses induced by dry turning of difficult-to-machine materials. CIRP annals 57(1):77–80
17. Yanda H, Ghani JA, Rodzi MNAM, Othman K, Haron CHC (2010) Optimization of material removal rate, surface roughness and tool life on conventional dry turning of FCD700. International Journal of Mechanical and Materials Engineering 5(2):182–190

SteelChain—Blockchain-Based Transparent Supply Chain Framework for the Steel Industry

Sreemana Datta and Parichay Bhattacharjee

1 Introduction

One of the foremost challenges the steel industry is facing is rising operational costs due to inefficiencies in the supply chain. Steel manufacturing is a complex and extremely labor-intensive process. Multiple stages of the production and supply processes bring about a unique complexity in the supply chain. Every year, countless stakeholders across the global logistics chain have to depend on a complex system including fleet of cargo vehicles over land, air, and seas to transfer almost 92 billion tons of goods physically. It has been estimated that by the year 2023, the value of the global logistics industry will stand at $15.5 trillion approximately [1]. The present disintegrated and localized brackets in operation demand processes that are pragmatic and prudent. The solutions need to lay supreme importance to ease of use and the staggering volumes in the supply chain industry call for shifting to digital counterparts of presently manual processes. SteelChain aims to offer a pragmatic solution to the problem.

The remainder of this paper is structured as follows: Sect. 2 presents a brief overview and benefits of Blockchain Technology; Sect. 3 focuses on the present scenario of the steel industry supply chain and its shortcomings. Section 4 enumerates the challenges and motivation. Section 5 details about the SteelChain framework elaborating each level of the SteelChain protocol suite; Sect. 6 concludes this work and provides an outlook on the future scope of work in related fields.

S. Datta (✉)
National Institute of Technology, Patna, India

P. Bhattacharjee
Steel Authority of India Limited, Bokaro Steel City, India

S. K. Ghosh et al. (eds.), *Advances in Thermal Engineering, Manufacturing, and Production Management*, Lecture Notes in Mechanical Engineering,
https://doi.org/10.1007/978-981-16-2347-9_30

2 Blockchain Technology

A Blockchain can be understood as a cryptographically linked chain of digital blocks that contains the records of transactions. The data is secured by the blocks in the blockchain [8]. It is difficult for hackers or intruders to alter the data of blockchain as all the blocks are connected to each other and are updated continuously. In typical blockchain architecture, every individual node maintains a local copy of the blockchain. Thus, there are multiple nodes in the network that are connected with each other, and are maintaining a local copy of the global data sheet. The distributed ledger is made up of blocks of data that are chained together with Cryptography that makes it almost impossible to modify once something is recorded [9]. Blockchain technology thus ensures that the information contained in the blocks remains decentralized, yet unreplicated thereby forming an enhanced manifestation of the internet we use [5]. The benefits of blockchain have been used in this work to design a transparent supply chain for the steel industry.

3 Benefits Offered by Blockchain Technology [5]

1. Data Immutability
2. Greater Transparency.
3. Enhanced Security.
4. Traceability is updated.
5. Low costs compared to similar technologies.

3.1 Blockchain Architecture [5, 15]

A Block consists of two components: Block Header and List of Transactions. The header of a block connects the transactions; any change in any transaction will eventually result in the Block Header getting changed. For the remaining blocks, each block header is linked to its preceding block's header, thereby forming a 'chain'. Block header contains the metadata about a block (Fig. 1).

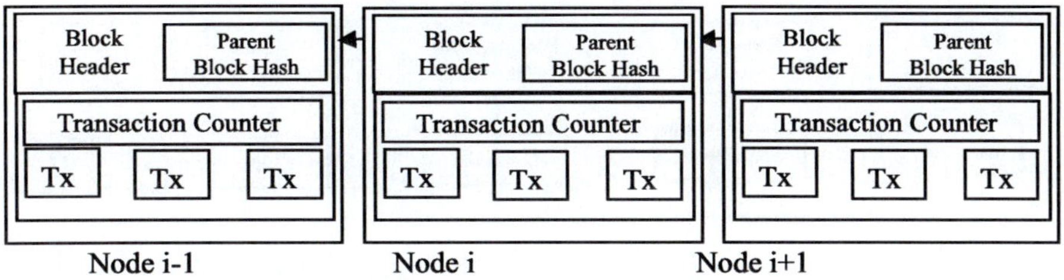

Fig. 1 Blockchain architecture [15]

4 Present Scenario of the Steel Industry Supply Chain

4.1 Global Supply Chain Scenario

As on date, the entire system of global logistics depends entirely on manual, inefficient, and repetitive cycles which can be affected human mistakes and fraudulent intrusions. Apart from incurring huge costs on companies owing to repetitive processing, any breakdown in the present scheme results in a failure to transfer items including critical medicines which has serious implications on mankind. Another parameter exercising stress on costly assets is friction which affects more than 160 million road vehicles, 92,000 sea cargo vessels, and 700,000 miles of rail transport [2]. Recent developments in this field have been proposed by companies such as SAP [11], Oracle [10] and SalesForce. They have proposed cloud-based techniques to improve activities such as freight management, physical transfer handling, optimizing costs, data analysis, and generating reports. A common shortfall of each of these methods is their inability to address the manual parameter in the process, which is affected by friction and deficiency of mutual faith. Carrying agents do not believe middlemen because they do not provide any clarity in freight details and related costs. Shipping authorities consider carriers as options, transacting with the cheapest ones. Broking agents and carrying parties often fight over claim settlements during a damaged delivery. These things contribute to shrinking profits [3], increasing operational costs and diminishing driver supplies [4].

4.2 Present Steel Industry Supply Chain

Steel manufacturing industry possesses a critical and complex supply chain which is both vulnerable to human errors and even has the potential to claim human lives as it consists of stages where the material being handled and transferred involves molten metal at temperatures above 1500 °C [12]. As a part of our research, we studied the supply chain of Steel Authority of India Limited's Integrated Steel Plant at Bokaro Steel City, India. The Bokaro Steel Plant situated in Jharkhand is an integrated steel

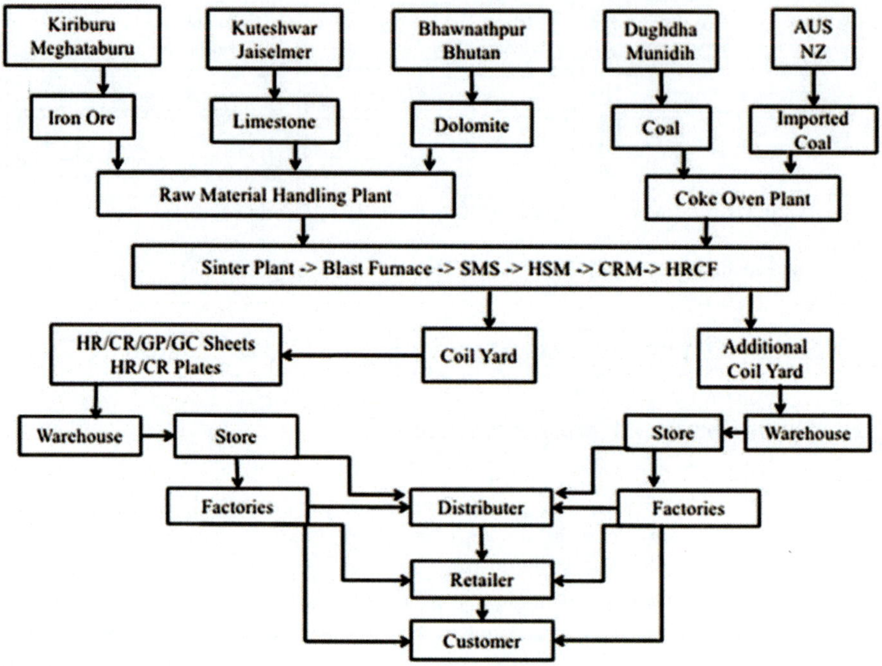

Fig. 2 Steel industry supply chain (example: bokaro steel plant)

plant having a 4.5 MTPA capacity [13]. It manufactures finished cold rolled steel from constituent raw materials within a single facility. It manufactures Hot Rolled, Cold Rolled and Galvanized Plates, Sheets, and Coils.

The supply chain of the plant is depicted in Fig. 2 as follows.

5 Shortcomings in the Present System

At present, there is practically no physical tracking of an entity until it reaches the mills zone in the form of a coil. Even in the mills zone, at each stage, the same coil gets newly renumbered and every stakeholder in the value chain performs this re-nomenclature to suit his own application. On the other hand, the customer has no idea of his order status until the consignment has been dispatched. En-route tracking is done with a different set of parameters which were not related to the production process. Unique identifiers at various stages such as Heat Numbers (in BF or SMS), Slab Numbers (in SM or HSM), HR Coil Numbers (in HRCF or CRM), CR Coil Numbers (in CRM), Consignment Numbers (during transport) are not related to one another. This engenders ambiguity in the system and necessitates repetitive data processing of the same item. For example, after a customer places an order, he has no idea at what stage of the manufacturing process his present order currently is; whether it is still in the 'heat' stage, casting stage, or has assumed an identity as a hot or cold rolled coil. It is only after the coil is dispatched, that the customer gets hold of a Tracking ID to trace the consignment. This process is entirely dependent on the carriers and varies from case to case. There is no standardized procedure to bring transparency to the system. This necessitates the development of an end-to-end communication system which is easy to implement and universal in appeal.

6 Motivation

Stakeholders at various levels of the supply chain in the steel industry operate on quaint and decentralized techniques. This leads to

i. Destitute Conveyance
ii. Constrained Visibility
iii. Lack of Transparency
iv. Repetition in Data Processing
v. Higher Costs
vi. Risk to health of Man, Machines
vii. Hazards to Ecosystem.

7 SteelChain Framework

In order to resolve the pertinent issues affecting the steel industry supply chain at both Intra-Plant, and Inter-Plant levels, the SteelChain Framework is proposed. It aims to bring about simplicity, transparency and reduce redundancy in the existing systems. It efficiently extracts the benefits of the Blockchain Technology and creates a robust, standardized, and easy to use system of tracking and delivering consignments pertaining to the steel industry. This may be extended to other industries bringing about minor modifications in the implementation. SteelChain ensures immaculate delivery of data throughout the steel industry value chain allowing stakeholders at a particular level access to authentic data based on the validation performed and their authorization permissive.

7.1 SteelChain Protocol Suite

The SteelChain protocol suite is given in Fig. 3 as follows.

The SteelChain Protocol Suite is designed to implement the SteelChain Protocol as a connected stack. Each level in the protocol suite performs a specific task. The Physical level resides at the bottom of the stack and houses all the information related

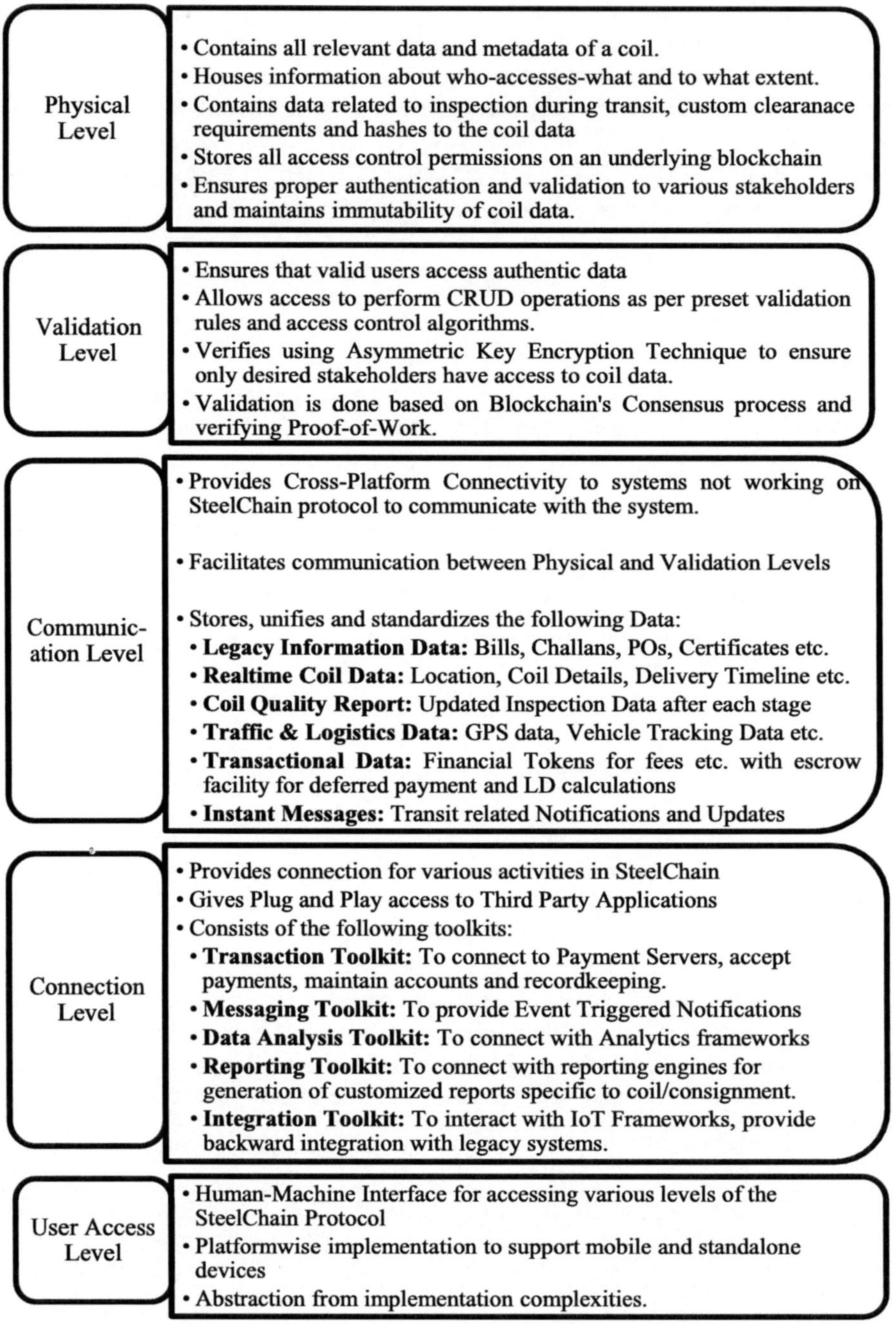
Physical Level
• Contains all relevant data and metadata of a coil.
• Houses information about who-accesses-what and to what extent.
• Contains data related to inspection during transit, custom clearanace requirements and hashes to the coil data
• Stores all access control permissions on an underlying blockchain
• Ensures proper authentication and validation to various stakeholders and maintains immutability of coil data.
Validation Level
• Ensures that valid users access authentic data
• Allows access to perform CRUD operations as per preset validation rules and access control algorithms.
• Verifies using Asymmetric Key Encryption Technique to ensure only desired stakeholders have access to coil data.
• Validation is done based on Blockchain's Consensus process and verifying Proof-of-Work.
Communic-ation Level
• Provides Cross-Platform Connectivity to systems not working on SteelChain protocol to communicate with the system.
• Facilitates communication between Physical and Validation Levels
• Stores, unifies and standardizes the following Data:
• Legacy Information Data: Bills, Challans, POs, Certificates etc.
• Realtime Coil Data: Location, Coil Details, Delivery Timeline etc.
• Coil Quality Report: Updated Inspection Data after each stage
• Traffic & Logistics Data: GPS data, Vehicle Tracking Data etc.
• Transactional Data: Financial Tokens for fees etc. with escrow facility for deferred payment and LD calculations
• Instant Messages: Transit related Notifications and Updates
Connection Level
• Provides connection for various activities in SteelChain
• Gives Plug and Play access to Third Party Applications
• Consists of the following toolkits:
• Transaction Toolkit: To connect to Payment Servers, accept payments, maintain accounts and recordkeeping.
• Messaging Toolkit: To provide Event Triggered Notifications
• Data Analysis Toolkit: To connect with Analytics frameworks
• Reporting Toolkit: To connect with reporting engines for generation of customized reports specific to coil/consignment.
• Integration Toolkit: To interact with IoT Frameworks, provide backward integration with legacy systems.
User Access Level
• Human-Machine Interface for accessing various levels of the SteelChain Protocol
• Platformwise implementation to support mobile and standalone devices
• Abstraction from implementation complexities.

Fig. 3 .

to a particular coil/rod/consignment. It also ensures integrity of the data by updating it in a decentralized ledger; the underlying blockchain. The Validation level allows for data modification and views by authenticated stakeholders through the blockchain consensus process using Proof-of-Work [14] technique. The communication level ensures that the Physical and Validation levels are connected. It transacts on detailed and specific data of each entity. The Connection level provides connector toolkits for plug and play access to the SteelChain framework. The User Access Level sits at the top of the stack and ensures that data is presented constantly, immediately, and continuously to all the stakeholders of the steel industry supply chain. The SteelChain protocol suite serves as a generic framework for the steel industry and is independent of implementation or platform restrictions.

8 Conclusion

The SteelChain Protocol Suite works atop an underlying blockchain to maintaining integrity and authenticity of a coil data and its consignment counterpart. Information attached to a coil remains constant throughout the value chain and is maintained by the SteelChain Protocol Suite. The information can assume a multitude of formats, including documents, coil tracking details, sensor parameters, and even geographical tracking data. For each coil, these voluminous data is banked on a database and the transaction is mined on the underlying blockchain. This saves much of the storage overheads and yet extracting the immutability offered by blockchain technology using the cryptographic transactional hash to bind together with the data. This ensures exactness and prevents tampering of the data which remains constant throughout the supply chain.

9 Future Scope and Research Challenges

As a future scope, it is envisaged to generalize the SteelChain framework for diverse industries beyond the metals and mining fields. We wish to perform a detailed analysis and refine the algorithm by reducing complexity and enhancing universality in implementation. We also plan to extend the framework utilizing popular blockchain manifestations such as BitCoin [6] and Ethererum [7] and undertake a comparative analysis in order to further reduce implementation costs and overheads.

Another crucial challenge to be addressed in the future scope is integrating the SteelChain framework with different IoT applications in fulfillment of Industry 4.0 requirements. This would open up an end-to-end information exchange protocol which may be universally implemented and even support future developments in the field of Industrial IoT.

Table 1 Table of abbreviations

Abbreviation	Meaning
TX	Transaction
MTPA	Million tonnes per annum
AUS	Australia
NZ	New Zealand
SMS	Steel melting shop
HSM	Hot strip mill
CRM	Cold rolling mill
HRCF	Hot rolled coil finishing section
HR	Hot rolled
CR	Cold rolled
GP	Galvanized plain
GC	Galvanized corrugated

Appendix

See Table 1.

References

1. Logistics Market Report. https://www.transparencymarket research.com/logistics-market.html. Accessed 24 Feb 2019
2. Rail lines. https://data.worldbank.org/indicator/IS.RRS.TOTL.KM? end = 2016 Accessed 24 Feb 2019. &start = 2016&view = bar
3. Branch J (2016). Hanjin: A timeline of actions and disruptions. https://www.supplychaindive.com/news/hanjin-timeline-bankruptcy-ship ping-effects/430694/. Accessed 24 Feb 2019
4. Keitz A (2018). America's Massive Truck Driver Shortage May Triple by 2026: Experts. https://www.thestreet.com/markets/truck-driver-shortage- may-triple-by-2026-analysts-say-14650452. Accessed 23 Feb 2019
5. Datta S, Das A K, Kumar A, Khushboo, Sinha D, "Authentication and Privacy Preservation in IoT based Forest Fire detection by using Blockchain – A Review", in Fourth International Conference on Internet of Things and Connected Technologies (ICIoTCT), Jaipur, India
6. Nakamoto, S. (2009). Bitcoin: A Peer-to-Peer Electronic Cash System. Cryptography Mailing list at https://metzdowd.com
7. Gavin, D., & Cto, W. (2014). Ethereum: A Secure Decentralised General ised Transaction Ledger EIP-150 REVISION (1e18248 - 2017-04-12)
8. Narayanan A, Bonneau J, Felten E, Miller A, Goldfeder S (2016) Bitcoin and cryptocurrency technologies: a comprehensive intro duction. Princeton University Press, Princeton. ISBN 978-0-691-17169-2
9. Haber, S; Stornetta, W. S (January 1991). "How to time-stamp a digital document". Journal of Cryptology. 3 (2): 99–111. CiteSeerX 10.1.1.46.8740. https://doi.org/10.1007/bf00196791
10. Simon Ellis, John Santagate, Matthew Marden (2018) Making Supply Chain Operations More Effective and Efficient by Moving to the Cloud with Oracle SCM Cloud. IDC Whitepaper, Oracle

11. Supply Chain Innovation Fuels Success, SAP ERP and Oracle Supply Chain Management: A Case for Coexistence, Oracle Whitepaper, https://www.oracle.com/assets/wp-sap-coexistence-2490975.pdf
12. Campbell John, Melting, Remelting, and Casting for Clean Steel, steel research international, vol 88, issue 1, January 2017, https://doi.org/10.1002/srin.201600093
13. Bokaro Steel Plant - A Partner in Nation Building. https://sail.co.in/plants/about-bokaro-steel-plant. Accessed 23 Feb 2019
14. Porat A, Pratap A, Shah P, Adkar V, Blockchain Consensus: An analysis of Proof-of-Work and its applications, https://www.scs.stanford.edu/17au-cs244b/labs/projects/porat_pratap_shah_adkar.pdf
15. Zheng Z, Xie S, Dai H, Chen X, Wang H (2017) An over view of blockchain technology: architecture. Consensus Future Trends. https://doi.org/10.1109/BigDataCongress.2017.85

Multi-objective Optimization of CNC Drilling Parameters on Machining of HcHcr Steel Using Taguchi's Technique and Grey Relational Analysis

Sandeep Kumar, Jagadish, and Amitava Ray

1 Introduction

The high carbon high chromium steel (HcHcr) is widely used in manufacturing of tool, die and mould for manufacturing and production industry which exhibits outstanding abrasion, good dimensional stability, hardenability, wear resistance hot hardness and high compressive strength to endure cyclic and compressive loads applied on the punches and dies during operation [1].

Amongst all machining operation in manufacturing and production industry, drilling is most useful method to producing cylindrical hole for riveting and fastening structural assembles [2].

Drilling is material removal machining process in which multi-point tools are used to remove the undesired material and produce desired holes and drilling process is recognized as major machining process that approximately 50–70% production time is used; therefore, it is require a specialized technique to achieved optimal drilling conditions for productive drilling that interpreted as efficient material removal in drilling and quality drilling represents satisfactory product attributes as preferred and desired by customers [3–6]. Hence, CNC drilling machine is extremely used in manufacturing industry for material removal operations with the various twist drill bits and the material removal process in drilling involve fracture, plastic deformation

S. Kumar (✉)
Department of Mechanical Engineering, Bengal College of Engineering and Technology, Durgapur 713212, West Bengal, India

Jagadish
Department of Mechanical Engineering, National Institute of Technology, Raipur, Chhattisghar 492010, India

A. Ray
Department of Mechanical Engineering, Jalpaiguri Government Engineering College, Jalpaiguri, West Bengal 735102, India

S. K. Ghosh et al. (eds.), *Advances in Thermal Engineering, Manufacturing, and Production Management*, Lecture Notes in Mechanical Engineering,
https://doi.org/10.1007/978-981-16-2347-9_31

of work material, impact on work material continuously and intermittently, multi-contact point and friction force between work material and twist drill bits which induced error such as delamination, roundness error, interlaminar crack propagation, surface roughness, splintering, swelling and burrs height must have affect the performance characteristics of drilling. Delamination is a most important concern in drilling of the holes that need to be minimizing the size of delamination zone which correlated with thrust force develops during drilling process. Thus, there are no damages occur below the critical thrust force [7–9]. Hence the response characteristics like delamination, thrust force, torque, interlaminar crack propagation, surface roughness, splintering, swelling and burrs height fibre pullout of materials, thermal damages, power consumption, holes circularity and holes diametric error are extensively influenced by process parameter like spindle speeds, feed rates, drill bit diameters, temperature, tool point geometry and others condition and parameters. In other hand several researchers worked carried out on thrust force, material removal rate, torque, surface roughness, thermal damages, holes circularity and holes diametric errors and correlated various process parameters of drilling to produced delamination [10–14] such as the investigation of affects of the drilling parameters like spindle speed, interface temperature, feed rate, drill bit geometry and drilling conditions on quality of holes and drilling tool life [15]; the influence of spindle speed on machining of CFRP [16]; the investigation of effect of processing variable on CNC drilling damage of CFRP [17]; the investigation of effects of drilling process parameters like spindle speed, feed rate, drill bit geometry tool wear rate and drill bit material on response parameters of composite materials such as delamination and surface roughness [18–22]; the effect of drill diameter, and feed rate on machining force of composites laminates[23]; the analysis of the feed rate, cutting speed(CS) and drill-material on thrust force and delamination of CFRP composites laminates [24]; the investigation of CS, feed rate and drill diameters on delamination of CFRP Composite [25]; the effect of spindle speed and feed rate on material damage mechanism such as delamination, SR, fibres pulling or pullout, thermal damages, circularity of holes, and holes diameter errors of woven composites [26]; the investigation of effect of CNC drilling process parameter such as spindle speed(SS), and feed rate(FR) on thrust force, torque and delamination that occurs at entrances and exits of drills bit in CNC machining composite in different fraction of fibre volume [27]; analysis and optimization of CNC drilling process parameters like drill bit diameters feed, drilling speed, slurry concentration on surface finish and material removal of AISI D3steel using HSS drill bit [28]; investigation and optimization of CNC drilling process variables on drilling thrust force, and power consumption of mild steel [29]; the investigation of machining parameters on SR of AISI B1113 using M2 HSS drill [30]; the study of CNC drilling process parameters on response parameters of AISI 304 steel [31]. However, none of the researchers in past literature have been worked on drilling of HcHcr Steel to determine the effects of CNC drilling process parameters like point angles, spindle speeds, feed rates and drill bit diameters on performance parameters.

Table 1 Composition of HcHcr

%C	%Si	%Mn	%Cr
1.83	0.25	0.61	12.45

Additionally, To reduced damage or drilling error, get a desire drill holes quality and improve the performance characteristics, the selection of optimal drilling controllable factors are a vital assignment that influence directly or indirectly the performance characteristics which directly affects the efficiency and quality of drilling. Hence a careful selection of optimal drilling parameters is essential to improve the response characteristics for drilling of HcHcr. While several researchers have used various optimizations methods to optimized drilling process parameters during drilling of different materials such as Taguchi method on drilling of OHNS [32]; Taguchi's with grey relational analysis on drilling of GFRP composites [33]; Artificial neural network and genetics algorithm on drilling of CFRP laminates [34]; ANOVA and grey relational analysis on drilling of SS316 [35]; regression analysis on drilling of Al 2024 Alloy [36]; RSM-based Taguchi's method on drilling of composites material[37], Taguchi's with grey relational analysis on drilling of MMCAl-$TiBr_2$ [38]; on the other hand none of research works are found in past on analysis and optimization of CNC drilling process parameters on HcHcr steel with single objective Taguchi's optimization method and grey relational analysis method(GRA).

The objective of present paper is estimation of optimal combination of CNC drilling process parameters and its significance reduce the response parameters like surface roughness (SR) and delamination using single objective optimization of Taguchi's method and multi-objective optimization grey relational analysis method on HcHcr with point angles, spindle speeds, drill bit diameters and feed rates as input process parameters.

2 Methods and Material

2.1 Work Material

In the present investigation, HcHcr of dimension 20 mm x 30 mm × 10 mm is used as work materials. The chemical composition and properties of work material are illustrated in Tables 1 and 2 respectably.

2.2 Design of Experiment

Taguchi's method is a useful statistical tool for experimental design, and estimation of the outcome of drilling parameters on performance parameters. In the present

Table 2 Properties of HcHcr

Density (Kg/m^3)	Melting point (°C)	Hardness	Poisson ratio	Elastic modulus (GPa)	Thermal expansion (°C)	Thermal conductivity (W/m–K)
7.7 × 1000	1421	62	0.27–0.3	190–210	10.4 × 10^{-6}	392

Table 3 Drilling process parameter and their levels

Input parameter	Symbols	Units	Level-1	Level-2	Level-3
point angle (*A*)	θ	Degree	118	127	136
spindle speed (*D*)	*N*	Rpm	900	1800	2700
Feed rate (*C*)	*f*	mm/min	80	120	160
Drill diameter (*D*)	*d*	mm	8	10	12
Helix angle	Φ		30°	30°	30°
Leap angle (Ψ)	Ψ		7°–10°	7°–10°	7°–10°
Shank type	Cylindrical				
Tool material	HSS				
Coolant	Cut60EP				

work, the experimental design has performed using Taguchi's design of experiment with L_{27} Orthogonal array. Since all drilling inputs process parameters considered for investigation are varied at three levels as shown in Table 3 and other parameters consider being constant [39, 40].

Hence, Taguchi's suggested to obtain the experimental response data by using L27-OA, and investigate the response characteristics from the experimental data to find the optimal drilling input process variables. Finally, L27 recorded the twenty-seven trail conditions with input drilling parameters, and output responses are shown in Table 4.

2.3 Taguchi's Optimization Method

The optimization of drilling process parameters is important step in Taguchi's technique used to obtain the high quality of performance characteristics with no raising the cost, and it is suitable for optimization single response characteristics. Taguchi suggested *S*/*N* ratios that are the logarithmic function of desired objectives for optimization where signal represents desired value and noise represent undesired value. The levels of factor at which the higher *S*/*N* ratios are the optimal parametric setting of CNC drilling machining process parameters [41]. The signal-to-noise ratios of response performance thrust, surface roughness and diametric error are calculated using Eq. (1) that transformed the experimental results into *S*/*N* ratios are desired to

Table 4 Experimental results and S/N ratio of surface roughness and delamination

Exp. No	Inputs drilling parameters				Output drilling parameters		Corresponding S/N ratio	
	Drill diameter	Feed rate	Spindle speed	Point angle	Surface roughness (SR)	Delamination	S/N ratio of SR	S/N ratio of delamination
1	8	0.8	900	118	2.12	0.903	– 3.263	0.443
2	8	1.2	1800	118	2.77	0.881	– 4.424	0.550
3	8	1.6	2700	118	2.64	0.786	– 4.216	1.045
4	10	1.6	900	118	2.36	0.940	– 3.729	0.268
5	10	1.2	1800	118	2.38	0.804	– 3.765	0.947
6	10	0.8	2700	118	1.82	0.743	– 2.600	1.290
7	12	1.6	900	118	3.20	0.923	– 5.051	0.347
8	12	0.8	1800	118	2.72	0.754	– 4.345	1.226
9	12	1.2	2700	118	2.14	0.675	– 3.304	1.706
10	8	0.8	900	127	2.88	0.923	– 4.593	0.347
11	8	1.2	1800	127	2.56	0.788	– 4.082	1.034
12	8	1.6	2700	127	2.64	0.762	– 4.216	1.180
13	10	1.6	900	127	2.81	0.897	– 4.487	0.472
14	10	1.2	1800	127	2.6	0.802	– 4.149	0.958
15	10	0.8	2700	127	1.83	0.649	– – 2.624	1.877
16	12	1.6	900	127	2.98	0.912	– 4.742	0.400
17	12	0.8	1800	127	2.56	0.735	– 4.082	1.337
18	12	1.2	2700	127	2.52	0.654	– 4.014	1.844
19	8	0.8	900	136	2.96	0.824	– 4.713	0.840
20	8	1.2	1800	136	2.03	0.792	– 3.074	1.012

(continued)

Table 4 (continued)

Exp. No	Inputs drilling parameters				Output drilling parameters		Corresponding S/N ratio	
	Drill diameter	Feed rate	Spindle speed	Point angle	Surface roughness (SR)	Delamination	S/N ratio of SR	S/N ratio of delamination
21	8	1.6	2700	136	2.52	0.732	− 4.014	1.354
22	10	1.6	900	136	2.48	0.931	− 3.944	0.310
23	10	1.2	1800	136	2.28	0.823	− 3.579	0.846
24	10	0.8	2700	136	1.98	0.639	− 2.966	1.944
25	12	1.6	900	136	3.31	0.912	− 5.198	0.401
26	12	0.8	1800	136	2.53	0.732	− 4.031	1.354
27	12	1.2	2700	136	1.97	0.698	− 2.944	1.561

Table 5 Mean *S/N* ratio of surface roughness

Input parameters	Level-1	Level-2	Level-3
Point angle (*A*)	– 3.8555	– 4.1099	**– 3.82931**
Spindle speed (*B*)	– 4.4134	– 3.9481	**– 3.43318**
Feed rate(*C*)	**– 3.6909**	– 3.7040	– 4.39970
Drill diameter (*D*)	– 4.0660	**– 3.5380**	– 4.19025

Table 6 Mean S/N ratio for delamination

Input parameters	Level-1	Level-2	Level-3
Point angle (*A*)	0.8695	1.0501	**1.0693**
Spindle speed (*B*)	0.4255	1.0296	**1.5337**
Feed rate (*C*)	**1.1844**	1.1624	0.6421
Drill diameter (*D*)	0.8677	0.9903	**1.1309**

be minimized the performance parameters that are shown in Table 4, thus lower the better characteristic is used to determine the S/N ratios.

ith trails ($i = 1, 2, 3, \ldots, m$; $j = 1, 2, 3, \ldots, p$) are dissimilar for different type of response characteristics and are represented as

$$\eta_{ij} = -10\log(\frac{1}{n}\sum_{i=1}^{n}(yij)^2 \tag{1}$$

The process of calculation of the main effects of all factors is called as analysis of mean (ANOM) and effects of factor levels are deviation from the total mean response parameters. The mean value of multi S/N ratios for SR, and delamination for four factors and three levels are shown in Tables 5 and 6 respectably.

From Fig. 1 the optimum drilling parameters that would yield the minimum surface finish is achieved at 136° point angle (A_3); 2700 RPM spindle speed (B_3); 80 mm/min feed rate (C_1) and 10 mm drill diameter (D_2), i.e., $\boldsymbol{A_3B_3C_1D_2}$.

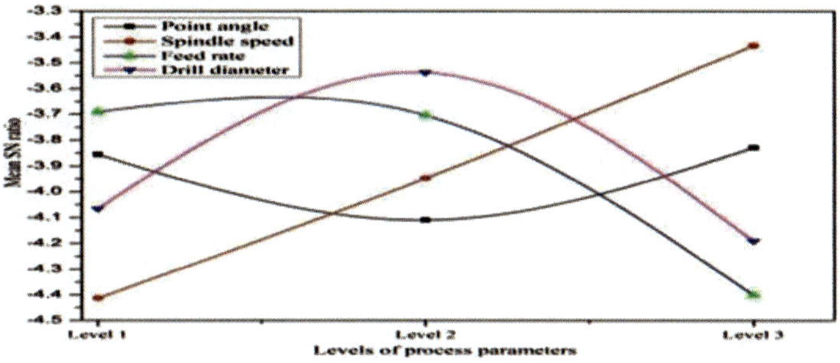

Fig. 1 Main effect plots on means of SN ratios for surface roughness

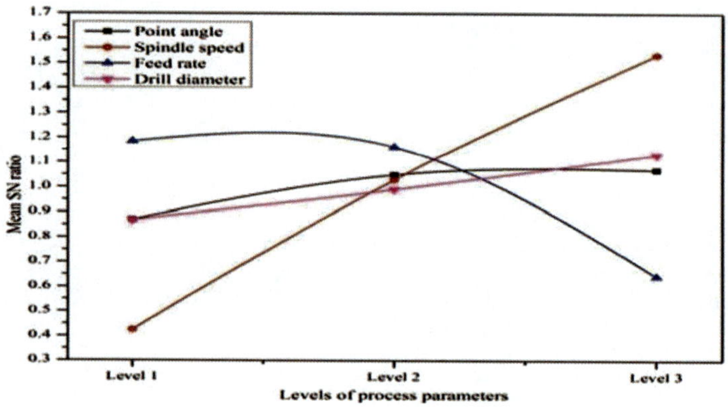

Fig. 2 Main effect plots on means of S/N ratio for delamination

From Fig. 2 the optimum drilling parameters that would yield the delamination are obtained at 136° point angle (A_3); 2700 RPM spindle speed (B_3); 80 mm/min feed rate (C_1) and 10 mm drill diameter (D_3), i.e., $\boldsymbol{A_3B_3C_1D_3}$.

2.4 Grey Relational Analysis(GRA)

Grey system concept is suitable technique to solve and optimize the machining process parameters which are associated with various response parameters [42, 43]. In GRA method, the multi-response optimization problem is transferred into single response optimization problem with GRA, and determine the grey relational grades(GRG) values are tabulated in Table 7. The Various steps are used in GRA as follow:

1. Step-1. Grey relational normalization: normalize the observed value of response parameters between 0 and 1.
2. Step-2. Grey relational gathering: The grey relational gathering values are determined from grey normalized value that are calculated with required response characteristics. While all response parameters can be optimizing for minimum value and lower, the better characteristics are used to get the minimum surface roughness and delamination respectably.
3. Step-3. Determination of grey relational coefficients: the grey coefficient is calculated to signify the correlation between the desired values and actual value.
4. Step-4. Calculation of grey relational grade(GRG): average value of all grey coefficients is known as overall grey relational grade.
5. Step-5. Select the optimal levels
6. Step-6. Perform the confirmation experiments

From Fig. 3 that is main effect plots on means of GRG values for *SR*, and delamination the optimum drilling parameters level that would be yield minimized surface

Table 7 Grey relational grade values of response parameter

Exp. No	GR normalize value		GR gathering values		GR coefficient		GRG
	SR	Delamination	SR	Delamination	SR	Delamination	
1	0.2553	0.8959	0.7447	0.1041	0.4017	0.8277	0.6876
2	0.7023	0.8319	0.2977	0.1681	0.6268	0.7484	0.5441
3	0.6220	0.5362	0.3780	0.4638	0.5695	0.5188	0.7346
4	0.4346	1.0000	0.5654	0.0000	0.4693	1.0000	0.5140
5	0.4487	0.5949	0.5513	0.4051	0.4756	0.5524	0.3920
6	0.0000	0.3904	1.0000	0.6096	0.3333	0.4506	0.9061
7	0.9436	0.9527	0.0564	0.0473	0.8986	0.9135	0.5352
8	0.6719	0.4285	0.3281	0.5715	0.6038	0.4666	0.3875
9	0.2710	0.1415	0.7290	0.8585	0.4068	0.3681	0.7979
10	0.7671	0.9527	0.2329	0.0473	0.6822	0.9136	0.5301
11	0.5706	0.5428	0.4294	0.4572	0.5380	0.5223	0.5242
12	0.6220	0.4560	0.3780	0.5440	0.5695	0.4789	0.7255
13	0.7263	0.8786	0.2737	0.1214	0.6463	0.8047	0.5509
14	0.5962	0.5884	0.4038	0.4116	0.5532	0.5485	0.3389
15	0.0092	0.0400	0.9908	0.9600	0.3354	0.3425	0.8023
16	0.8245	0.9216	0.1755	0.0784	0.7402	0.8645	0.4887
17	0.5704	0.3623	0.4296	0.6377	0.5379	0.4395	0.4352
18	0.5443	0.0597	0.4557	0.9403	0.5232	0.3471	0.6613
19	0.8133	0.6590	0.1867	0.3410	0.7281	0.5945	0.4545
20	0.1824	0.5559	0.8176	0.4441	0.3795	0.5296	0.4793
21	0.5443	0.3517	0.4557	0.6484	0.5232	0.4354	0.7309
22	0.5173	0.9753	0.4827	0.0247	0.5088	0.9530	0.5186
23	0.3768	0.6554	0.6232	0.3446	0.4452	0.5920	0.3506
24	0.1411	0.0000	0.8589	1.0000	0.3679	0.3333	0.9322
25	1.0000	0.9216	0.0000	0.0784	1.0000	0.8645	0.4811
26	0.5509	0.3517	0.4491	0.6483	0.5268	0.4354	0.3794
27	0.1324	0.2284	0.8676	0.7716	0.3656	0.3932	0.6147

and delamination simultaneously are obtained at 118° point angle (A_1); 900 RPM spindle speed (B_1); 120 mm/min feed rate (C_3) and 12 mm drill diameter (D_3), i.e., $\boldsymbol{A_1B_1C_3D_3}$.

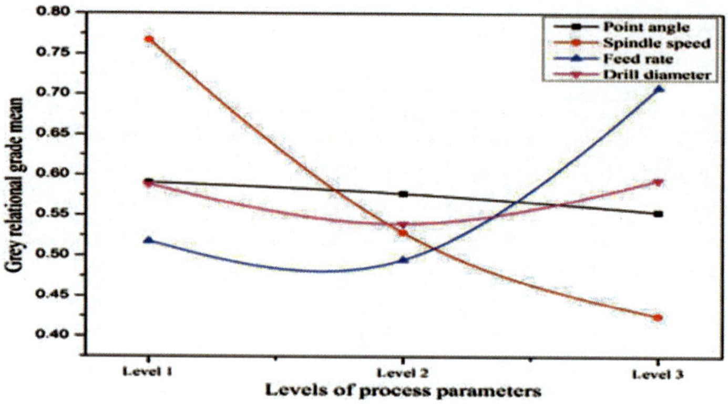

Fig. 3 Main effect plot on mean grey relational grade

2.5 *Analysis of Variance (ANOVA)*

Analysis of variance (ANOVA) is statistical tool use to study the significant effects of drilling process parameters on response parameters. Based on analysis of variance, the relative significance of drilling process parameter by means of reference to output parameters are studied to find out more accurate optimal combinations of drilling parameter and ANOVA established the comparative importance of machining parameters in terms of percentage contribution. This analysis of variance is performed which is based on *S/N* ratios to determine the contribution of CNC drilling process parameter. ANOVA is required to determine errors of variance in the drilling to estimate the effects the drilling process factor with highest percentage contribution ranked highest significance in analysis and have important effect to control the whole performance characteristics, and result of analysis of variance of surface roughness in Table 9 that represents drilling parameters spindle speed most significant parameters influencing process parameters subsequently feed, drill diameter and point angle. From the results, analysis of variance spindle speed has highest contribution about 42.296% and feed rate about 28.930%.

The analysis of variance results of delamination in Table 10 that represents drilling parameters spindle speed most significant process parameter subsequently feed rate, drill diameter, and point angle. From the results, analysis of variance spindle speed has highest contribution about 64.4247% and feed rate about 19.832%.

The analysis of variance results of grey grade values in Table 11 that represents drilling parameters spindle speed most significant parameters influencing process parameters subsequently feed rate, drill diameter, and point angle. From the result, analysis of variance spindle speed has highest contribution about 60.095% and feed rate about 26.94%.

3 Result and Discussion

Multi-objective response parameters like SR and delamination are determined in this investigation. The observed experiment of SR and delamination is revealed in Table 4. Taguchi's method is used to understand the experimental observation based on *S*/*N* ratios and higher values of mean *S*/*N* ratios are optimal parametric combinations and GRA is used to optimize the multi-objective performence parameters, and it is required to minimize *SR*, and delamination. Hence, lower the best type of characteristics has been suggested.

The analysis of means of *S*/*N* ratio for SR in Table 5 is used to find out optimal parametric combination of CNC drilling process parameters to minimized surface roughness (SR), and the higher values of means of *S*/*N* ratios are optimal levels in various levels of process parameters. Hence in this study optimum drilling process parameters are 136° point angle (A_3); 2700 RPM spindle speed (B_3); 80 mm/min feed rate (C_1) and 10 mm drill diameter (D_2), i.e., $\boldsymbol{A_3B_3C_1D_2}$. Similarly, the analysis of means of S/N ratios for delamination in Table 6 is used to determine optimal parametric combination of CNC drilling process parameters to minimized delamination and the higher value of means of *S*/*N* ratios are optimal levels in various levels of process parameters. Hence in this study optimal drilling process parameters are 136° point angle (A_3); 2700 RPM spindle speed (B_3); 80 mm/min feed rate (C_1) and 10 mm drill diameter (D_3), i.e., $\boldsymbol{A_3B_3C_1D_3}$. These combinations of parametric setting are used to achieve minimum delamination. The analysis of means of grey relational grade for multi-objective response like delamination and surface roughness in Table 8 are used to find out optimal parametric combination of CNC drilling process parameters to minimized SR, and delamination. The higher values of means of GRG values are optimal levels in various levels of process parameters. hence in this study optimal drilling process parameters are 118° point angle (A_1); 900 RPM spindle speed (B_1); 120 mm/min feed rate (C_3), and 12 mm drill diameter (D_3), i.e., $\boldsymbol{A_1B_1C_3D_3}$.

The ANOVA is used to determine the % contribution of all drilling process parameters on multi-objective and most significance drilling parameters.

The results of analysis of variance (ANOVA) for surface roughness in Table 9 that represents drilling parameters spindle speed most significant parameters influencing process parameters subsequently feed, drill diameter and point angle. From the results, analysis of variance spindle speed has highest contribution about 42.296% and feed rate about 28.930%.

Table 8 Means of grey relational grade(GRG) values

Process parameters	Level-1	Level-2	Level-3
Point angle (A)	**0.5906**	0.5770	0.5542
Spindle speed (B)	**0.7672**	0.5289	0.4256
Feed rate (C)	0.5178	0.4953	**0.7088**
Drill diameter (D)	0.5882	0.5395	**0.5941**

Table 9 Results of ANOVA for surface roughness

Process parameters	Sum of square	DoF	Mean of square	F_{val}	%Contribution
Point angle (A)	0.4880	2	0.244	0.8171	4.770
Spindle speed (B)	4.3265	2	2.16325	7.2446	42.296
Feed rate (C)	2.9593	2	1.47965	4.9552	28.930
Drill diameter (D)	2.1575	2	1.07875	3.6126	21.090
Error	0.2986	18	0.01659		2.9190
Total	10.229				

Table 10 Results of ANOVA for delamination

Process parameters	Sum of square	DOF	Mean of square	Fval	%contributions
Point angle (A)	0.2185	2	0.10925	0.1326	2.5399
Spindle speed (B)	5.5422	2	2.7711	3.3650	64.4247
Feed rate (C)	1.7061	2	0.85305	1.0358	19.832
Drill diameter (D)	0.3123	2	0.15615	0.1896	3.6300
Error	0.8235	18	0.04575		9.5726
Total	8.6026				

Table 11 Results of ANOVA grey relational grade(GRG) values

Process parameters	Sum of square	Dof	Mean of square	F_{val}	% contribution
Point angle (A)	0.00609	2	0.003045	0.031	0.6624
Spindle speed (B)	0.55246	2	0.27623	5.735	60.095
Feed rate (C)	0.24768	2	0.12384	2.571	26.940
Drill diameter (D)	0.01676	2	0.00838	0.087	1.8231
Error	0.09632	18	0.00535		10.477
Total	0.91931				

The analysis of variance results for delamination in Table 10 that represents drilling parameters spindle speed most significant process parameters subsequently feed rate, drill diameter, and point angle. From the results, analysis of variance spindle speed has highest contribution about 64.4247% and feed rate about 19.832%.

The results of ANOVA for multi-response parameters like surface roughness and delamination in Table 11 that represents drilling parameters spindle speed most significant parameters influencing response parameters followed by feed rate, drill diameters and point angle. From the results, spindle speed has highest contribution about 60.095%, and feed rate about 26.94%.

4 Conclusion

In this analysis of performance of CNC drilling on high carbon, high chromium steel has been analysed for surface roughness and delamination. The main conclusion of present research are as follows:

- Taguchi's based grey relational optimization method is used to find out the key effects, influencing factors and optimal drilling conditions to obtained the better response parameters.
- The optimal parametric combination of drilling process parameters obtained using Taguchi's technique are 136° point angle (A_3); 2700RPM spindle speed (B_3); 80 mm/min feed rate (C_1) and 10 mm drill diameter (D_2), i.e., $\boldsymbol{A_3B_3C_1D_2}$ for surface roughness and 136° point angle (A_3); 2700 RPM spindle speed (B_3); 80 mm/min feed rate (C_1) and 10 mm drill diameter (D_3), i.e., $\boldsymbol{A_3B_3C_1D_3}$ for delamination.
- ANOVA results for surface roughness show that drilling parameters spindle speed most influencing process parameter subsequently feed rate, drill diameters and point angle and spindle speed has maximum contribution about 42.296% and feed rate about 28.930%.
- ANOVA results for delamination shows that spindle speed most influencing parameters subsequently feed rate, drill diameters and point angle and spindle speed has maximum contribution about 64.4247% and feed rate about 19.832%.
- The optimal parametric combinations of drilling process parameters obtained using grey relational multi-objective optimization are 118° point angle (A_1); 900 RPM spindle speed (B_1); 120 mm/min feed rate (C_3) and 12 mm drill diameter ($D3$), i.e., $\boldsymbol{A_1B_1C_3D_3}$.
- ANOVA results for GRA show that the spindle speed is most significant parameter that influencing response parameter subsequently feed rate, drill diameters, and point angle. From the ANOVA results, spindle speed has highest contribution about 60.095% and feed rate about 26.94%.

References

1. Arsecularatne JA, Zhang LC, Montross C, Mathew P (2005) On machining of hardened AISI D2 Steel with PCBN tools. J Mater Process Technol 171:244–252
2. Ravishankar SR, Murthy CRL (1996) Ultrasonic imaging for evaluation of drill induced delaminations in composite laminates. In: Proceedings of 14th conference on non-destructive testing. Oxford and IBH Publishing, pp 8–13
3. Benes J (2000) Hole making trends run deep, fast and dry. Am Mach 144(5):10–97
4. Kumar D, Singh LP, Singh G (2012) Operational modeling for optimizing surface roughness in mild steel drilling using Taguchi technique. Int J Res Manage 2(3):66–67
5. Tyagi Y, Chaturvedi V, Vimal J (2012) Parametric optimization of drilling machining process using Taguchi design and ANOVA approach. J Emerg Technol Adv Eng 2(7):339–347
6. Jindal A, Singla VK (2011) Experimental investigation of process parameters in drilling operation using different software technique. Int J Eng Sci 1(1):135–154

7. Koenig W, Wulf C, Grass P, Willerscheid H (1985) Machining of fiber reinforced plastics. Ann CIRP 34:537–548
8. Davim JP, Reis P (2003) Drilling carbon fiber reinforced plastic manufactured by auto clave experimental and statistical study. Mater Design 24(5):315–324
9. Riza Motorcu R (2010) The optimization of machining parameters using the taguchi method for surface roughness of AISI 8660 hardened alloy steel. J Mech Eng 56(6):391–401
10. Cicek A, Kivak T, Samtas G (2012) Application of Taguchi's method for surface roughness and roundness error in drillingof AISI 316 stainless steel. J Mech Eng 58(3):122–129
11. Galloway DF (1957) Some experiments on the influence of various factors on drill performance. Trans ASME 79:191–223
12. Caprino G, Tagliaferri V (1985) Development in drilling glass fiber reinforced plastics. Int J Mach Tools Manuf 35(6):817–829
13. Haggerty WA, Ernst H (1958) The spiral point drills—self-centering drill point geometry. ASTE 101:58
14. Tsao CC, Hocheng H (2008) Evaluation of thrust force and surface roughness in drilling composite material using Taguchi analysis and neural network. J Mater Process Technol (2031–3):342–348
15. Lin SC, Chen IK (1996) Drilling carbon fiber-reinforced composite material at high speed Wear. J Compos Mater 194:156–162
16. Koenig W, Wulf C, Grass P, Willerscheid H (1985) Machining of fiber reinforced plastics. Ann CIRP 34(2):538–548
17. Hocheng H, Puw HY (1992) On drilling characteristics of fiber-reinforced thermoset and thermoplastics. Int J Mach Tool Manuf 32(4):583–592
18. Chen WC (1997) Some experimental investigations in the drilling of carbon fiber-reinforced plastic (CFRP) composite laminates. Int J Mach Tool Manuf 37(8):1097–1108
19. Doran JH, Maikish (1937) Machining boron composite. In: Noton BR (ed) Composite materials in engineering design. ASM Press, pp 242–250
20. Veniali F, Di Ilio A, Tagliaferri V (1995) An experimental study of the drilling of aramid composites. Trans ASME J Energy Resour Technol 117:271–278
21. Koplev A, Lystrup A, Vorm P (1983) The cutting process, chips and cutting forces in machining CFRP. Composites 14(4):371–376
22. Singh I, Bhatnagar N (2006) Drilling of uni-directional glass fiber reinforced plastics composites laminates. Int J Adv Manuf Technol 27:870–76
23. Azmir MA, Nair Sivasankaran P, Hamedon Z (2013) Experimental study on drilling process of CFRP composite laminate. Mater Sci Forum 638–42:927–932
24. Krishnamoorthy A, Rajendra Boopathy S, Palanikumar K (2009) Delamination analysis in drilling of CFRP composites using response surface methodology. J Compos Mater 43(24):2885–2902
25. Rawata S, Attia H (2009) Characterization of the dry high speed drilling process of woven composites using machinability maps approach. CIRP Ann Manuf Technol 58:105–108
26. Khashaba UA, Seif MA, Elhamid MA (2007) Drilling analysis of chopped composites. Compos Appl Sci Manuf 38(1):61–70
27. Singh P, Bhambri K (2016) Optimization of Process Parameters of AISI D3 steel with abrasive assisted drilling. Int Res J Eng Technol 31446–31452
28. Chouhan YS, SalodaMA JS, Agarwal C (2016) Optimization of drilling process parameters for thrust force: a review. Int J Fract Damage Mech 1:1–7
29. Kalita B, Nath T (2016) An experimental investigation and optimization of cutting parameter in drilling AISI B1113 usingM2HSS drill bit. Int Conf Explor Innov Eng Tech (ICEIET)
30. Kurt M, Bagei E, Kaynak Y (2009) Application of taguchi method in the optimization of cutting parameters for surface finish and hole diameter accuracy in dry drilling process. Int J Adv Manuf Technol 40:458–469
31. Tsao CC, Hocheng H (2004) Taguchi analysis of delamination associated with various drill bits in drilling of composite material. Int J Mach Tools Manuf 44:1403–1416

32. Krishnaraj V, Prabukarthi A, Ramanathan A, Davim JP (2012) Optimization of machining parameters at high speed drilling of carbon fiber reinforced plastic (CFRP) laminates. Composites Parts B 43(4):1791–1799
33. Prajapati, Navneet K, Patel SM (2013) Optimization of process parameters for surface roughness and material removal rate for SS 316 on CNC turning machine. Int J Res Modern Eng Emerg Technol **1**(3):40–47
34. Palanikumar K (2011) Experimental investigation andoptimization in drilling of GFRP composites. Measurement 44:2138–2148
35. Kurt M, Bagci E, Kaynak Y (2009) Application of Taguchi methods in the optimization of cutting parameters for surface finish and hole diameter accuracy in dry drilling processes. Int J Adv Manuf Technol 40(5–6):458–469
36. Tsao CC (2012) Evaluation of the drilling-induced delamination of compound core-special drills using response surface methodology based on the Taguchi method. Int J Adv Manuf Technol 62:241–247
37. Shivapragash B, Chandrasekharan K, Parthasarathy C, Samuel M (2013) Multiple response optimizations in drilling using taguchi and grey relational analysis. Int J Modern Eng Res 3(2):765–768
38. Sathiya P, Abdul Jaleel MY (2010) Grey based taguchi method for optimization of bead geometry in laser bead-on-plate welding. Adv Prod Eng Manage 5(4):225
39. Kumar S, Singh I (2016) The influence of process parameters on cutting speed of WEDM using Taguchi's technique. Int J Emerg Technol. **7**(2):332–337. (Special Issue on ICRIET-2016)
40. Kumar S, Ramola IC, Kumar R (2015) Analysis of surface roughness and material removal rate for high carbon high chromium steel on die sinking EDM using Taguchi technique. SSRG Int J Mech Eng (SSRG-IJME)—EFES. ISSN: 2348–8360
41. Gaitonde SRK, Paulo Davim J (2009) Multi performance optimization in turning of free-machining steel using Taguchi method and utility concept. J. Mater. Eng. Perform 18:231–236
42. Durairaj M, Gowri S (2012) Optimization of inconel 600 alloy micro turning process using grey relation analysis. Adv Mater Res 576:548
43. Kumar S, Jagadish, Singh AK, Kumar N (2020) Multi objective optimization of CNC drilling parameters on HcHcr steel using Taguchi's based utility concept & GRA-PCA methods. Int J Adv Sci Technol Res 2394–9627, 55–63. (Special issue ICAME-2020)

Supplier Selection Problem by Applying Additive Ratio Assessment (ARAS) Methodology

Shankha Shubhra Goswami, Dhiren Kumar Behera, and Soupayan Mitra

1 Introduction

In this highly competitive market [1], proper selection of suppliers plays an important role in industries. The most effective supplier selection [2] is one of the most important factors behind the success of every manufacturing firm [1]. day by day more attention is given to the suppliers and logistics management related problems by the industrial concern [1] just to improve the product quality and to enables JIT manufacturing [3]. The market demand and the customers satisfaction depend largely on the quality and the delivery time of the raw materials. The most important thing is the price at which the companies sold their products in the market, mostly depends on the price and availabilities of the raw materials at which the suppliers supply it to the industries. So, it can be concluded from the present market strategies that the selection of an efficient supplier [4] helps to improve the business policies as well as to improve the product quality and efficiency of manufacturing.

Venkata Rao [5] performed an analysis on supplier selection problem by using TOPSIS [6] methodology in 2007 based on five criteria. Venkata Rao [5] found out the weightages of the criteria's by using AHP [7, 8] process and proposed the possible ranking of the alternatives by TOPSIS method. The same problem was carried forward by Madic et al. [1] in 2014 by implementing COPRAS [9–11] method, using the same weightages, and a new ranking of the alternatives was given. Madic et al. [1] found out that the results do not show much difference between the COPRAS and TOPSIS method. Table 6 shows that the first three and the last preference of the supplier are exactly the same for both methodologies, although there are some changes in the ranking of middle-order alternatives [1].

S. S. Goswami (✉) · D. K. Behera
Indira Gandhi Institute of Technology, Sarang, Odisha 759146, India

S. Mitra
Jalpaiguri Government Engineering College, Jalpaiguri, West Bengal 735102, India

S. K. Ghosh et al. (eds.), *Advances in Thermal Engineering, Manufacturing, and Production Management*, Lecture Notes in Mechanical Engineering,
https://doi.org/10.1007/978-981-16-2347-9_32

Hence, there is a scope to analyze the same problem by introducing other MCDM methods [12, 13] and compare the results with the outcomes of the previous researchers. The same problem is examined by performing additive ratio assessment (ARAS) method [14, 15], using the same weightages determined by Venkata Rao [5] through AHP analysis. All the three outcome results by TOPSIS [16, 17], COPRAS [10] and ARAS [14] methodologies are compared and a final ranking order is also proposed by combining the three-ordering hierarchy as shown in Table 7.

2 Literature Review

In the last few years, supplier selection have become a trending subject in the field of decision making and since then several researchers analyzed several supplier selection problems considering different conflicting criteria [18, 19] by using different types of MCDM tools. Various optimization techniques have been developed and proposed for the selection process and several literatures have been made considering COPRAS, ARAS, and TOPSIS.

Bayazit [20] adopted ANP for the supplier selection problems. Simunovic et al. [21] uses AHP process for the selection of suppliers. Venkata Rao [4] has developed a paper that uses AHP and genetic algorithm methods for vendor selection in the supply chain. Shyur and Shih [22] presented a combined MCDM model of ANP and TOPSIS for solving strategic vendor selection problem. Kumar and Roy [23] uses the hybrid approach of AHP and ANN for vendor selection. Kwang et al. [24] proposed a combined strategy of scoring method with fuzzy expert systems for vendor assessment. Stevic [10] presented a supplier selection analysis by using COPRAS method where the weightages are found out by using AHP. Ho et al. [25] gives an extensive literature review on MCDM approaches for selection and supplier evaluation.

The above literature presented that different researchers adopted different MCDM tools for supplier selection problem but the applications of COPRAS, ARAS, and TOPSIS method is not limited to only these field of area. Kaklauskas et al. [26] applied COPRAS method for the renovation and design of buildings. Das et al. [27] uses the integrated COPRAS and fuzzy AHP methodology to measure relative performance of IITs. Andruskevicius [28] selected the best contractor by this method and Kaklauskas et al. [29] also used this method to evaluate contractors for windows replacement. Chatterjee and Chakraborty [30] selected the gear material using ARAS and COPRAS methodology. Esbouei and Ghadikolaei [31] uses an integrated hybrid approach of ARAS with FAHP for evaluating the financial performance and again later on, Ghadikolaei and Esbouei [32] incorporated fuzzy ARAS with FAHP for evaluating the same. Dadelo et al. [33] adopted ARAS method for the personnel selection. Streimikiene and Balezantis [34] have done the sustainability assessment by applying ARAS and TOPSIS method in Lithuania. Ertugrul and Karakasoglu [35] used TOPSIS and FAHP to assess the efficiency of Turkish cement companies. Degdeviren et al. [36] introduced methods of AHP and TOPSIS in military services for weapon selection. Simsek et al. [37] followed a Taguchi-based TOPSIS approach for maximizing the mixing proportions of high-strength self-compacting concrete. Srikrishna et al. [38] applied TOPSIS technique for new car selection by analyzing under five different criteria. Secme et al. [39] performed an evaluation in Turkish banking sector using AHP and TOPSIS under fuzzy environment.

The above literature indicates that many researchers have implemented various MCDM tools for the supplier selection process and applied these three approaches in a wide variety of applications. However, an attempt is being made to incorporate the newly developed ARAS method with the supplier selection process to choose the right efficient supplier, since very few works have been recorded and the ARAS method has very limited applications in the transport and logistics field.

3 Materials and Methods

This section includes all the discussions and calculation details about ARAS method. The method was firstly developed by Zavadkas and Turkis [14] in 2010 and refined since then. This paper presents the supplier selection problem under five different conflicting criteria by ARAS method which was initially examined by Liu et al. [3] and Madic et al. [1] using DEA and COPRAS methodology. The steps of ARAS are as follows [11].

3.1 Additive Ratio Assessment (ARAS)

Step 1: Create a decision (evaluation) matrix *X* having *n* numbers of criteria's and *m* numbers of alternatives as shown by Eq. 1.

$$X = \left[x_{ij}\right]_{m\times n} = \begin{bmatrix} x_{11} & x_{12} & \ldots & x_{1n} \\ x_{21} & x_{22} & \ldots & x_{2n} \\ \ldots & \ldots & \ldots & \ldots \\ x_{m1} & x_{m2} & \ldots & x_{mn} \end{bmatrix} \quad (1)$$

where, x_{ij} is the performance value of the *ith* alternative and *jth* criteria. $i = 1, 2, 3\ldots., m; j = 1, 2, 3\ldots.,n.$

Step 2: The decision matrix is normalized using Eqs. 2 and 4 based on the nature of the criteria. For beneficial criteria, the normalization is done by using Eq. 2.

$$N_{ij} = \frac{x_{ij}}{\sum_{i=1}^{m} x_{ij}} \quad (2)$$

where, N_{ij} is the normalized value of *ith* alternative and *jth* criteria. $i = 1, 2, 3\ldots., m; j = 1, 2, 3\ldots., n$. Normalization of the non-beneficial (or cost) criteria is done in two stages as shown by Eqs. 3 and 4.

$$x_{ij}^{*} = \frac{1}{x_{ij}} \tag{3}$$

$$N_{ij} = \frac{x_{ij}^{*}}{\sum_{i=1}^{m} x_{ij}^{*}} \tag{4}$$

For normalizing the non-beneficial criteria, the reciprocal of all the elements is found out by using Eq. 3 and further, the normalization is done by using Eq. 4.

Step 3: Find out the weighted normalized matrix *W* as shown in Eq. 5 by multiplying each column with their respective criteria weights by using Eq. 6.

$$W = \left[\mathrm{d}_{ij}\right]_{m \times n} = \begin{bmatrix} \mathrm{d}_{11} & \mathrm{d}_{12} & \ldots & \mathrm{d}_{1n} \\ \mathrm{d}_{21} & \mathrm{d}_{22} & \ldots & \mathrm{d}_{2n} \\ \ldots & \ldots & \ldots & \ldots \\ \mathrm{d}_{m1} & \mathrm{d}_{m2} & \ldots & \mathrm{d}_{mn} \end{bmatrix} \tag{5}$$

$$\mathrm{d}_{ij} = N_{ij} \times w_j \tag{6}$$

where, d_{ij} and N_{ij} are the weighted normalized values and the normalized values of the *ith* alternatives and *jth* criteria and w_j is the weightages of the criterias respectively. $i = 1, 2, 3\ldots., m; j = 1, 2, 3\ldots., n.$

Step 4: Determine the optimality value (S_i) for each alternative by using Eq. 7.

$$S_i = \sum_{j=1}^{n} \mathrm{d}_{ij} \tag{7}$$

where, $i = 1, 2, 3\ldots., m; j = 1, 2, 3\ldots., n$

Step 5: The degree of utility (U_i) is determined for each alternative by using Eq. 8.

$$U_i = \frac{S_i}{S_0} \tag{8}$$

where, S_O is the optimality value of the optimal alternative. The relative efficiency of each alternative with respect to the best alternative [40] is judged by this utility function (U_i) and this utility function value ranges from 0 to 100%. The alternative with the highest utility function is termed as the best alternatives [14] and ranking have been made in ascending order based on these utility values [11].

3.2 Experimental Details and Research Analysis

This section consists of the research methodology and all the calculation details step by step. The supplier selection problem is taken from Liu et al. [3] which was solved by using DEA method and later, Venkata Rao [5] and Madic et al. [1] adopted the same problem and analyzed by using TOPSIS and COPRAS methodology using the weightages determined by Venkata Rao [5] through AHP analysis. There are five criteria's and 18 suppliers are considered for this analysis and the weightages of all the criteria are determined by AHP method in [5] which is as follows: $w_{\text{price}} = 0.1361$, $w_{\text{quality}} = 0.4829$, $w_{\text{delivery performance}} = 0.2591$, $w_{\text{distance}} = 0.0438$ and $w_{\text{supply variety}} = 0.0782$. Now proceeding with the same decision matrix [1] according to Eq. 1 as shown in Table 1. The suppliers are denoted by S1, S2, S3….., S18.

In Table 1 above, *A0* values are the most desirable values for each and every criterion. For example, in case of price criteria the minimum value is desirable so 80 is the most minimum value in the price column so it is taken as *A0* value, similarly for quality, maximum value is desirable so the max value i.e., 100 is taken as *A0* value.

Table 1 Evaluation matrix

Goal	Min	Max	Max	Min	Max
Suppliers	Price ($)	Quality (%)	Delivery performance (%)	Distance (Miles)	Supply variety
A0	80	100	100	238	53
S1	100	100	90	249	2
S2	100	99.79	80	643	13
S3	100	100	90	714	3
S4	100	100	90	1809	3
S5	100	99.83	90	238	24
S6	100	96.59	90	241	28
S7	100	100	85	1404	1
S8	100	100	97	984	24
S9	100	99.91	90	641	11
S10	100	97.54	100	588	53
S11	100	99.95	95	241	10
S12	100	99.85	98	567	7
S13	100	99.97	90	567	19
S14	100	91.89	90	967	12
S15	80	99.99	95	635	33
S16	100	100	95	795	2
S17	80	99.99	95	689	34
S18	100	99.36	85	913	9

(*Source* Madic et al. [1])

The optimum performance values [11] and the goal for every criterion is shown in Table 1 above.

The evaluation matrix is shown in Table 1 are normalized by using Eq. 2 for beneficial criteria and by using Eqs. 3 and 4 for non-beneficial criteria. Here, in this case, quality, delivery performance, and supply variety are the beneficial criteria whose larger values are desired whereas, price and distance are the cost (non-beneficial) criteria whose smaller values are desired. The normalization matrix is shown in Table 2.

Table 3 shows the weighted normalized decision matrix according to Eq. 5 where the weighted values in each cell are calculated by multiplying the criteria weightages with each of the elements of Table 2 of their respective column by using Eq. 6. All the elements in each row are summed up to find out the optimality function values of each and every alternative by using Eq. 7 as shown in Table 3. The sum values are shown in Table 3 is nothing but the optimality values of the alternatives where *0.0659* is the optimality value of the optimal alternative which is denoted by S_0. The

Table 2 Normalization of the evaluation matrix

Weights	0.1361	0.4829	0.2591	0.0438	0.0782
Suppliers	Price ($)	Quality (%)	Delivery performance (%)	Distance (Miles)	Supply variety
A0	0.0633	0.0531	0.0573	0.1073	0.1554
S1	0.0506	0.0531	0.0516	0.1026	0.0059
S2	0.0506	0.0529	0.0458	0.0397	0.0381
S3	0.0506	0.0531	0.0516	0.0358	0.0088
S4	0.0506	0.0531	0.0516	0.0141	0.0088
S5	0.0506	0.0530	0.0516	0.1073	0.0704
S6	0.0506	0.0513	0.0516	0.1060	0.0821
S7	0.0506	0.0531	0.0487	0.0182	0.0029
S8	0.0506	0.0531	0.0556	0.0260	0.0704
S9	0.0506	0.0530	0.0516	0.0398	0.0323
S10	0.0506	0.0518	0.0573	0.0434	0.1554
S11	0.0506	0.0530	0.0544	0.1060	0.0293
S12	0.0506	0.0530	0.0562	0.0450	0.0205
S13	0.0506	0.0530	0.0516	0.0450	0.0557
S14	0.0506	0.0488	0.0516	0.0264	0.0352
S15	0.0633	0.0531	0.0544	0.0402	0.0968
S16	0.0506	0.0531	0.0544	0.0321	0.0059
S17	0.0633	0.0531	0.0544	0.0371	0.0997
S18	0.0506	0.0527	0.0487	0.0280	0.0264

(*Source* Author himself)

Table 3 Weighted normalized evaluation matrix

Suppliers	Price ($)	Quality (%)	Delivery performance (%)	Distance (Miles)	Supply variety	Sum
A0	0.0086	0.0256	0.0148	0.0047	0.0122	0.0659
S1	0.0069	0.0256	0.0134	0.0045	0.0005	0.0508
S2	0.0069	0.0256	0.0119	0.0017	0.0030	0.0491
S3	0.0069	0.0256	0.0134	0.0016	0.0007	0.0481
S4	0.0069	0.0256	0.0134	0.0006	0.0007	0.0472
S5	0.0069	0.0256	0.0134	0.0047	0.0055	0.0560
S6	0.0069	0.0247	0.0134	0.0046	0.0064	0.0561
S7	0.0069	0.0256	0.0126	0.0008	0.0002	0.0462
S8	0.0069	0.0256	0.0144	0.0011	0.0055	0.0536
S9	0.0069	0.0256	0.0134	0.0017	0.0025	0.0501
S10	0.0069	0.0250	0.0148	0.0019	0.0122	0.0608
S11	0.0069	0.0256	0.0141	0.0046	0.0023	0.0535
S12	0.0069	0.0256	0.0146	0.0020	0.0016	0.0506
S13	0.0069	0.0256	0.0134	0.0020	0.0044	0.0522
S14	0.0069	0.0235	0.0134	0.0012	0.0028	0.0477
S15	0.0086	0.0256	0.0141	0.0018	0.0076	0.0577
S16	0.0069	0.0256	0.0141	0.0014	0.0005	0.0485
S17	0.0086	0.0256	0.0141	0.0016	0.0078	0.0578
S18	0.0069	0.0255	0.0126	0.0012	0.0021	0.0483

(*Source* Author himself)

last step is to find out the degree of utility (U_i) by using Eq. 8. The degree of utility is found out for every alternative and shown in Table 4.

4 Result and Discussion

The utility values (U_i) for every alternative are calculated and depicted in Table 4. From the Table 4 it can be clearly seen that the degree of utility is highest for supplier 10 with 92.19% followed by supplier 17 and supplier 15 with 87.60 and 87.46%, respectively. Now, arranging the supplier in ascending order according to the utility values and giving a preference ranking order of the suppliers. Table 5 below shows the ranking order of the suppliers.

Hence, the preference order of the suppliers from best to worst can be given as follows:

Table 4 Degree of utility of the alternatives

Suppliers	S_i	U_i	%
A0	0.0659	1	100
S1	0.0508	0.7708	77.08
S2	0.0491	0.7440	74.40
S3	0.0481	0.7299	72.99
S4	0.0472	0.7156	71.56
S5	0.0560	0.8498	84.98
S6	0.0561	0.8503	85.03
S7	0.0462	0.7001	70.01
S8	0.0536	0.8122	81.22
S9	0.0501	0.7601	76.01
S10	0.0608	0.9219	92.19
S11	0.0535	0.8120	81.20
S12	0.0506	0.7674	76.74
S13	0.0522	0.7916	79.16
S14	0.0477	0.7235	72.35
S15	0.0577	0.8746	87.46
S16	0.0485	0.7353	73.53
S17	0.0578	0.8760	87.60
S18	0.0483	0.7319	73.19

(*Source* Author himself)

$$S10 > S17 > S15 > S6 > S5 > S8 > S11 > S13 > S1 >$$
$$S12 > S9 > S2 > S16 > S18 > S3 > S14 > S4 > S7$$

4.1 Comparisons with Other Researchers Work

The outcome ranking order presented by Venkata Rao [5] by analyzing through TOPSIS methodology is as follows:

$$S10 > S17 > S15 > S6 > S5 > S8 > S13 > S11 > S12 >$$
$$S9 > S2 > S1 > S16 > S14 > S3 > S18 > S4 > S7$$

Similarly, the ranking order proposed by Madic et al. [1] by implementing COPRAS methodology is as follows:

$$S10 > S17 > S15 > S8 > S6 > S5 > S13 > S11 > S12 >$$

Table 5 Preference ranking order of the suppliers

Supplier	Utility values (U_i)	Ranking
S10	92.19	1
S17	87.60	2
S15	87.46	3
S6	85.03	4
S5	84.98	5
S8	81.22	6
S11	81.20	7
S13	79.16	8
S1	77.08	9
S12	76.74	10
S9	76.01	11
S2	74.40	12
S16	73.53	13
S18	73.19	14
S3	72.99	15
S14	72.35	16
S4	71.56	17
S7	70.01	18

(*Source* Author himself)

$$S9 > S2 > S18 > S16 > S14 > S3 > S4 > S1 > S7$$

The preference ranking order of the suppliers by the three methods are compared in Table 6 and also shown graphically in Fig. 1

From Table 6 it is observed that the suppliers in the first, second, third and the last position are exactly same in all the three cases. Supplier 10 is coming out to be the best efficient supplier followed by supplier 17 and supplier 15 whereas, supplier 7 is the most inefficient supplier by all three methods. However, there is a bit alteration in ranking of the middle-order alternatives [1]. The ranking of the suppliers by all the three methods are also compared graphically which is shown in Fig. 1 and a modified final ranking of the suppliers is also proposed by combining all the three ranking hierarchical order which is shown in Table 7.

A modified final ranking of the suppliers is proposed which is shown in the above Table 7. Now the question arises how the final ranking has been made by combining these three orders of preferences. Let's explain it clearly, the first three and the last position must have supplier 10, 17, 15, and 7, respectively since all the three methods are giving the same supplier rating, now supplier 6 and supplier 5 are allotted in the 4th position and 5th position since two methods are giving the same results. Likewise, the suppliers are allotted in the position which are given by at least two of the methods. Now the confusion arises while allotting the 12th position and the 16th position since

Table 6 Comparisons of the ranking order

Ranking	By TOPSIS	By COPRAS	By ARAS
1	S10	S10	S10
2	S17	S17	S17
3	S15	S15	S15
4	S6	S8	S6
5	S5	S6	S5
6	S8	S5	S8
7	S13	S13	S11
8	S11	S11	S13
9	S12	S12	S1
10	S9	S9	S12
11	S2	S2	S9
12	S1	S18	S2
13	S16	S16	S16
14	S14	S14	S18
15	S3	S3	S3
16	S18	S4	S14
17	S4	S1	S4
18	S7	S7	S7

(*Source* Venkata Rao [5], Madic et al. [1] and Author himself)

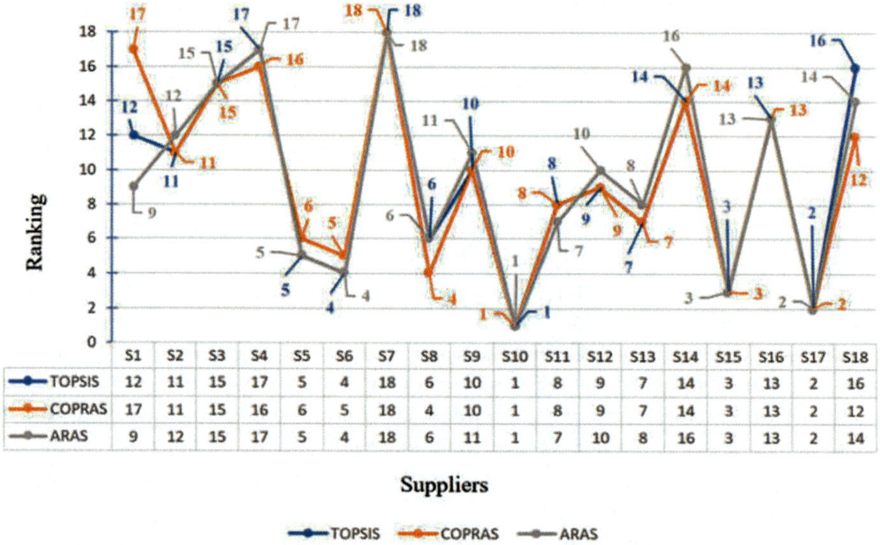

Fig. 1 Ranking order comparisons of TOPSIS, COPRAS and ARAS (*Source* Madic et al. [1] and Author himself)

Table 7 Modified final ranking of the suppliers

Ranking	By TOPSIS	By COPRAS	By ARAS	Final ranking
1	S10	S10	S10	S10
2	S17	S17	S17	S17
3	S15	S15	S15	S15
4	S6	S8	S6	S6
5	S5	S6	S5	S5
6	S8	S5	S8	S8
7	S13	S13	S11	S13
8	S11	S11	S13	S11
9	S12	S12	S1	S12
10	S9	S9	S12	S9
11	S2	S2	S9	S2
12	S1	S18	S2	S1
13	S16	S16	S16	S16
14	S14	S14	S18	S14
15	S3	S3	S3	S3
16	S18	S4	S14	S18
17	S4	S1	S4	S4
18	S7	S7	S7	S7

(*Source* Venkata Rao [5], Madic et al. [1] and Author himself)

three different suppliers are proposed by the three different methods, for this case Copeland and Borda voting rule [41] is followed for allotting these two positions. However, both the voting rule are giving the same results that supplier 1 must come ahead of supplier 18, so supplier 1 is allotted in position 12 and supplier 18 is allotted in position 16 respectively, thus giving the final order ranking of the suppliers which is shown in Table 7.

5 Conclusion

From the above analysis, it is concluded that supplier 10 is the most efficient supplier followed by supplier 17 and supplier 15 based on the five criteria considered for this analysis. In addition to these, other criteria's e.g., lead time, behavior of the supplier, etc. can also be considered to make the selection process more precise and accurate. It can also be concluded that the outcome results from the TOPSIS methodology is more accurate since the final ranking is exactly same as the ranking obtained from TOPSIS as shown in Table 7.

The same MCDM approaches can be applied to other decision-making areas, e.g., waste management, green energy, water management, etc., and the same issue of supplier selection can be expanded by embedding other decision-making tools such as SAW, VIKOR, ELECTRE, PROMETHEE, etc.

Acknowledgements Our heartfelt thanks to all the authors of the research papers from whose references this paper has been prepared, special thanks to our friends, colleagues and to all the faculty members and professors of our institutions whose direct and indirect involvements enables us to complete this research work on time.

References

1. Madic M, Markovic D, Petrovic G, Radovanovic M (2014) Application of COPRAS method for supplier selection. In: The fifth international conference transport and logistics, TIL. http://til2014.masfak.ni.ac.rs/elementi/15.pdf
2. Taherdoost H, Brard A (2019) Analyzing the process of supplier selection criteria and methods. Proc Manuf 32:1024–1034. https://doi.org/10.1016/j.promfg.2019.02.317
3. Liu F, Ding FY, Lall V (2000) Using data envelopment analysis to compare vendors for vendor selection and performance improvement. Supply Chain Manage: An Int J 5(3):143–150. https://doi.org/10.1108/13598540010338893
4. Rao RV (2007) Vendor selection in a supply chain using analytic hierarchy process and genetic algorithm methods. Int J Serv Operations Manage 3(3):355–369. https://doi.org/10.1504/IJSOM.2007.013097
5. Rao RV (2007) Decision making in the manufacturing environment: using graph theory and fuzzy multiple attribute decision making methods. Springer, London. https://epdf.pub/decision-making-in-the-manufacturing-environment-using-graph-theory-and-fuzzy-mu.html
6. Hwang CL, Yoon K (1981) Multiple attributes decision making methods and applications. Springer, Berlin. https://www.springer.com/gp/book/9783540105589
7. Saaty TL (1980) The analytic hierarchy process. McGraw-Hill, New York
8. Saaty TL (2008) Decision making with the analytic hierarchy process. Int J Serv Sci 1(1). http://www.rafikulislam.com/uploads/resourses/197245512559a37aadea6d.pdf
9. Zavadskas EK, Kaklauskas A, Turskis Z, Tamošaitien J (2008) Selection of the effective dwelling house walls by applying attributes values determined at intervals. J Civil Eng Manage 14(2):85–93 https://www.tandfonline.com/doi/pdf/10.3846/1392-3730.2008.14.3?needAccess=true
10. Stevic Z (2016) Supplier selection using AHP and COPRAS method. In: Strategic management and decision support systems in strategic management, 21th international scientific conference. https://www.researchgate.net/profile/Zeljko_Stevic/publication/303462041_SUPPLIER_SELECTION_USING_AHP_AND_COPRAS_METHOD/links/5744498908ae9ace841d0236/SUPPLIER-SELECTION-USING-AHP-AND-COPRAS-METHOD.pdf
11. Adali EA, Tusisik A (2016) Air conditioner selection problem with COPRAS and ARAS methods. Manas J Soc Stud 5(2):124–138. http://journals.manas.edu.kg/mjsr/archives/Y2016_V05_I02/5e1edf45b0f32a68a1d91bae2b7d874b.pdf
12. Bhole GP, Deshmukh T (2018) Multi-criteria decision making (MCDM) methods and its applications. Int J Res Appl Sci Eng Technol (IJRASET) 6(5):899–915. https://www.ijraset.com/fileserve.php?FID=17056
13. Velasquez M, Hester PT (2013) An analysis of multi-criteria decision making methods. Int J Operations Res 10(2):56–66. http://www.orstw.org.tw/ijor/vol10no2/ijor_vol10_no2_p56_p66.pdf

14. Zavadskas EK, Turskis Z (2010) A new additive ratio assessment (ARAS) method in multicriteria decision-making. Technol Econ Develop Econ 16(2):159–172. https://doi.org/10.3846/tede.2010.10
15. Zavadskas EK, Turskis Z, Vilutiene T (2010) Multiple criteria analysis of foundation instalment alternatives by applying additive ratio assessment (ARAS) method. Archiv Civil Mech Eng 10(3):123–141. https://doi.org/10.1016/S1644-9665(12)60141-1
16. Yoon K (1987) A reconciliation among discrete compromise situations. J Operational Res Soc 38(3):277–286. https://link.springer.com/article/10.1057/jors.1987.44
17. Hwang CL, Lai YJ, Liu TY (1993) A new approach for multiple objective decision making. Comput Operations Res 20(8):889–899. https://doi.org/10.1016/0305-0548(93)90109-V
18. Chatterjee P, Athawale VM, Chakraborty S (2011) Materials selection using complex proportional assessment and evaluation of mixed data methods. Mater Des 32(2):851–860. https://doi.org/10.1016/j.matdes.2010.07.010
19. Madurika HKGM, Hemakumara GPTS (2017) GIS based analysis for suitability location finding in the residential development areas of greater matara region. Int J Sci Technol Res 6(2):96–105. http://www.ijstr.org/final-print/feb2017/Gis-Based-Analysis-For-Suitability-Location-Finding-In-The-Residential-Development-Areas-Of-Greater-Matara-Region.pdf
20. Bayazit O (2006) Use of analytic network process in vendor selection decisions. Benchmarking: An Int J 13(5):566–579. http://www.rafikulislam.com/uploads/resourses/14167927475909bd7d234d5.pdf
21. Šimunović K, Draganjac T, Lujić R (2011) Supplier selection using a multiple criteria decision making method. Strojarstvo 53:293–300. https://hrcak.srce.hr/80442
22. Shyur HJ, Shih HS (2006) A hybrid MCDM model for strategic vendor selection. Math Comput Modell 44(7–8):749–761. https://doi.org/10.1016/j.mcm.2005.04.018
23. Kumar J, Roy N (2010) Hybrid method for vendor selection using neural network. Int J Comput Appl 11(12):35–40. https://pdfs.semanticscholar.org/4352/ab5b1a591022edc2b595a98dbfaa76488c1b.pdf
24. Kwong CK, Ip WH, Chan JWK (2002) Combining scoring method and fuzzy expert systems approach to vendor assessment: A case study. Integr Manuf Syst 13:512–519. https://doi.org/10.1108/09576060210442671
25. Ho W, Xu X, Dey PK (2010) Multi-criteria decision making approaches for supplier evaluation and selection: a literature review. Europ J Operational Res 202(1):16–24. https://doi.org/10.1016/j.ejor.2009.05.009
26. Kaklauskas A, Zavadskas EK, Raslanas S (2005) Multivariant design and multiple criteria analysis of building refurbishments. Energy Build 37(4):361–372. http://www.raslanas.lt/uploads/5.pdf
27. Das MC, Sarkar B, Ray S (2012) A framework to measure relative performance of Indian technical institutions using integrated fuzzy AHP and COPRAS methodology. Socio-Econ Plann Sci 46(3):230–241. https://doi.org/10.1016/j.seps.2011.12.001
28. Andruškevicius A (2005) Evaluation of contractors by using COPRAS: the multiple criteria method. Technol Econ Develop Econ 11(3):158–169. https://doi.org/10.3846/13928619.2005.9637694
29. Kaklauskas A, Zavadskas EK, Raslanas S, Ginevicius R, Komka A, Malinauskas P (2006) Selection of low e-windows in retrofit of public buildings by applying multiple criteria method COPRAS: a Lithuanian case. Energy Build 38(5):454–462. https://doi.org/10.1016/j.enbuild.2005.08.005
30. Chatterjee P, Chakraborty S (2013) Gear material selection using complex proportional assessment and additive ratio assessment based approaches: a comparative study. Int J Mater Sci Eng 1(2):104–111. http://www.ijmse.net/uploadfile/2014/0519/20140519020917152.pdf
31. Esbouei SK, Ghadikolaei AS (2013) An integrated approach based on FAHP and ARAS methods for financial performance evaluation. ARPN J Syst Softw 3(4):53–56. https://fdocuments.in/document/an-integrated-approach-based-on-fahp-and-aras-scientific-integrated-approach.html

32. Ghadikolaei AS, Esbouei SK (2014) Integrating Fuzzy AHP and Fuzzy ARAS for evaluating financial performance. Boletim da Sociedade Paranaense de Matemática 32(2):163–174. https://doi.org/10.5269/bspm.v32i2.21378
33. Dadelo S, Turskis Z, Zavadskas EK, Dadeliene R (2012) Multiple criteria assessment of elite security personal on the basis of ARAS and expert methods. Econ Comput Econ Cybernet Stud Res 46(4):65–87. http://www.ecocyb.ase.ro/20124pdf/Edmund%20Zavadskas%20(T).pdf
34. Štreimikienė D, Baležentis A (2013) Integrated sustainability index: tThe case study of Lithuania. Intelektinė Ekonomika 7(3):289–303. https://doi.org/10.13165/IE-13-7-3-02
35. Ertuğrul I, Karakaşoğlu N (2009) Performance evaluation of Turkish cement firms with fuzzy analytic hierarchy process and TOPSIS methods. Expert Syst Appl 36(1):702–715. https://doi.org/10.1016/j.eswa.2007.10.014
36. Değdeviren M, Yavuz S, Kliniç N (2009) Weapon selection using the AHP and TOPSIS methods under Fuzzy environment. Expert Syst Appl 36(4):8143–8151. https://doi.org/10.1016/j.eswa.2008.10.016
37. Şimşek B, Yusuf T, Şimşek EH (2013) A TOPSIS based Taguchi optimization to determine optimal mixture proportions of the high strength self-compacting concrete. Chemometrics Intell Laboratory Syst 125:18–32. https://doi.org/10.1016/j.chemolab.2013.03.012
38. Sri Krishna S, Sri Nivasulu Readdy A, Vani S (2014) A new car selection in the market using TOPSIS. Int J Eng Res General Sci 2(4):177–181. http://ijergs.org/files/documents/A-NEW20.pdf
39. Seçme NY, Bayrakdaroglu A, Kahraman C (2009) Fuzzy performance evaluation in Turkish banking sector using analytic hierarchy process and TOPSIS. Expert Syst Appl 36(9):11699–11709. https://doi.org/10.1016/j.eswa.2009.03.013
40. Chatterjee P, Chakraborty S (2014) Flexible manufacturing system selection using preference ranking methods: A comparative study. Int J Indus Eng Comput 5:315–338. http://www.growingscience.com/ijiec/Vol5/IJIEC_2013_47.pdf
41. Işık AT, Adalı EA (2016) A comparative study for the agricultural tractor selection problem. Decision Sci Lett 5(4):569–580.. http://www.growingscience.com/dsl/Vol5/dsl_2016_10.pdf

Green Business in Blue Economy: Quality Management and Development of Innovative Products

Nataliia Kraus, Kateryna Kraus, and Valerii Osetskyi

1 Introduction

Ukraine ranked 60th in Global and Digital Competitiveness Rankings IMD-2017 [1, 2] of the International Institute for Management Development (IMD).

However, in digital competitiveness ranking for the implementation and study of digital technologies that are transformative in government practice, business models and society as a whole, Ukraine is in the last positions alongside Indonesia, Mongolia, Peru and Venezuela. Innovation has become a major factor in the success of states and businesses today.

Countries and corporations seeking to become world economic leaders are investing heavily in research and development, new manufacturing and quality management. Among the countries that invest the largest share of their GDP in research and development (R&D), South Korea leads by a large margin of 4.3% and Israel of 4.1%. Japan invests 3.6% of GDP, Austria, Germany and Switzerland—about 3%. Whereas, by comparison, Ukraine invests only 0.2% of its already insignificant GDP in scientific development.

In terms of money in R&D investment, in purchasing power parity, the absolute leader is the US—$ 480 billion and China—$ 371 billion a year. For comparison, in

N. Kraus
Department of Finance and Economics, Borys Grinchenko Kyiv University, Kiev, Ukraine
e-mail: k2205n@ukr.net

K. Kraus (✉)
Department of Management, Borys Grinchenko Kyiv University, Kiev, Ukraine
e-mail: k23k@ukr.net

V. Osetskyi
Department of Economic Theory, Macro-and Microeconomics, Taras Shevchenko National University of Kyiv, Kiev, Ukraine
e-mail: val_osetski@ukr.net

S. K. Ghosh et al. (eds.), *Advances in Thermal Engineering, Manufacturing, and Production Management*, Lecture Notes in Mechanical Engineering,
https://doi.org/10.1007/978-981-16-2347-9_33

Ukraine—$ 2.5 billion. Expenditure on R&D 1000 corporations, the world's largest R&D investor, increased 3.2% in 2017 to $ 702 billion.

In 2017, the Government of Ukraine presented a National Report on "Sustainable Development Goals: Ukraine" [3], which set the benchmarks for achieving them. However, there are now many concepts that are based on utility estimates such as energy, water, waste management, transportation and complicate long-term smart community planning.

The European Strategy for Intellectual, Sustainable and Inclusive Development by 2020 [4] has declared the achievement of objectives as a common interest on the basis of three complementary economic policy priorities: smart growth (economic development based on knowledge and innovation); sustainable development (promoting a more resource-saving, green and competitive economy); inclusive growth (stimulating the employment economy, ensuring social and territorial cohesion).

2 Green Business

2.1 The Essence of Implementing Green Business

Valuable in the scientific sense of the study of green entrepreneurship are scientific works and practical achievements of such well-known scientists as Gryunwald [5], Merts et al. [6], Shchulst. The theory concerning the future modern development of the economy, the so-called "blue economy" theory, by a scientist economist, a member of the Roman Club of G. Pauli, deserves due attention.

Economists from Ukraine are also actively involved in research and development in the field of doing business [7], which seeks to combine environmental concerns, fair treatment of employees and economic success and the issues of quality management of environmental business. Among them are the names of V. Heits, V. Galasyuk, A. Grytsenko, O. Lanovenko, O. Ostapishyna, I. Yurinova. But at the same time, there are a number of pressing issues of type:

- The particularities of problems and risk management at the various stages of green business in blue economy;
- Previous recommendations regarding the control and quality management of green production at the stage of its growth;
- Ranking of the most promising green business ideas that allow us to join green market and lay the foundations for the development of blue economy in Ukraine remains poorly understood.

The purpose of the article is to study the features of green business in blue economy. Presentation of the core of VI-th and VII-th technological structures that lay foundations for the establishment of green business in blue economy.

Finding out the specificities of problems and risk management in the various stages of green business in blue economy. Substantiation and disclosure of previous recommendations regarding the control and quality management of green production at the stage of its growth.

Green business (sustainable business) or circular (non-exhaustive) business is an activity that has minimal adverse impact on the global or local environment, society or economy—a business that seeks to combine environmental, equitable treatment and economic success [8]. Most green companies have progressive environmental principles and policies on human issues. A business is generally considered green if it meets the following four criteria:

- Is guided by the principle of "circulation" in making all its business decisions;
- Supplies environmental goods or services that replace the demand for non-environmental goods or services [9];
- "Greener" than typical competitors;
- Has long been committed to environmental standards.

The Brundtland Commission [10] emphasized that the understanding of inexhaustibility rests on three pillars: people, the planet and profit. Circular business in its content aims to balance these three factors, using circular production and distribution in order to influence the environment, economic development and society [11].

In general, all types of businesses affect circulation in the context of world market and our planet in some sense. The development of a circular principle in business can be significant for the buyer, the investor, and it is also environmentally friendly. The circular business must meet the needs of the buyer and be environmentally friendly.

A scientist economist, a member of the Roman Club G. Pauli put forward his theory about the future of the latest economic development, the so-called blue economy theory. Her content was revealed in her research paper entitled "Blue Economy: 10 Years, 100 Innovations, 100 Million Jobs".

In her research, the scientist is devoted to the aspects of becoming an innovative economy in the context of the transition of society to sustainable environmentally sound development. G. Pauli argues that economic development and the restoration of environmental balance must be interrelated processes [12].

2.2 *Blue Economy*

The author has found innovative solutions to the pressing problems of humanity. G. Pauli's scientific postulate is that "there is nothing superfluous in nature", so the rational use of resources at every stage of humanity's economic activity is a necessary condition for humanity's competitiveness.

"Blue economy" demonstrates how many environmental and environmental degradation problems can be avoided through the conservation of material resources and the adaptation of production processes to the laws of nature. G. Pauli gives examples of the interrelationship between nature and economy and points to the ways of harmonious environmental and economic development of society.

Based on the existing concept of green business, which is based on the principle of "circulation" and theory, regarding the future modern development of the economy, the so-called blue economy theory, it can be stated that their implementation in practice pursues the same goal, namely: a consistent and reasonable combination of economic and environmental development on a circular basis in business with complete satisfaction of the customer's needs.

For these reasons, we consider it appropriate to present the core of VI and VII technological frameworks that lay foundations for green business formation in blue economy, which are presented in Table 1.

Regarding the functions performed by green business, most of them relate to all possible lines of business. For example, setting prices for goods and services sold, conducting research and development work and creating innovative goods, methods and technologies, creating new jobs.

2.3 Problems of Green Business in Blue Economy

Functions specific only to green business entities include: meeting public needs for environmentally friendly products, protecting the environment, minimizing environmental-destructive environmental impacts, shaping the environmental awareness of society and its environmental culture, and preserving natural capital. It is the fulfilment of these functions that distinguishes green business in blue economy, among other areas of activity.

It should be noted that the functions that are inherent in all areas of activity in the green business are manifested in a more environmentally rational way. For example, an efficient use of resources not only minimizes the amount of their use, but also is based on the choice of those resources, and the use of which will have the least possible environmental impact [8].

Specific features of problems and risk management at the various stages of green business in blue economy are presented in Table 2.

2.4 Green Products in Blue Economy

Previous recommendations regarding the control and quality management of green production at the stage of its growth are presented in Table 3. The forms of manifestation of green production today are: eco-company; direct green entrepreneurship; production of environmental goods and services.

The first form of green production is the introduction of environmental management, which is mainly related to the administrative regulation of the enterprises in

Table 1 Core of VI and VII technological structures that lay for green business in blue economy [15, p. 57 and own experience]

Direction	Problems solved	Possible implementation options
Characteristics of VI technological way		
Non-traditional energy	Reducing the burden on the environment, saving natural resources	Hydrogen energy, synthetic fuel, solar energy converters, closed-loop nuclear power plants, fast reactors, vortex heat generators
Information systems	Globalization of the world economy through partnership	Bioenergy, optics, quantum-vacuum computers, artificial intelligence, torsion communication systems
Biotechnology	A new level of well-being	Water purification, seafood desalination, modified agro-culture, disease treatment, cloning
Transport	Environmental safety, speed, efficiency	Underwater superliners, string transport, electric vehicles, aerospace transport systems
Ecology	Sustainable development	Waste-free and closed technological "circuits"
Materials	Durability, safety, reliability, efficiency	Nanotechnology, amorphous metals, memory materials, high-temperature superconductivity, torsional materials processing technologies
Characteristics of VII technological way		
Cognitive and socio-human technologies. The main production factor is creative intelligence	"World vacuum control". New forms of life on the planet. Constructing a new social reality	Fusion technologies, psi-technologies (advances in modern psychology that include new tools for human management), bioenergy, technologies related to morality and responsibility. This way is realized by means of hyper intelligence, hyper knowledge, hyper information, hyper communication. "Subconscious and mind games". It is projected that there are five cognitive technologies: neuroimaging, cognotropic drugs, cognitive assistants, brain-machine interfaces and artificial sensory organs

Table 2 Characteristic features of problems and risk management at different stages of green business in blue economy [16, p. 102–104; 10]

Possible life-cycle options "green business (WB)—innovation market (RI)—new products (NP)"	Preliminary diagnosis for questions major problems and risks of GB
1	2
1. WB growth is the embryonic state of RI—the technical idea of creating an NP	The problem is related to the need for rapid implementation of the technical idea and the emergence of NP in RI. The most significant risk is the mismatch of the expected and actual consumer response to the NP
2. WB growth—crystallization of RI—technical idea of creation of state of emergency	The problem is the same, but time is limited. Main risks are events that can slow down the process of RI production (counterparties, internal inconsistencies)
3. WB growth—RI growth—technical idea of creating an NP	The problem is the same, but the time resource is almost exhausted. Main risks are the same
4. WB growth—RI saturation—technical idea of creating a NP	Technical idea was "late"
5. WB growth—RI maturity—technical idea of creating a NP	
6. WB growth—decline of RI—technical idea of creation of state of emergency	
7. ST growth—embryonic state of RI—development of NP	There are no visible problems. There is a good chance for the successful development of the ST in the future. Main risk is the mismatch of the expected and actual consumer response to the NP
8. WB growth—crystallization of RI—development of NP	There are no visible problems. However, the time resource for RI output is limited. Main risk is the same + events that can slow down the output of RI (counterparties, internal discrepancies)
9. WB growth—RI growth—NP development	Main problem is the limited time resource for the NP to enter RI. Main risks are the same
10. WB growth—RI saturation—NP development	Development is "late". It is obvious that there are problems in management: marketing "sleeping", organization of the basic processes "limping". Main risk is the deterioration of financial condition
11. WB growth—RI maturity—NP development	
12. WB growth—decline of RI—development of NP	
13. ST growth—embryonic state of RI—exit of NP to RI	Has a chance for successful development of the WB in the future. Main risk is the mismatch of the expected and actual consumer response to the NP
14. WB growth—crystallization of RI—output of NP on RI	

(continued)

Table 2 (continued)

Possible life-cycle options "green business (WB)—innovation market (RI)—new products (NP)"	Preliminary diagnosis for questions major problems and risks of GB
15. WB growth—RI growth—NP output to RI	Unbeatable/wonderful! There are no problems. Of paramount importance are the risks of an internal nature: will the company "drive" such rapid/rapid growth?
16. WB growth—RI saturation—NP output at RI	Exit to RI NP "a little late". Underestimation of management. Difficulties with the financial state of the ST are possible
17. WB growth—RI maturity—NP output to RI	The same, but financial problems can be more serious
18. WB growth—decline of RI—exit of NP on RI	The exit of the NP at RI was too late. Obvious problems in management and marketing. Financial losses can be critical to the ST
19. WB growth—crystallization RI—sales growth of innovation	The situation is quite promising for the WB. Main problem is to maintain a position in the emerging market. Main risks are actions by competitors
20. WB growth—RI growth—NP sales growth	Fantastic situation! The problem is to save it. Risks of internal character are most clearly seen: in the part of the analysis of trends of development of RI, as well as the organization of the basic processes of ST
21. WB growth—RI saturation—stable sale of NP	There are no obvious problems. Main risk is the financial loss that can occur if the ST "hangs" on RI with "old" product
22. WB growth—RI maturity—stability of NP sales	The problem is the limited time available for "production upgrades". Most important are marketing risks as well as risks related to reorganization of internal processes

the country. As a result, the enterprise contributes to the reduction of environmental pollution, but as a rule, not at will, but in accordance with the requirements of laws and regulations or regulations of world organizations.

The peculiarity of the second form of green production is that, in addition to minimizing the harmful environmental externalities from production activities, the enterprise has managers and employees forming an ecological consciousness, which is based on providing environmental needs not only for the modern generation, but also for creating favourable living conditions for the future by minimizing the environmental impact of production over the long term.

The third form of green production is manifested when the company, at its own discretion, switches to energy-efficient, environmentally friendly technologies for the

Table 3 Previous recommendations regarding the control and quality management of green production at the stage of its growth [16, p. 102–104; 11]

Possible options for the life-stage ratio "green production (DR)—innovation market (RI)—new products (NP)"	Previous recommendations
1	2
1. WB growth is the embryonic state of RI—the technical idea of creating an NP	Develop a rigorous timetable of activities to ensure a rapid and efficient transition from a technical idea to a serial issue of an NP. Clearly motivate managers to implement the plan in terms of time and quality
2. WB growth—crystallization of RI—technical idea of creation of state of emergency	
3. WB Growth—RI growth—technical idea of Creating an NP	Mobilize all resources for organizing activities to bring the NP to RI
4. WB growth—RI saturation—technical idea of creating a NP	Use a technical idea for a second product that addresses the needs of a more promising sector of RI (second market niche)
5. WB growth—RI maturity—technical idea of creating a NP	
6. WB growth—decline of RI—technical idea of creation of state of emergency	
7. ST growth—embryonic state of RI—development of NP	Perform continuous analysis of the state of RI. Develop the programme for the fastest and most effective product promotion
8. WB growth—crystallization of RI—development of NP	Develop a timetable for activities that provide a quick and effective transition from product development to production. Clearly motivate managers to implement the plan
9. WB growth—RI growth—NP development	Mobilize all resources to accelerate the exit of the NP in RI
10. WB growth—RI saturation—NP development	Use the development for a second NP focused on the needs of a more promising sector of RI (other market niche). Reorganize the main processes
11. WB growth—RI maturity—NP development	
12. WB growth—decline of RI—development of NP	
13. ST growth—embryonic state of RI—exit of NP to RI	Continuous monitoring of the consumer response to the state of emergency and the state of RI. Develop a programme for the fastest and most effective promotion of NP
14. WB growth—crystallization of RI—output of NP on RI	
15. WB growth—RI growth—NP output to RI	Analyse and refine basic processes. Pay more attention to discipline, responsibilities and responsibilities
16. WB growth—RI saturation—NP output at RI	
17. WB growth—RI maturity—NP output to RI	The same NP + modified to more promising sectors RI (other niche market)

(continued)

Table 3 (continued)

Possible options for the life-stage ratio "green production (DR)—innovation market (RI)—new products (NP)"	Previous recommendations
18. WB growth—decline of RI—exit of NP on RI	Reorganize the management system and main processes of the EE. Modify NP for a more promising sector of RI (other market niche)
19. WB growth—crystallization RI—sales growth of innovation	To carry out constant analysis of the state of RI, and especially, the actions of competitors. Ensure the readiness of the EE (in terms of organizing major processes) for its potential growth
20. WB growth—RI growth—NP sales growth	Constantly analyse the state of RI. Ensure that the VP is ready to move to another product line or to another sector (niche) of RI
21. WB growth—RI saturation—stable sale of NP	Develop a production upgrade programme: move to another product line or to other sectors (niches) of RI
22. WB growth—RI maturity—stability of NP sales	Implement the production upgrade programme

production of goods and services and promotes environmental improvements through the use in the process of activity of methods and raw materials that minimize the environmental-destructive impact on the environment and lead to improved living conditions [13].

Green technologies are production processes that are nature-friendly and produce a product that meets high quality standards. For example, the largest vehicle companies have turned their attention to hybrid and electric cars, ecotires and ecogasoline, which are still in high demand today. One should not ignore the fact that in these areas there is an active cooperation between business and science.

Scientists from different universities in the world are opening up new opportunities in the use of solar panels, growing organic crops and creating an "eco-culture" [14]. Rating of the most promising green business ideas that allow you to join the environmental market and a basis for the formation of a blue economy in Ukraine is presented in Table 4.

3 Conclusions

Key features of the future patterns of development of the blue economy in Ukraine should be:

- Institutional order of the cluster structure based on innovative hubs;
- Direct communication between its members;
- A collective way of creating innovation;

Table 4 Rating of the most promising green business ideas that allow you to join the green market and lay the foundations for becoming a blue economy in Ukraine (compiled by authors based on *Source* 16)

Green business idea of green direction	The content and overall characteristics of green business idea that underlies blue economy
Eco-consulting services	Ideal for those who are well versed in green technology. Companies operating in the sector provide environmental expertise for residential buildings, office space, and are developing options for improving eco-equipment, ranging from energy-efficient technologies to waste management programmes. In order to gain the trust of potential clients, you need to get a certificate, better international, that allows such activities
Bicycle repair	This service is in dire demand in a number of CIS countries. Not only the desire to preserve the environment, but also more rational arguments (high prices for gasoline, endless traffic jams and accidents) make citizens abandon cars. In addition, having ridden a bicycle about ten kilometres away, you do not have to go to the fitness room, so you can save money and burn calories, which will not be unnecessary. For these reasons, there is a need to open bicycle repair centres
Cleaning of ventilation systems	Provides services that improve the energy efficiency of buildings. The fact is that dust and other small particles in any case enter the ventilation systems, but you can get rid of them. This may come in handy for a company that will show you how to save money on gas and electricity. Potential clients can certainly be allergy sufferers: they are always willing to pay a lot of money just to get rid of the allergens
Growing vegetables and fruits	This industry clearly demonstrates the desire of many to eat properly, favouring those fruits and vegetables offered not by large companies but by small farms. It is important that there are certificates that confirm that no chemicals were used when growing the crops. You should also be prepared for the approval of such activities to be agreed with the local authorities
Release of organic cosmetics	A real hit in today's market. Nowadays, girls and women are striving not only to be beautiful, using a huge amount of products every day, but also to want cosmetics to be safe for the skin. At the same time, most beauties simply do not have the time to prepare their own scrub or balm, so it is much easier to buy the finished product
Furniture restoration	A service that is not cheap but is of interest to many. The content of the service is to "revive" old furniture, which may have been inherited from grandparents and which are just so easy to say goodbye to

(continued)

Table 4 (continued)

Green business idea of green direction	The content and overall characteristics of green business idea that underlies blue economy
Eco-cleaning of the house	Actual service for people who are constantly employed. Cleaning up with clean, non-chemical cleaners is a difficult job, but it is essential for families with small children or allergic family members

- An effectively functioning harmonization institute;
- Structure of production of V and VI technological structures.

The strategic priorities for the development of green production of innovative type in the course of R&D include: development of new technologies of energy transportation, introduction of resource-saving technologies, modernization of power plants and technological updating of the agro-industrial complex, high-tech updating of machine and instrument making, systematic modernization of health care.

References

1. Create your own infographics (2019). Available at http://www.liga.net/infografica/336825_ukraina-opustilas-v-reytinge-samykh-konkurentosposobnykh-ekonomik.htm (Accessed 19 January 2019)
2. Global innovation index. Analysis (2018). Explore the interactive database of the gii 2018 indicators. Available at https://www.globalinnovationindex.org/analysis-indicator (Accessed 19 December 2018)
3. National report of Ministry of Economic Development and Trade of Ukraine (2017) *Tsili Staloho Rozvytku: Ukraina* [Sustainable Development Goals: Ukraine]. Kyiv, Ukraine
4. EUROPE2020 (2019) A strategy for smart, sustainable and inclusive growth. Available at http://eur-lex.europa.eu/legal-content/en/ALL/?uri=CELEX%3A52010DC2020 (Accessed 12 January 2019)
5. Gryunvald M, Merts A, Rikhter K, Shults A (2018) Technological transfer. *Tekhnolohichnyi transfer,* Available at http://www.nw-innovations.ru/analitika/orgformi/list1/text/tema2.shtml (Accessed 19 May 2018)
6. Fronshtein NM (1999) *Transfer tekhnolohii ta efektyvne vprovadzhennia innovatsii* [Technology transfer and effective implementation of innovations]. Russia, Moscow
7. Mukhopad VI (2002) *Litsenzovana torhivlia: marketynh, tsinoutvorennia, upravlinnia* [Licensed trade: marketing, pricing, management], 2nd edn. INITs, Moscow, Russia
8. Green Business (2019). Available at https://uk.wikipedia.org/wiki/Зелений_бізнес (Accessed 15 January 2019)
9. Kyiv International Economic Forum (2018) Eight state incentives for innovation are world-wide experience. Available at http://forumkyiv.org/uk/analytics/visim-derzhavnih-stimuliv-dlya-rozvitku-innovacij-svitovij-dosvid (Accessed 8 January 2019)
10. Our Common Future (2019). Available at https://en.wikipedia.org/wiki/Our_Common_Future (Accessed 14 January 2019)
11. Triple bottom line (2019). Available at https://en.wikipedia.org/wiki/Triple_bottom_line (Accessed 18 January 2019)
12. Umbrella patenting (2019) Umbrella patenting. Available at http://forum.triz-guide.com/viewtopic.php?p=858 (Accessed 10 January 2019)

13. Lanovenko OG, Ostapishina OO (2013) Ecological business. Dictionary—a guide to ecology: a teaching method. Kherson: PP Vyshemirsky VS, S. 78–79
14. Yurinova I (2019) Green production: best ideas for eco-business development in the world. Available at http://prostir.co.ua/vyrobnytstvo-najkraschi-rozvytku-4466/ (Accessed 12 January 2019)
15. Krasnoshchekov PA (2008) Structure of technological style [Innovations]. 2(112):S.56–59
16. Medvedeva AM (2011) Blitz analysis of a growing company and risk management [MIR: Modernization. Innovation Development], 3(7):S.100–104

New Heuristics to Minimize Makespan of Permutation Flowshop Scheduling Problem with Uniformly Distributed Processing Times

Rose Dhar, Alok Mukherjee, Kingshuk Chatterjee, and Partha Haldar

Nomenclatures

N, n	Number of Jobs
M, m	Number of Machines
$t_p\ (j, i)$	Processing Time of Job i on Machine j
$t_c\ (j, i)$	Completion Time of Job i on Machine j
π_x	x^{th} Sequence of all jobs
$C(\pi_i)$	Completion Time of a sequence π_i
PFSP	Permutation Flowshop Scheduling Problem
NEH	Nawaz-Enscore-Ham
FLM	Modified Framinan and Leisten
PH	Proposed Heuristic
H1	The Stochastic Method, proposed by Chakraborty et al.
PRE	Percentage Relative Error

R. Dhar · K. Chatterjee
Department of Computer Science and Engineering, Government College of Engineering and Ceramic Technology, Kolkata, India

A. Mukherjee
Department of Electrical Engineering, Government College of Engineering and Ceramic Technology, Kolkata, India

P. Haldar (✉)
Department of Mechanical Engineering, Government College of Engineering and Ceramic Technology and Ceramic Technology, Kolkata, India

S. K. Ghosh et al. (eds.), *Advances in Thermal Engineering, Manufacturing, and Production Management*, Lecture Notes in Mechanical Engineering,
https://doi.org/10.1007/978-981-16-2347-9_34

1 Introduction

Flowshop scheduling problem has great importance in manufacturing systems, production planning to maintain the quality as well as quantity [1]. In flowshop, scheduling a continuous flow of jobs is fed following the same route. The global optimal solution is not possible to find so such problems are known as NP-Hard problem [2]. In permutation flowshop, scheduling problem (PFSP) the final sequence of all jobs remains same for all the machines. PFSP has extensive application in several industries such as steel and iron production, pharmaceutical, food processing and automobile industries and many different fields in real life [3]. In solving PFSPs, minimization of makespan is one of the exoteric and most important objectives [4]. Makespan is the time interval between the starting of the first operation in the first machine and the completion of the last operation in the last machine, i.e. maximum total flow time. Makespan is commonly studied in literature as minimization of the total completion time, for example Chakraborty and Laha [5]; Dong et al. [6]; Kalcynski and Kamburowski [7]; Rad et al. [8]; Ribas, Companys, and Tort-Martorell [9].

PFSPs are majorly of three different types [10]. The first kind of PFSP is the precision category, applicable for very small-scale problem since they have high computational complexity and use large memory. For execution of such algorithm, enumeration method, cutting plane method, branch and bound method [11], etc. are used. The second category uses constructive algorithms, e.g. algorithms used by Gupta et al. [12], Nawaz-Enscore-Ham (NEH) algorithm [13], Rajendran et al. [14], Johnson et al. [15], Palmer [16], Campleu-Dudek-Smith [17] and so on. The third one is the metaheuristic algorithms. They do not follow strict rules and deterministic process. They are more flexible and universal compared to deterministic algorithms in number of movements in their searches. With the help of some nature inspired and other accumulated work experiences by human beings, metaheuristic is sketched. For example, use of genetic algorithm (GA) [18–20], simulated annealing (SA) algorithms [21, 22] etc. to solve scheduling problems. To solve bigger problems, researchers have focused on heuristics. On the basis of dispatching rule heuristic algorithms are constructive and improvement heuristics. In case of constructive heuristics through the list of unscheduled jobs, a series of passes of the jobs is maintained. During each pass, one or more jobs are selected and then it is added to that schedule. Once a sequence is prepared, it remains fixed i.e. cannot be altered. Nawaz et al. (NEH) heuristic [13] is commonly regarded as the most effective constructive heuristic. It is a core heuristic for most of the scheduling problems and was mainly proposed for the PFSPs for minimization of makespan. Three stochastic versions were proposed by Laha et al. based on the ideas from the NEH algorithm of Nawaz et al. [13] and simulated annealing (SA) of Kirkpatrick et al. [23] are H1, H2, H3 [5]. Framinan and Leisten proposed an heuristic [24]. In the heuristic procedure proposed by Nawaz et al. [13] to optimize the partial schedules this mechanism is used. Further, a modification of this heuristic was made by Laha and Sarin [25]. This modification enhances the performance of the previous heuristic significantly. Sometimes ties occur for NEH heuristic during the initial ordering of jobs as well as during the choosing of the best or optimal partial sequence among the sequences which have the same makespan. For this, two tie-breaking rules called NEHMSWG and NEHMinS-PS are proposed by Rajendran et al. [26]. On the other hand, in case of improvement heuristics, it is started with an existing solution and after that it will apply some tricks to improve the result [27].

In this paper, we focus on PFSP. The main claim of the paper is to minimization of makespan. The two assumptions of PFSP are as follow: (i) Same priority of importance in all jobs. (ii) Uninterrupted operations and one machine can perform one operation at a time.

Consider a small two machines two job problem. The processing times are given:

Job	Machine 1	Machine 2
J1	10	2
J2	5	7

The makspan for the sequence 1–2 is 22. For the 1st job, the weightage in 1st machine is greater than 2nd.

Now, considering the processing times are well distributed for 1st job.

Job	Machine 1	Machine 2
J1	6	6
J2	5	7

The makespan is for the same sequence is 19, i.e. reduced. In our work, we apply this strategy, i.e. a well-distributed machining time in all machines. It is obvious that if processing times are large then the corresponding machines will be occupied by the job. Hence, it will delay the completion. If the processing times are nearly same, then all machines will be occupied for nearly same time. In the previously worked algorithms by the researchers, they are not focused on well-distributed processing times. Here, on considering this a significant improvement is obtained on the results.

2 A Description of Flowshop Scheduling

In flowshop scheduling system the n jobs, denoted as $\{J_1, J_2, J_3, \ldots, J_n\}$. Each job is processed under m machines $\{M, M_2, M_3, \ldots, M_m\}$ where the Jth operation needs to be processed by J^{th} machine and each machine can process only one operation at a time interval without preemption. After completing J operation at J^{th} machine then it only go to the next $(J+1)^{\text{th}}$ machine.

3 General Notation and Problem Description of Permutation Flowshop Scheduling

The mathematical model of PFSP is described as follow:

(1) $t_p(j, i)$ denotes the time required to process the job i on machine j
(2) $t_c(j, i)$ denotes the completion or execution time of job i on machine j.

$\{\pi_1, \pi_2, \pi_3, \ldots, \pi_{n!}\}$ referred to all possible sequence of n jobs (a total n! is possible).

(3) The completion time of a sequence π_i is represented as $C(\pi_i)$

For $j = 1, 2, \ldots, m$ and $i = 1, 2, \ldots, n$.

$$t_c(M_1, J_1) = t_p(M_1, J_1)$$
$$t_c(M_j, J_1) = t_c(M_{j-1}, J_1) + t_p(M_j, J_1)$$
$$t_c(M_1, J_i) = t_c(M_1, J_{i-1}) + t_p(M_1, J_i)$$
$$\vdots$$
$$\therefore t_c(M_j, J_i) = \max\{t_c(M_{j-1}, J_i), t_c(M_j, J_{i-1})\} + t_p(M_j, J_i)$$

Now, the makespan is defined as the completion time of the last job, $t_c(M_m, J_n)$ of the sequence.

Therefore, for each sequence $(\pi_i) = t_c(M_m, J_n)$.

Among all π_i for a specific i, $C(\pi_i)$ is minimum.

4 Proposed Algorithm

The flowchart of the proposed constructive heuristic is given in Fig. 1.

5 Performance Evaluations

The proposed heuristic is compared against three most efficient constructive heuristics. These are: (a) the modified Framinan and Leisten [25], hereinafter referred to as FLM. (b) The stochastic method, H1 proposed by Chakraborty et al. [5] and (c) NEH with 2nd tie braking rule NEHMinS-PS (here it has been referred as NEH1) [26].

All the standard reference algorithms (FLM, H1, NEH1) and the proposed heuristic were coded in C programming language and executed on a AMD Turion, 3 GB, 1.90 GHz PC. The quality of the solution is judged by the generated makespan by an algorithm (smaller makespan means better result). Here, two phases of experiment are taken to compare the performances of the heuristics. Small In the first

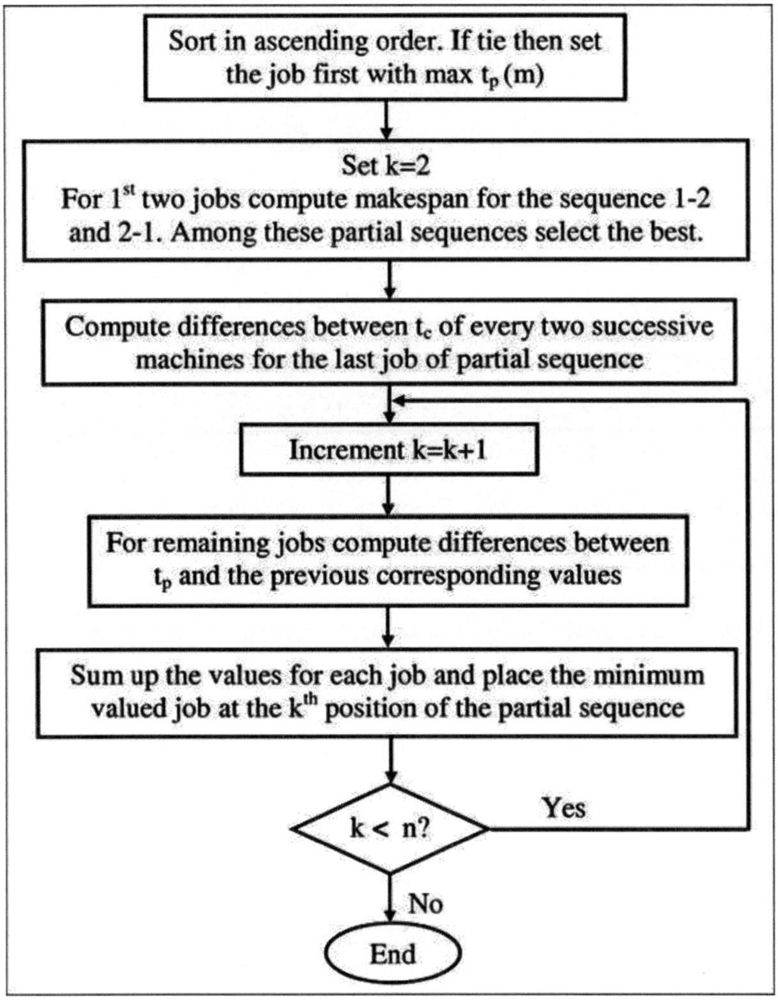

Fig. 1 Flowchart of the proposed constructive heuristic

phase, small-size problems with $n = 6, 7$, and 8, and $m = 5, 10, 15, 20$ and 25 are taken. The second phase, $n = 10, 20, 30$ and 40, and $m = 5, 10, 15, 20$ and 25 are considered. For each set of jobs and machines 20 instances are chosen. Therefore, 300 problems are considered in first phase and 400 problems in second phase. To generate the problem dataset, uniform random discrete distribution of numbers from 1 to 99 are selected.

To evaluate the performance of the coded algorithms, percentage relative error (PRE) and number of optimal or best solutions for a particular problem set are calculated.

The Percentage Relative Error is given by,

$$\text{PRE} = \frac{\begin{array}{c}(\text{makespan by a respective}\\ \text{heuristic} - \text{optimal makespan(or)}\\ \text{best upper bound on the makespan})\end{array}}{\begin{array}{c}\text{optimal makespan(or)}\\ \text{best upper bound on the makespan}\end{array}} \times 100$$

For small-size problem, complete enumeration by constituting the optimal makespan from all possible values is taken in order to find PRE. In the second phase, the optimal solution is taken as the best upper bound on makespan generated by all heuristics.

To compute the experiment, we take three processes in heuristic.

(i) Proposed Heuristic1 (PH1): Here, standard deviation(SD) is used to sort the jobs initially. For each job, the SD of all processing times are calculated. Then for each job, processing times are divided by corresponding SD.
(ii) Proposed Heuristic2 (PH2): Here, the mean of processing time of each job is calculated. Next, for each job, all processing times are divided by the corresponding mean.
(iii) Proposed Heuristic3 (PH3): SD and Mean both are calculated for each job. Then we take a ratio $x = 0.3 \times$ standard deviation $+ 0.7 \times$ Mean. Then the processing times are divided by x for the corresponding job.

The comparative PRE for the small-sized and large-sized problems are shown in Tables 1 and 2, respectively. For each combination of job and machine set, 20

Table 1 Comparison of heuristics for small problems with respect to PRE ($n = 6$, 7 and 8)

	m	Instances	FLM	NEH1	H1	PH1	PH2	PH3
6	5	20	0.2051	0.2326	0.1922	0.1078	0.0990	**0.0958**
	10	20	0.2313	0.2475	0.2821	**0.0945**	0.1296	0.1213
	15	20	0.1668	0.1748	0.1745	**0.0750**	0.0772	0.0871
	20	20	0.4032	0.4537	0.4840	0.3557	**0.3313**	0.3554
	25	20	0.1361	0.1405	0.1748	0.0875	**0.0720**	0.0902
7	5	20	0.2536	0.3577	0.1891	**0.1150**	0.1156	0.1215
	10	20	0.1995	0.2067	0.2078	**0.0933**	0.1098	0.0998
	15	20	0.1760	0.2011	0.2395	0.0898	**0.0805**	0.0964
	20	20	0.1693	0.1819	0.1929	**0.1019**	0.1040	0.1042
	25	20	0.1482	0.1514	0.1556	**0.0938**	0.1011	0.0970
8	5	20	0.2757	0.2677	0.2208	0.1499	**0.1267**	0.1283
	10	20	0.2166	0.2481	0.3272	0.1310	**0.1241**	0.1286
	15	20	0.2113	0.2293	0.3092	0.1047	0.1121	**0.0961**
	20	20	0.1585	0.1769	0.2379	0.0908	0.1071	**0.0988**
	25	20	0.1619	0.1665	0.1313	0.1069	0.1028	**0.1008**

Table 2 Comparison of heuristics for large problems with respect to PRE ($n = 10, 20, 30$ and 40)

	m	Instances	FLM	NEH1	H1	PH1	PH2	PH3
10	5	20	0.1252	0.1623	0.0920	**0.0168**	0.0295	0.0247
	10	20	0.1231	0.1201	0.0780	0.0305	**0.0230**	0.0374
	15	20	0.1069	0.1289	0.0815	**0.0199**	0.0233	0.0219
	20	20	0.0716	0.0726	0.0426	**0.0180**	0.0184	0.0201
	25	20	0.0869	0.0878	0.0711	0.0246	0.0186	**0.0155**
20	5	20	0.1619	0.1179	0.0766	0.0110	0.0158	**0.0105**
	10	20	0.1449	0.1505	0.0884	0.0341	0.0278	**0.0192**
	15	20	0.0871	0.1114	0.0562	**0.0180**	0.0183	0.0269
	20	20	0.1035	0.1010	0.0687	**0.0056**	0.0216	0.0124
	25	20	0.0802	0.0982	0.0556	0.0188	0.0258	**0.0139**
30	5	20	0.1555	0.0988	0.0770	0.0187	0.0068	**0.0052**
	10	20	0.1081	0.1058	0.0759	**0.0127**	0.0364	0.0168
	15	20	0.1076	0.1223	0.0815	0.0262	**0.0115**	0.0234
	20	20	0.0840	0.0996	0.0677	0.0190	0.0197	**0.0127**
	25	20	0.0959	0.1119	0.0692	**0.0158**	0.0230	0.0239
40	5	20	0.1666	0.1141	0.0773	0.0131	**0.0032**	0.0130
	10	20	0.1534	0.1379	0.0985	0.0343	0.0157	**0.0104**
	15	20	0.1058	0.1132	0.0745	**0.0182**	0.0157	**0.0182**
	20	20	0.0890	0.0943	0.0586	0.0093	0.0152	**0.0081**
	25	20	0.0816	0.1013	0.0590	**0.0118**	0.0159	**0.0118**

instances are selected and evaluate the PRE for each heuristic. Out of these PRE values which are less is the best solution and are marked bold in the table. The result shows that the proposed heuristic performs better than the existing standard reference algorithms.

From Tables 3 and 4, it is evident that, in case of small-sized problem, the number of optimal solution out of 300 problems are 0, 0, 1, 8, 5, 7 for FLM, NEH1, H1, PH1, PH2 and PH3, respectively. For large-sized problems, these are 1, 3, 22, 147, 151, 167 respectively out of 400 problem set. For large-sized problem, the normalized version of the heuristic performs significantly well among all heuristics.

6 Time Complexity

The time complexity of FLM heuristic is $O(n^4m)$ [25] and NEH1 is $O(n^2m)$ [26]. In case of proposed heuristic initially the jobs are shorted using bubble sort in order to the sum of processing times up to $(m - 1)^{th}$ machines. It takes order $O(n^2)$ times. The complexity is associated with generating the best partial sequence. Then the

Table 3 Number of problems in favour of optimal solution for small problems ($n = 6, 7$ and 8)

	m	Instances	FLM	NEH1	H1	PH1	PH2	PH3
6	5	20	0	0	0	1	2	2
	10	20	0	0	0	1	0	1
	15	20	0	0	0	3	1	2
	20	20	0	0	0	0	0	0
	25	20	0	0	1	0	0	1
7	5	20	0	0	0	0	0	0
	10	20	0	0	0	1	1	0
	15	20	0	0	0	2	1	1
	20	20	0	0	0	0	0	0
	25	20	0	0	0	0	0	0
8	5	20	0	0	0	0	0	0
	10	20	0	0	0	0	0	0
	15	20	0	0	0	0	0	0
	20	20	0	0	0	0	0	0
	25	20	0	0	0	0	0	0
Sum		300	0	0	1	8	5	7

processing times of the remaining jobs are subtracting performing an order of O(nm) work, which is performed for $n = 3$ to $n$jobs. So, the overall order is $O(n \cdot nm + n^2)$, i.e. $O(n^2m)$. This is significantly lesser than FLM.

7 Conclusions

Here, a new heuristic for permutation flowshop scheduling problems has been proposed and compared with some existing standard algorithm with an objective of minimization of the makspan. Specifically, we have compared the new algorithm with three existing algorithms with two size of problems. The proposed algorithm has consistently shown better computational result for both the large and small-sized problem with respect to others. For the large-sized problem, it works much better. The computational complexity of the proposed algorithm is also less than the existing standards.

Table 4 Number of problems in favour of the best solution for large problems ($n = 10, 20, 30$ and 40)

	m	Instances	FLM	NEH1	H1	PH1	PH2	PH3
10	5	20	1	0	1	11	14	13
	10	20	1	1	1	12	8	9
	15	20	2	0	1	13	9	12
	20	20	0	1	4	10	8	8
	25	20	0	0	0	11	12	8
20	5	20	0	2	4	10	10	10
	10	20	0	0	5	9	8	11
	15	20	0	0	1	11	11	8
	20	20	0	0	0	9	17	13
	25	20	0	0	1	10	9	12
30	5	20	0	1	3	13	9	13
	10	20	0	0	1	7	11	14
	15	20	0	0	0	14	8	9
	20	20	1	0	0	10	7	11
	25	20	0	0	1	8	10	9
40	5	20	0	0	1	14	12	10
	10	20	0	0	2	8	5	11
	15	20	0	0	0	11	12	12
	20	20	0	0	3	6	10	11
	25	20	0	0	0	7	12	13
Sum		400	1	3	22	147	151	167

References

1. Li X, Yin M (2013) A hybrid cuckoo search via Lévy flights for the permutation flow shop scheduling problem. Int J Prod Res 51(16):4732–4754
2. Pan QK, Ruiz R (2012) An estimation of distribution algorithm for lot-streaming flow shop problems with setup times. Omega 40(2):166–180
3. Fernandez-Viagas V, Ruiz R, Framinan JM (2017) A new vision of approximate methods for the permutation flowshop to minimise makespan: state-of-the-art and computational evaluation. Eur J Oper Res 257(3):707–721
4. Reza Hejazi S, Saghafian S (2005) Flowshop-scheduling problems with makespan criterion: a review. Int J Prod Res 43(14):2895–2929
5. Chakraborty UK, Laha D (2007) An improved heuristic for permutation flowshop scheduling. Int J Inf Commun Technol 1(1):89–97
6. Dong X, Huang H, Chen P (2008) An improved NEH-based heuristic for the permutation flowshop problem. Compu Oper Res 35(12):3962–3968
7. Kalczynski PJ, Kamburowski J (2009) An empirical analysis of the optimality rate of flow shop heuristics. Eur J Oper Res 198(1):93–101
8. Rad SF, Ruiz R, Boroojerdian N (2009) New high performing heuristics for minimizing makespan in permutation flowshops. Omega 37(2):331–345

9. Ribas I, Companys R, Tort-Martorell X (2010) Comparing three-step heuristics for the permutation flow shop problem. Compu Oper Res 37(12):2062–2070
10. Qiu CH, Wang C (2010) An immune particle swarm optimization algorithm for solving permutation flowshop problem. In: Key engineering materials. vol 419. Trans Tech Publications. pp 133–136
11. Takano MI, Nagano MS (2017) A branch-and-bound method to minimize the makespan in a permutation flow shop with blocking and setup times. Cogent Engg 4(1):1389638
12. Gupta JN, Shanthikumar JG, Szwarc W (1987) Generating improved dominance conditions for the flowshop problem. Compu oper Res 14(1):41–45
13. Nawaz M, Enscore EE Jr, Ham I (1983) A heuristic algorithm for the m-machine, n-job flowshop sequencing problem. Omega 11(1):91–95
14. Gangadharan R, Rajendran C (1993) Heuristic algorithms for scheduling in the no-wait flowshop. Int J Prod Econ 32(3):285–290
15. Garey MR, Johnson DS, Sethi R (1976) The complexity of flowshop and jobshop scheduling. Math of oper Res 1(2):117–129
16. Palmer DS (1965) Sequencing jobs through a multi-stage process in the minimum total time-a quick method of obtaining a near optimum. J Oper Res Soc 16(1):101–107
17. Campbell HG, Dudek RA, Smith ML (1970) A heuristic algorithm for the n job, m machine sequencing problem. Manage Sci 16(10):B-630
18. Iyer SK, Saxena B (2004) Improved genetic algorithm for the permutation flowshop scheduling problem. Compu Oper Res 31(4):593–606
19. Reeves CR (1995) A genetic algorithm for flowshop sequencing. Compu oper Res 22(1):5–13
20. Zhang Y, Li X, Wang Q (2009) Hybrid genetic algorithm for permutation flowshop scheduling problems with total flowtime minimization. Eur J Oper Res 196(3):869–876
21. Osman IH, Potts CN (1989) Simulated annealing for permutation flow-shop scheduling. Omega 17(6):551–557
22. Tian P, Ma J, Zhang DM (1999) Application of the simulated annealing algorithm to the combinatorial optimisation problem with permutation property: an investigation of generation mechanism. Eur J Oper Res 118(1):81–94
23. Kirkpatrick S, Gelatt CD, Vecchi MP (1983) Optimization by simulated annealing. Science 220(4598):671–680
24. Framinan JM, Leisten R (2003) An efficient constructive heuristic for flowtime minimisation in permutation flow shops. Omega 31(4):311–317
25. Laha D, Sarin SC (2009) A heuristic to minimize total flow time in permutation flow shop. Omega 37(3):734–739
26. Rajendran S, Rajendran C, Leisten R (2017) Heuristic rules for tie-breaking in the implementation of the NEH heuristic for permutation flow-shop scheduling. Int J Oper Res 28(1):87–97
27. Ye H, Li W, Abedini A (2017) An improved heuristic for no-wait flow shop to minimize makespan. J Manuf Syst 44:273–279